STUDENT STUDY GUIDE WITH SELECTED SOLUTIONS

JOSEPH BOYLE

Miami-Dade College

Sixth Edition

PHYSICS

GIANCOLI

Volume 2

PEARSON
Prentice Hall

Upper Saddle River, NJ 07458

Associate Editor: Christian Botting
Senior Editor: Erik Fahlgren
Editor-in-Chief, Science: John Challice
Vice President of Production & Manufacturing: David W. Riccardi
Executive Managing Editor: Kathleen Schiaparelli
Assistant Managing Editor: Becca Richter
Production Editor: Dana Dunn
Supplement Cover Manager: Paul Gourhan
Supplement Cover Designer: Joanne Alexandris
Manufacturing Buyer: Ilene Kahn
Cover Photo Credit: Art Wolfe/Getty Images, Inc.

© 2005 Pearson Education, Inc.
Pearson Prentice Hall
Pearson Education, Inc.
Upper Saddle River, NJ 07458

Printed in the United States of America

20 19 18 17 16 15

ISBN 0-13-146557-0

Pearson Education Ltd., *London*
Pearson Education Australia Pty. Ltd., *Sydney*
Pearson Education Singapore, Pte. Ltd.
Pearson Education North Asia Ltd., *Hong Kong*
Pearson Education Canada, Inc., *Toronto*
Pearson Educación de Mexico, S.A. de C.V.
Pearson Education—Japan, *Tokyo*
Pearson Education Malaysia, Pte. Ltd.

FOR   M. F. S.

PREFACE

This study guide was written to accompany PHYSICS: PRINCIPLES WITH APPLICATIONS, Sixth Edition, by Douglas C. Giancoli. It is intended to provide additional help in understanding the basic principles covered in the textbook and add to the student's problem solving skills.

Each chapter begins with a list of Course Objectives based on the information covered in the chapter. This section is followed by a list of Key Terms and Phrases and a list of the basic mathematical equations used in the textbook. The Concept Summary section of the study guide summarizes the main topics covered in the corresponding chapter of the textbook. The concept summary section also includes the answers to three or four End-Of-Chapter questions from the textbook as well as the step-by-step process to the solution to example problems similar to the type of problems found in the textbook. Each chapter of the study guide concludes with the step-by-step process to the solution to six representative problems taken from the textbook.

Because beginning physics courses emphasize problem solving, hints on problem solving skills are placed just before the representative problems taken from the textbook. The problems are solved using a programmed problem approach. A suggestion for the proper use of the programmed problem method to solving problems is included in Chapter 1. The section of the textbook to which each problem corresponds is included as part of the solution.

The student should be aware that the study guide is meant to complement the textbook, not to replace the textbook as a learning tool. Because of this, it is suggested that the student carefully read the chapter in the textbook before using the study guide.

I wish to acknowledge the help given by Dr. Bob Davis for reviewing the solution to each problem and to Dr. Erik Hendrikson for reviewing the answers to the textbook questions. Also, I wish to acknowledge the help given by Christian Botting, Associate Editor, Physics and Astronomy, Prentice Hall, Inc. for his assistance. Every effort has been made to avoid errors; however, I alone have responsibility for any errors which remain and corrections and comments are most welcome.

Joseph J. Boyle
Professor Emeritus
Miami-Dade College

CONTENTS

PREFACE

CHAPTER

16 ELECTRIC CHARGE AND ELECTRIC FIELD
17 ELECTRIC POTENTIAL
18 ELECTRIC CURRENTS
19 DC CIRCUITS
20 MAGNETISM
21 ELECTROMAGNETIC INDUCTION AND FARADAY'S LAW
22 ELECTROMAGNETIC WAVES
23 LIGHT: GEOMETRIC OPTICS
24 THE WAVE NATURE OF LIGHT
25 OPTICAL INSTRUMENTS
26 THE SPECIAL THEORY OF RELATIVITY
27 EARLY QUANTUM THEORY AND MODELS OF THE ATOM
28 QUANTUM MECHANICS OF ATOMS
29 MOLECULES AND SOLIDS
30 NUCLEAR PHYSICS AND RADIOACTIVITY
31 NUCLEAR ENERGY; EFFECTS AND USES OF RADIATION
32 ELEMENTARY PARTICLES
33 ASTROPHYSICS AND COSMOLOGY

CHAPTER 16

ELECTRIC CHARGE AND ELECTRIC FIELD

OBJECTIVES

After studying the material of this chapter, the student should be able to:

- state from memory the magnitude and sign of the charge on an electron and proton and also state the mass of each particle.
- apply Coulomb's law to determine the magnitude of the electrical force between point charges separated by a distance r and state whether the force will be one of attraction or repulsion.
- state from memory the law of conservation of charge.
- distinguish between an insulator, a conductor, and a semi-conductor and give examples of each.
- explain the concept of electric field and determine the resultant electric field at a point some distance from two or more point charges.
- determine the magnitude and direction of the electric force on a charged particle placed in an electric field.
- sketch the electric field pattern in the region between charged objects.
- use Gauss's law to determine the magnitude of the electric field in problems where static electric charge is distributed on a surface which is simple and symmetrical.

KEY TERMS AND PHRASES

electrostatics is the study of interaction between electric charges which are not moving.

electric charge is a fundamental property of matter. Electric charge appears as two kinds, arbitrarily designated **positive** and **negative**. The negative charges are carried by particles called **electrons** while the positive charge carriers are known as **protons**.

law of conservation of electric charge states that the net amount of electric charge produced in any process is zero. Another way of saying this is that in any process electric charge cannot be created or destroyed; however, it can be transferred from one object to another.

force of attraction or repulsion between charged particles can be summarized as unlike charges attract and like charges repel. A negative charge and a positive charge will attract one another while two negative charges or two positive charges repel.

coulomb (C) is the SI unit of charge where $1 C = 6.25 \times 10^{18}$ electrons or protons. The charge carried by the electron is represented by the symbol -e, and the charge carried by the proton is +e. $1 e = 1.6 \times 10^{-19}$ coulomb, $m_{electron} = 9.11 \times 10^{-31}$ kg and $m_{proton} = 1.672 \times 10^{-27}$ kg. The neutron carries no electrical charge. The neutron has a mass of 1.675×10^{-27} kg.

insulators are materials in which the electrons are tightly held by the nucleus and are not free to move through the material. There is no such thing as a perfect insulator, however, examples of good insulators include substances such as glass, rubber, plastic, and dry wood.

conductors are materials through which electrons are free to move. Just as in the case of the insulators, there is no such thing as a perfect conductor. Examples of good conductors include metals, such as silver, copper, gold, and mercury.

semiconductors are materials where there are a few free electrons and the material is a poor conductor of electricity. As the temperature rises, electrons break free and move through the material. Examples of elements which are semiconductors are silicon, germanium, and carbon.

charging by contact occurs when electric charge is transferred from a charged object to an uncharged object.

charging by induction occurs when an electrically charged object is brought near but does not touch an uncharged second object. The negative and positive charges on the second object separate. Overall, the second conductor is still electrically neutral and if the first object is removed, a redistribution of the negative charge occurs.

Coulomb's law states that two point charges exert a force (F) on one another that is directly proportional to the product of the magnitudes of the charges (Q) and inversely proportional to the square of the distance (r) between their centers.

electric fields exist in the space surrounding a charged particle or object. Electric fields are represented by lines of force that start on a positive charge and end on a negative charge.

SUMMARY OF MATHEMATICAL FORMULAS

electric charge	$q = n e$ where $e = 1.6 \times 10^{-19}$ C	Relates the total charge on an object (q) to the fundamental unit of charge (e) and the total number of charges on the object (n).
Coulomb's law	$F = k Q_1 Q_2 / r^2$ where $k = 9 \times 10^9$ N m^2/C^2	Coulomb's law states that two point charges exert a force (F) on one another that is directly proportional to the product of the magnitudes of the charges (Q) and inversely proportional to the square of the distance (r) between their centers.

electric field (E)	$\vec{E} = \vec{F}/q$ or $\vec{F} = q\,\vec{E}$	The magnitude of the electric field (E) at any point in space can be determined by the ratio of the force (F) exerted on a test charge placed at the point to the magnitude of the charge on the test particle (q).
electric field due to a point charge	$\vec{E} = k\,Q/r^2$	The magnitude of the electric field (E) a distance (r) from a single point charge (q) is related to the magnitude of the charge (q) and is inversely proportional to the square of the distance (r) from the charge.

CONCEPT SUMMARY

Electric Charge

There are two types of **electric charge**, arbitrarily called **positive** and **negative**. Rubbing certain electrically neutral objects together (e.g., a glass rod and a silk cloth) tends to cause the electric charges to separate. In the case of the glass and silk, the glass rod loses negative charge and becomes positively charged while the silk cloth gains negative charge and therefore becomes negatively charged. After separation, the negative charges and positive charges are found to attract one another.

If the glass rod is suspended from a string and a second positively charged glass rod is brought near, a force of electrical repulsion results. Negatively charged objects also exert a repulsive force on one another. These results can be summarized as follows: **unlike charges attract and like charges repel.**

TEXT QUESTION 2. Why does a shirt or blouse taken from a clothes dryer sometimes cling to your body?

ANSWER: The air in a clothes dryer is hot and dry. As the clothes tumble they rub against one another and acquire a net charge. When you remove the clothes from the dryer they tend to cling to one another (i.e., static cling) and also to you.

Conservation of Electric Charge

In the process of rubbing two solid objects together, electrical charges are **not** created. Instead, both objects contain both positive and negative charges. During the rubbing process, the negative charge is transferred from one object to the other and this leaves one object with an excess of positive charge and the other with an excess of negative charge. The quantity of excess charge on each object is exactly the same. This is summarized by the **law of conservation of electric charge**: the net amount of electric charge produced in any process is zero. Another way of saying this is that in any process electric charge **cannot** be created or destroyed, however, it

16-3

can be transferred from one object to another.

During the past century, the negative charges have been shown to be carried by particles which are now called **electrons** while the positive charge carriers are known as **protons**. Both particles, as well as many others, are found in the atoms which make up a substance. Note: in a later chapter, the wave properties of electrons, protons, and other particles will be discussed. However, for ease of visualization and discussion, the term particle will be applied for the present.

The SI unit of charge is the coulomb (C). The amount of charge transferred when objects like glass or silk are rubbed together is in the order of microcoulombs (μC).

$1 \text{ C} = 6.25 \times 10^{18}$ electrons or protons and $1 \text{ }\mu\text{C} = 10^{-6} \text{ C}$

The charge carried by the electron is represented by the symbol -e, and the charge carried by the proton is +e.

$e = 1.6 \times 10^{-19}$ coulomb and $m_{electron} = 9.11 \times 10^{-31}$ kg while $m_{proton} = 1.672 \times 10^{-27}$ kg

A third particle, which carries no electrical charge, is the **neutron**. The neutron has a mass of 1.675×10^{-27} kg.

Experiments performed early in the 20th century led to the conclusion that protons and neutrons are confined to the nucleus of the atom while the electrons exist outside of the nucleus. When solids are rubbed together, it is the electrons that are transferred from one object to the other. The positive charges, which are located in the nucleus, do not move.

Insulators, Semiconductors, and Conductors

An **insulator** is a material in which the electrons are tightly held by the nucleus and are not free to move through the material. There is no such thing as a perfect insulator, however, but examples of good insulators include substances such as glass, rubber, plastic, and dry wood.

A **conductor** is a material through which electrons are free to move. Just as in the case of the insulators, there is no such thing as a perfect conductor. Examples of good conductors include metals such as silver, copper, gold, and mercury.

A few materials, such as silicon, germanium, and carbon, are called **semiconductors.** At ordinary temperatures, there are a few free electrons and the material is a poor conductor of electricity. As the temperature rises, electrons break free and move through the material. As a result, the ability of a semiconducting material to conduct improves with temperature.

Charging by Induction and Charging by Contact

As shown in the diagram at the top of the next page, if a negatively charged rod is brought near an uncharged electrical conductor, the negative charges in the conductor travel to the far end of the conductor.

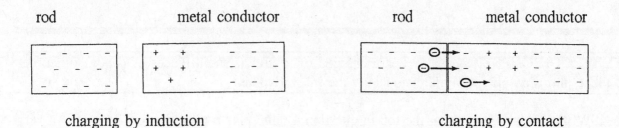

rod   metal conductor   rod   metal conductor

charging by induction   charging by contact

The positive charges are not free to move through a solid object and a charge is temporarily **induced** at the two ends of the conductor. Overall, the conductor is still electrically neutral and if the rod is removed, a redistribution of the negative charge would occur.

If the metal conductor is touched by a person's finger or a wire connected to ground, it is said to be "grounded." The negative charges would flow from the conductor to ground. If the ground is removed and then the rod is removed, a permanent positive charge would be left on the conductor. The electrons would move until the excess positive charge was uniformly distributed over the conductor.

If the rod touches the metal conductor, some of the negative charges on the rod transfer to the metal. This charge distributes uniformly over the metal. The metal has been charged by **contact** and a permanent charge remains when the rod is removed.

TEXTBOOK QUESTION 4. A positively charged rod is brought close to a neutral piece of paper, which it attracts. Draw a diagram showing the separation of charge and explain why attraction occurs.

ANSWER: Paper is a poor conductor, and only a slight displacement of the electric charge occurs when the positively charged rod is brought near the neutral piece of paper. The negative charges in the paper are attracted toward the rod while the positive charges are located in the atomic nucleus and cannot move. In the diagram shown below, the piece of paper remains neutral overall, but a temporary separation of electric charge is induced in the paper. Because the negative charges in the paper are closer to the rod than the positive charges, the force of attraction is greater than the force of repulsion between the rod and the positive charges in the paper. Thus the paper is attracted to the rod.

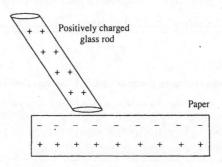

ANSWER: Let us assume that the rod in question 4 represents a plastic ruler. Plastic is a very poor conductor of electricity and paper is a poor conductor. However, if the paper touches the plastic it is possible for some of the negative charges in the paper to transfer to the plastic. If this should happen then both the plastic ruler and the paper would contain an excess of positive charge. The paper would then be repelled from the ruler.

Coulomb's Law

Coulomb's law states that two point charges exert a force (F) on one another that is directly proportional to the product of the magnitudes of the charges (Q) and inversely proportional to the square of the distance (r) between their centers. The formula relating the force to the charges and the distance is

$$\vec{F} = (k\, Q_1\, Q_2)/r^2 \quad \text{where} \quad k = \text{constant} = 9 \times 10^9 \text{ N m}^2/\text{C}^2$$

In the case of objects in which the electric charge is distributed in such a way that they can be considered to act like "point" charges, e.g., small metal spheres with uniform charge distribution, the distance r is measured from the center of one object to the center of the second object.

The value of k given above is for two point charges separated by a distance (r) in a vacuum and, to a good approximation, dry air. The constant may also be expressed in terms of the permittivity of free space (ϵ_o),

where $\epsilon_o = 1/(4\pi k) = 8.85 \times 10^{-12} \text{ C}^2/\text{N m}^2$

The proportionality constant (k) can only be used if the medium which separates the charges is a vacuum. If the region between the point charges is not a vacuum then the value of the proportionality constant to be used is determined by dividing k by the **dielectric constant** (K). For a vacuum K = 1, for distilled water K = 80, and for wax paper K = 2.25. Coulomb's law written to include the dielectric constant of the medium is

$$\vec{F} = (k/K)(Q_1\, Q_2)/r^2)$$

EXAMPLE PROBLEM 1. $Q_1 = +4.0 \ \mu\text{C}$ and $Q_2 = +1.0 \ \mu\text{C}$ are located on the x-axis at x = 0 and x = 6.0 m, respectively. Determine the point on the x-axis where a third charge $Q_3 = +1.0 \ \mu\text{C}$ can be placed and experience no net force.

Part a. Step 1. Determine the point where the net force equals zero.	Solution: (Sections 16-5 and 16-6) At the point in question, the vector sum of the two forces equals zero. The force exerted by Q_1 on Q_3 must be equal in magnitude but opposite in direction from the force exerted by Q_2 on Q_3. The only location that would satisfy this condition would be a location on the x-axis between Q_1 and Q_2.
Part a. Step 2. Draw a diagram locating each point charge.	Let r represent the distance from Q_1 to Q_3 and 6 - r represent the distance from Q_2 to Q_3. Q_1 $\rightarrow$ Q_3 $\rightarrow$ Q_2 $\oplus$ $\vec{F}_{2\ on\ 3}\Leftarrow \oplus \Rightarrow \vec{F}_{1\ on\ 3}$ $\oplus$ $\vdash\leftarrow$ r $\rightarrow\vdash\leftarrow$ 6 - r $\rightarrow\vdash$
Part a. Step 3. Use Coulomb's law to write an equation for the force that 1 exerts on 3 and 2 exerts on 3.	$\vec{F}_{1\ on\ 3} = (k\ Q_1\ Q_3)/r^2$ and $\vec{F}_{2\ on\ 3} = (k\ Q_2\ Q_3)/(6-r)^2$ $F_{1\ on\ 3} = F_{2\ on\ 3}$ $(k\ Q_1\ Q_3)/r^2 = (k\ Q_2\ Q_3)/(6-r)^2$ both k and Q_3 cancel algebraically, and rearranging $r^2/(6-r)^2 = Q_1/Q_2$ but $Q_1/Q_2 = +4.0\ \mu C/+1.0\ \mu C = 4$ Solving algebraically, $r^2/(6-r)^2 = 4$ and $r/(6-r) = 2$ $r = 2(6-r)$ and $r = 12 - 2r$ and $3r = 12$ $r = 4.0$ m Therefore, Q_3 is located on the x-axis 4.0 m to the right of Q_1.

EXAMPLE PROBLEM 2. Two small, identical metal spheres contain excess charges of -10.0 μC and +6.0 μC, respectively. The spheres are mounted on insulated stands and placed 0.40 m apart. a) Determine the magnitude and direction of the force between the spheres. b) The spheres are touched together and then returned to their original 0.40-m separation. Determine the magnitude and direction of the force between the spheres.

Part a. Step 1. Use Coulomb's law to determine the magnitude of the force between the charges.	Solution: (Sections 16-5 and 16-6) $Q_1\ Q_2 = (-10.0 \times 10^{-6}\ C)(+6.0 \times 10^{-6}\ C) = -6.0 \times 10^{-11}\ C$ $\vec{F} = (k\ Q_1\ Q_2)/\vec{r}^2$ $= (9.0 \times 10^9 N\ m^2/C^2)(-6.0 \times 10^{-11}C^2)/(0.40\ m)^2$

16-7

	$\vec{F} = -3.4$ N The negative value indicates that the force between the charges is one of attraction.
Part b. Step 1. Determine the magnitude of the charge on each sphere after contact.	When the spheres touch, electrons travel from the negatively charged sphere to the positively charged sphere. The magnitude of the negative charge is greater than that of the positive charge; therefore, the negative charge completely neutralizes the positive charge. The excess negative charge will distribute uniformly over the two spheres until they both contain equal amounts of negative charge. The magnitude of the charge on each sphere after separation can be determined as follows: the overall excess charge on the spheres is $- 10.0$ μC $+ 6.0$ μC $= - 4.0$ μC. This excess charge is shared equally between the two identical metal spheres. $Q_1' = Q_2' = - 4.0$ $\mu C/2 = - 2.0$ μC
Part b. Step 2. Apply Coulomb's law to determine the magnitude of the force between the charges.	$Q_1 Q_2 = (-2.0 \times 10^{-6}$ C$)(-2.0 \times 10^{-6}$ C$)$ $Q_1 Q_2 = +4.0 \times 10^{-12}$ C^2 $\vec{F} = k\, Q_1\, \vec{Q_2}/r^2$ $\quad = (9.0 \times 10^9$ N m^2/C$^2)(4.0 \times 10^{-12}$ C$^2)/(0.40$ m$)^2$ $\vec{F} = + 0.23$ N Note: the positive value indicates that the force between the charges is one of repulsion.

Electric Field

If an electric charge experiences an electric force at a particular point in space, it is in the presence of an **electric field**. The magnitude of the electric field (E) at any point in space can be determined by the ratio of the force (F) exerted on a test charge placed at the point to the magnitude of the charge on the test particle (q).

$\vec{E} = \vec{F}/q$ where the electric field (E) is measured in units of newton/coulomb (N/C).

Electric field is a vector quantity; the direction of the vector is in the same direction as the direction of the force vector if the test charge is positive and directed opposite from the force vector if the test charge is negative. For a single point charge (Q) the electric field a distance (r) from the charge is given by

$\vec{E} = k\, \vec{Q}/r^2$

In order to visualize the path taken by a charged particle placed in an electric field, **electric field lines,** also known as **lines of force**, are drawn. The diagrams shown at the top of the next page represent the electric field patterns for certain arrangements of charge.

Since a **positive test charge** is arbitrarily chosen to analyze the field, electric field lines are drawn away from an object with the excess of positive charge and toward an object with an excess of negative charge. In diagrams c, d, and e the electric field lines start on the positive charge and terminate on the negative charge.

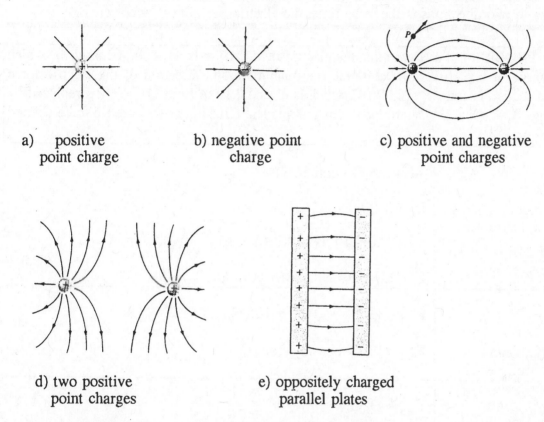

a) positive
    point charge

b) negative point
   charge

c) positive and negative
   point charges

d) two positive
   point charges

e) oppositely charged
   parallel plates

The electric field is strongest in regions where the lines are close together and weak where the lines are farther apart. Thus in diagrams (a) and (b), the field is strongest close to the point charges. In diagram (c) the field weakens as the lines diverge as they leave the positive charge and strengthens as the lines converge on the negative charge. In diagram (e), two parallel plates of opposite charge produce an electric field where the lines are parallel near the center of the plates. In this region the electric field is uniform and E is constant in magnitude and direction.

Coulomb's law can be used to predict that the electric field inside a closed conductor is zero. An example of a closed conductor is a hollow metal sphere which contains an excess of static electric charge. The charge on the conductor tends to reside on its outer surface. Inside the conductor, the electric field is zero. Outside the conductor, the electric field is not zero and the electric field lines are drawn perpendicular to the surface.

Electric charges tend to distribute throughout the volume of a non-conductor which contains an excess of static charge. It can be shown that a charged non-conductor has an electric field inside as well as outside the non-conductor.

TEXTBOOK QUESTION 18. Why can electric field lines never cross?

ANSWER: An electric field line indicates the direction of the force on a positive test charge and the path the test charge would follow. At a point where two electric field lines cross, the force on the test charge would be in two different directions at the same time. Since the test charge cannot move in two directions at the same time, the electric field lines cannot cross.

EXAMPLE PROBLEM 3. a) Determine the electric field 0.10 m from a point charge Q_1 which carries a charge of 0.20 µC. b) Determine the magnitude and direction of the electric force exerted on a second charge Q_2 = 0.80 µC which is placed 0.10 m from Q_1: c) Q_2 is replaced by a third charge Q_3 = -4.0 µC. Determine the magnitude and direction of the electric force exerted on a charge Q_3.

Part a. Step 1.	Solution: (Section 16-7)
Use the formula for the electric field due to a point charge.	$\vec{E}_1 = k\, Q_1/r^2$ $= (9.0 \times 10^9 \text{ N m}^2/\text{C}^2)(2.0 \times 10^{-7} \text{ C})/(0.10 \text{ m})^2$ $\vec{E}_1 = 1.8 \times 10^5$ N/C away from Q_1
Part b. Step 1. Calculate the magnitude and direction of the force.	$\vec{F}_{on\,2} = Q_2\,\vec{E}_1 = (0.80 \times 10^{-6} \text{ C})(1.8 \times 10^5 \text{ N/C})$ $\vec{F}_{on\,2} = 0.144$ N away from Q_1
Part c. Step 1. Calculate the magnitude and direction of the force.	$\vec{F}_{on\,3} = Q_3\,\vec{E}_1$ $= (-4.0 \times 10^{-6} \text{ C})(1.8 \times 10^5 \text{ N/C})$ $F_{on\,3} = -0.72$ N toward Q_1

EXAMPLE PROBLEM 4. Two point charges, Q_1 = + 5.0 µC and Q_2 = - 5.0 µC, are located on the y-axis at y = + 3.0 m and y = - 3.0 m, respectively. Determine the magnitude and direction of the electric field on the x-axis at x = 4.0 m.

Part a. Step 1.	Solution: (Section 16-7)
Draw a diagram locating the position of each charge.	

16-10

Part a. Step 2. Determine the distance r and the angle θ shown in the diagram.	$r = [(3.0 \text{ m})^2 + (4.0 \text{ m})^2]^{1/2} = 5.0 \text{ m}$ $\theta = \tan^{-1} (3.0 \text{ m})/(4.0 \text{ m}) = 37°$

Part a. Step 3. Determine the magnitude and direction of the electric field produced by each point charge at point p.	$\vec{E}_1 = (k\, Q_1)/\vec{r}^2$ $= [(9.0 \times 10^9 \text{ N m}^2/\text{C}^2)(5.0 \times 10^{-6} \text{ C})]/(5.0 \text{ m})^2$ $\vec{E}_1 = 1.8 \times 10^3 \text{ N/C} \angle 37°$ below the horizontal and away from Q_1 $\vec{E}_2 = (k\, Q_2)/\vec{r}^2$ $= (9.0 \times 10^9 \text{ N m}^2/\text{C}^2)(-5.0 \times 10^{-6} \text{ C})/(5.0 \text{ m})^2$ $\vec{E}_1 = -1.8 \times 10^3 \text{ N/C} \angle 37°$ below the horizontal and toward Q_2

Part a. Step 4. Use the trigonometric component method to determine the magnitude and direction of the resultant vector. Note: use the sign convention adopted in Chapter 3 in designating the direction of each component.	 $E_{1x} = E_1 \cos 37° = (1.8 \times 10^3 \text{ N/C}) \cos 37°$ $E_{1x} = + 1.4 \times 10^3 \text{ N/C}$ $E_{1y} = E_1 \sin 37° = (1.8 \times 10^3 \text{ N/C}) \sin 37°$ $E_{1y} = - 1.1 \times 10^3 \text{ N/C}$ $E_2 = E_2 \cos 37° = (1.8 \times 10^3 \text{ N/C}) \cos 37°$ $E_{2x} = - 1.4 \times 10^3 \text{ N/C}$ $E_{2y} = E_2 \sin 37° = (1.8 \times 10^3 \text{ N/C}) \sin 37°$ $E_{2y} = - 1.1 \times 10^3 \text{ N/C}$ $\Sigma X = E_{1x} + E_{2x}$ $\Sigma X = + 1.4 \times 10^3 \text{ N/C} + - 1.4 \times 10^3 \text{ N/C} = 0$ $\Sigma Y = E_{1y} + E_{2y} = -1.1 \times 10^3 \text{ N/C} + -1.1 \times 10^3 \text{ N/C}$ $\Sigma Y = - 2.2 \times 10^3 \text{ N/C} = -2200 \text{ N/C}$

The resultant electric field has a magnitude of 2200 N/C and is directed in the negative y direction at point p.

The **electric flux** (Φ_E) passing through a flat area A due to a uniform electric field (E) is given by $\Phi_E = E\,A\cos\theta$ where θ is the angle between the electric field vector and a line drawn perpendicular to the area.

Gauss's law states that the total flux passing through any closed surface is equal to the net charge (Q) enclosed by the surface divided by ϵ_0. The total flux through the closed surface also equals the sum of the flux through the incremental areas ΔA which make up the surface. Gauss's law is written as follows:

$$\Phi_E = \Sigma\, E\,\Delta A\cos\theta = Q/\epsilon_0 \quad \text{where} \quad \epsilon_0 = 8.85 \times 10^{-12}\ \text{N m}^2/\text{C}^2 \quad \text{and} \quad \Sigma \text{ means "sum of."}$$

EXAMPLE PROBLEM 5. A very long, thin cylindrical shell of length L and radius R carries a charge +Q which is uniformly distributed over the outer part of the shell. Use Gauss's law to determine the electric field at a) points inside the shell (r < R), b) at points outside the shell (r > R). Assume that the points are far from the ends but not too far from the shell.

Part a. Step 1.	Solution: (Section 16-10)
Draw a diagram showing the direction of the lines around the shell.	It is necessary to make assumptions based on symmetry. As shown in the diagram, we expect that field to be radially outward and away from the shell.

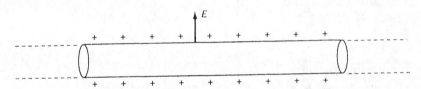

Part a. Step 2.	
Draw a gaussian surface for R < r and determine the charge enclosed by the gaussian surface.	

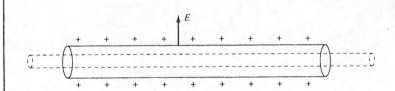

There is no charge enclosed by the gaussian surface.

Part a. Step 3.	$\Sigma\, E\,\Delta A = E\,A = Q/\epsilon_0$
Use Gauss's law to determine the electric field strength.	$E = Q/\epsilon_0\,A$ but $Q = 0$, therefore
	$E = 0$, the electric field inside the cylindrical shell equals zero.

Part b. Step 1.	The charge enclosed by the gaussian surface equals the total charge.
Draw a gaussian surface for r > R and determine the charge enclosed by the surface.	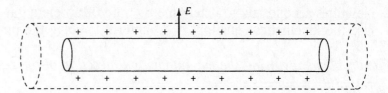

Part b. Step 2.	The cylinder is very long and the question asks for the electric field at points far from the ends. Therefore, only the lateral surface area of the gaussian surface need be considered ($A = 2\pi rL$).
Use Gauss's law to determine the magnitude of the electric field at r > R.	$\Sigma \ E \ \Delta A = E \ A \cos 0° = Q/\epsilon_o$
	$E = Q/(\epsilon_o A)$ but $A = 2\pi rL$
	$E = Q/(\epsilon_o \ 2\pi rL) = Q/(2\pi \epsilon_o rL)$
	alternate solution: The charge on the cylinder can be expressed in terms of the surface area charge density (σ) and the lateral surface area of the cylinder $A = 2\pi RL$.
	$E = Q/(2\pi \epsilon_o rL) = (\sigma \ A)/(2\pi \epsilon_o rL)$
	$E = (\sigma \ 2\pi RL)/(2\pi \epsilon_o rL)$
	$E = (\sigma \ R)/(\epsilon_o \ r)$
	Just above the surface of the cylinder r is approximately equal to R. At this point $E = \sigma/\epsilon_o$ which agrees with the solution to example 16-10 in the textbook.

PROBLEM SOLVING SKILLS

For problems involving Coulomb's law:

1. Complete a data table listing the charge on each object and the distance between the objects. If more than two charged objects are given, draw a diagram showing the position of each object.
2. If the objects touch then charge transfer occurs and the law of conservation of charge must be applied to determine the charge on each object.
3. If more than two charges are given it may be necessary to use the methods of vector algebra discussed in Chapter 3 to solve the problem.
4. Apply Coulomb's law and solve the problem.

For problems involving the motion of a charged particle in an electric field:

1. Draw an accurate diagram showing the motion of the particle in the field.

2. Complete a data table based on information both given and implied in the problem.
3. Determine the magnitude and direction of the electric force acting on the particle. Use Newton's second law to determine the rate of acceleration.
4. Use the kinematics equations of Chapter 2 to solve the problem.

For problems involving the magnitude and direction of the resultant electric field due to two or more point charges:

1. Draw an accurate diagram locating the position of each charge.
2. Determine the magnitude and direction of the electric field at the point in question due to each charge.
3. Use the methods of vector algebra discussed in Chapter 3 to solve for the resultant electric field.

For problems involving Gauss's law:

1. Draw a diagram showing the electric field pattern around the object in question.
2. Draw a gaussian surface to take advantage of the object's symmetry. The surface is such that the lines of electric flux are either perpendicular to or parallel to the gaussian surface.
3. If the flux lines are perpendicular to the surface, then $\theta = 0°$. If the flux lines are parallel to the surface, then $\theta = 90°$.
4. Determine the surface area of both the object and the gaussian surface. The lateral surface area of a cylinder is $2\pi rL$ while the surface area of each end cap $= \pi R^2$. The surface area of a sphere $= 4\pi r^2$.
5. If necessary, express the charge enclosed by the gaussian surface in terms of the charge density on the object. For example, a charge uniformly distributed along a line can be written as $Q = \lambda L$, where λ is the charge per unit length (coulombs/meter). A charge uniformly distributed across a surface of area A can be written as $Q = \sigma A$, where σ is the charge per unit area (coulombs/meter^2). A charge uniformly distributed throughout the object's volume can be written as $Q = \rho V$ where ρ is the charge per unit volume (coulombs/meter^3).
   Note: $V_{sphere} = 4/3 \pi R^3$ and $V_{cylinder} = \pi R^2L$.
6. Use Gauss's law to solve the problem.

SOLUTIONS TO SELECTED TEXTBOOK PROBLEMS

TEXTBOOK PROBLEM 20. A $+4.75 \mu C$ and a $-3.55 \mu C$ are placed 18.5 cm apart. Where can a third charge be placed so that it experiences no net force.

Part a. step 1.	Solution: (Sections 16-5 and 16-6)
Draw a diagram locating each charge. For convenience, place the charges on the x-axis.	Let $Q_1 = +4.75 \mu C$ at $x = 0$ and $Q_2 = -3.55 \mu C$ at $x = 0.185$ m. Let r represent the distance from Q_2 to Q_3 and $(0.185$ m $+ r)$ represent the distance from Q_1 to Q_3.

$$Q_1 \qquad\qquad Q_2 \qquad\qquad\qquad \overset{\rightarrow}{} \quad Q_3 \;\overset{\rightarrow}{}$$
$$\oplus \qquad\qquad\quad \ominus \qquad\qquad\qquad \mathbf{F}_{2\,on\,3}\Leftarrow \oplus \Rightarrow\mathbf{F}_{1\,on\,3}$$
$$|\leftarrow 0.185\text{ m} \rightarrow |\leftarrow \qquad\qquad r \qquad\qquad \rightarrow|$$

16-14

Part a. Step 2. Determine the point where the net force equals zero.	At the point in question, the vector sum of the two forces equals zero. The force exerted by Q_1 on Q_3 must be equal in magnitude but opposite in direction from the force exerted by Q_2 on Q_3. Since the magnitude of the charge on Q_1 is much greater than the charge on Q_2, the only location that would satisfy this condition would be a location on the x-axis to the right of charge Q_2.

Part a. Step 3. Use Coulomb's law to write an equation for the force that 1 exerts on 3 and 2 exerts on 3. Hint: the magnitude of the forces are equal; however, $F_{2\ on\ 3}$ has a negative value because $Q_2 = -3.55\ \mu C$.	$F_{1\ on\ 3} = (k\ Q_1\ Q_3)/(0.185\ m + r)^2$ and $F_{2\ on\ 3} = (k\ Q_2\ Q_3)/r^2$ $F_{1\ on\ 3} = -F_{2\ on\ 3}$ $(k\ Q_1\ Q_3)/(0.185\ m + r)^2 = - (k\ Q_2\ Q_3)/r^2$ Note: Both k and Q_3 cancel algebraically, and rearranging $(0.185\ m + r)^2/r^2 = - Q_1/Q_2$ but $Q_1/Q_2 = -(+4.75\ \mu C/-3.55\ \mu C) = 1.34$ $(0.185\ m + r)^2/r^2 = 1.34$ and $(0.185\ m + r)/r = \pm 1.16$ either $(0.185\ m + r)/r = +1.16$ or $(0.185\ m + r)/r = -1.16$ $0.185\ m + r = 1.16\ r$ or $0.185\ m + r = -1.16\ r$ either $r = 1.16\ m$ or $r = -0.086\ m$ Based on the diagram, r has a positive value. Therefore, Q_3 is located 1.16 m to the right of charge Q_2, i.e., at 1.35 m on the x-axis.

TEXTBOOK PROBLEM 28. What are the magnitude and direction of the electric field at a point midway between a -8.0 μC and a +7.0 μC charge 8.0 cm apart? Assume that no other charges are nearby.

Part a. Step 1. Draw a diagram locating each charge. For convenience the charges are placed on the x-axis.	Solution: (Sections 16-7 and 16-8) Place Q_1 = -8.0 μC at x = 0 and Q_2 = +7.0 μC at x = 0.080 m. A positive test charge is placed at point p which is located at x = 0.040 m. Q_1 p Q_2 ⊖ E_1⇐ ⊕ ⊕ E_2⇐ ⎸← 0.040 m →⎸← 0.040 m →⎸ Q_1 is a negative charge; therefore, E_1 is directed toward the left. Q_2 is a positive charge, E_2 is also directed toward the left.

Part a. Step 2. Determine the electric field at the midpoint between the charges.	At p, the resultant electric field is the vector sum of E_1 and E_2. $\vec{E}_{total} = \vec{E}_1 + \vec{E}_2$ $= \mid k\,Q_1/r^2\mid + \mid k\,Q_2/r^2 \mid = [k\,(Q_1 + Q_1)]/r^2$ where r = 0.040 m $= [(9 \times 10^9\ Nm^2/C^2)(\mid -8.0 \times 10^{-6}\ C\mid + 7.0 \times 10^{-6}\ C)]/(0.040\ m)^2$ $E_{total} = 8.4 \times 10^7$ N/C toward Q_1

TEXTBOOK PROBLEM 41. An electron (mass = 9.11×10^{-31} kg) is accelerated in the uniform E (E = 1.45×10^4 N/C) between two parallel charged plates. The separation of the plates is 1.10 cm. The electron is accelerated from rest near the negative plate and passes through a tiny hole in the positive plate, Fig. 16-60. (a) With what speed does it leave the hole? (b) Show that the gravitational force can be ignored.

Part a. Step 1. Use F = q E to determine the mag- nitude of the force.	Solution: (Section 16-7) $F = q\,E = (1.6 \times 10^{-19}\ C)(1.45 \times 10^4\ N/C)$ $F = 2.32 \times 10^{-15}$ N
Part a. Step 2. Use Newton's second law to determine the acceleration.	$a = F/m = (2.32 \times 10^{-15}\ N)/(9.11 \times 10^{-31}\ kg)$ $a = 2.55 \times 10^{15}$ m/s^2
Part a. Step 3. Complete a data table and use the kinematics equations from Chapter 2 to solve for the velocity as the electron leaves the hole.	$v_0 = 0$ m/s $a = 2.55 \times 10^{15}$ m/s^2 $v = ?$ $x - x_0 = 1.10$ cm $= 0.0110$ m $2\,a\,(x - x_0) = v^2 - v_0^2$ $2\,(2.55 \times 10^{15}\ m/s^2)(0.0110\ m) = v^2 - (0\ m/s)^2$ $v = 7.49 \times 10^6$ m/s

Part b. Step 1.	$F_{gravity} = m\,g$
Determine the magnitude of the gravitational force and compare this force to the electric force.	$F_{gravity} = (9.11 \times 10^{-31}\ kg)(9.8\ m/s^2) = 8.9 \times 10^{-30}\ N$
	$F_{electic}/F_{gravity} = (2.96 \times 10^{-15}\ N)/(8.9 \times 10^{-30}\ N) = 3.33 \times 10^{14}$
	The electric force is much, much greater than the gravitational force. Also, the time for the particle to pass between the plates is insufficient for the gravitational force to affect the electron's motion.

TEXTBOOK PROBLEM 54. A proton (m = 1.67×10^{-27} kg) is suspended at rest in a uniform electric field $\vec{E}$. Take into account gravity at the Earth's surface and determine $\vec{E}$.

Part a. Step 1.	Solution: (Section 16-7)

Draw a diagram locating the forces acting on the proton.	- - - - - - -
	      $\Uparrow$ $F_{electric}$
	      $\oplus$
	      $\Downarrow$ $F_{gravity}$
	 + + + + + +

Part a. Step 2.	The particle is suspended at rest; therefore, net F = 0.
The electric force balances gravity. Use F = q E to determine the magnitude of the electric field.	$F_{electric} - F_{gravity} = 0$
	$q\,E - m\,g = 0$
	$E = (m\,g)/q = [(1.67 \times 10^{-27}\ kg)(9.8\ m/s^2)]/(1.6 \times 10^{-19}\ C)$
	$E = 1.02 \times 10^{-7}\ N/C$ vertically upward

TEXTBOOK PROBLEM 61. An electron with speed $v_o = 21.5 \times 10^6$ m/s is traveling parallel to an electric field of magnitude E = 11.4×10^3 N/C. (a) How far will the electron travel before it stops? (b) How much time will elapse before it returns to its starting point?

Part a. Step 1. Use the work-energy theorem to solve for the distance traveled.	Solution: (Section 16-7) $W = \Delta KE$ $F\,d = \frac{1}{2}\,m\,v^2 - \frac{1}{2}\,m\,v_o^2$ but $v_o = 21.5 \times 10^6$ m/s $(-1.6 \times 10^{-19}$ C$)(11.4 \times 10^3$ N/C$)\,d =$ $\qquad\qquad 0 - \frac{1}{2}(9.1 \times 10^{-31}$ kg$)(21.5 \times 10^6$ m/s$)^2$ $d = 0.115$ m
Part b. Step 1. Use Newton's second law to solve for the acceleration.	$F = m\,a$ but $F = q\,E$ $a = (q\,E)/m = [(-1.6 \times 10^{-19}$ C$)(11.4 \times 10^3$ N/C$)]/(9.1 \times 10^{-31}$ kg$)$ $a = -2.00 \times 10^{15}$ m/s^2
Part b. Step 2. Use the kinematics equations to solve for the time to return to the starting point.	$x = \frac{1}{2}\,a\,t^2 + v_o t + x_o$ 0 m $= \frac{1}{2}(-2.00 \times 10^{15}$ m/s$^2)\,t^2 + (21.5 \times 10^6$ m/s$)\,t + 0$ m factoring gives $0 = [(1.00 \times 10^{15}$ m/s$^2)\,t + 21.5 \times 10^6$ m/s$]\,t$ either $t = 2.15 \times 10^{-8}$ s or $t = 0$ s (starting time)

TEXTBOOK PROBLEM 67. A point charge (m = 1.0 g) at the end of an insulating string of length 55 cm is observed to be in equilibrium in a uniform horizontal electric field of E = 12,000 N/C, when the pendulum position is as shown in Figure 16-66, with the charge 12 cm above the lowest (vertical) position. If the field points to the right in Fig. 16-66, determine the magnitude and sign of the point charge.

Part a. Step 1. Complete a data table.	Solution: (Section 16-7) m = 1.0 g = 0.0010 kg L = 55 cm = 0.55 m y = 0.43 m E = 12,000 N/C Q = ? k = 9 × 10^9 N m^2/C^2
Part a. Step 2. Draw an accurate diagram locating the forces acting on the point charge.	

16-18

Part a. Step 3. Determine the angle θ shown in Fig. 16-66.	$\cos \theta = (0.43 \text{ m})/(0.55 \text{ m})$ $\cos \theta = 0.782$ $\theta = 38.6°$ 0.43 m $\|\theta$ 0.55 m

---

Part a. Step 4. Apply the first condition of static equilibrium and solve for Q.	$\Sigma F_y = 0 \quad T_y - m\,g = 0$ $T \cos 38.6° - (0.0010 \text{ kg})(9.8 \text{ m/s}^2) = 0$ $T = (0.0098 \text{ N})/(\cos 38.6°) = 0.0125 \text{ N}$ $\Sigma F_x = 0 \quad T_x - QE = 0$ $(0.0125 \text{ N})(\sin 38.6°) - Q\,(12000 \text{ N/m}) = 0$ $Q = (7.80 \times 10^{-3} \text{ N})/(12000 \text{ N/m}) = 6.5 \times 10^{-7} \text{ C}$

Part a. Step 5. Determine the sign on the point charge.	Based on the diagram, the direction of the force on the point charge is in the same direction as the electric field. Since the electric field is toward the right, the charge must be positive.

CHAPTER 17

ELECTRIC POTENTIAL

OBJECTIVES

After studying the material of this chapter, the student should be able to:

- write from memory the definitions of electric potential and electric potential difference.
- distinguish between electric potential, electric potential energy, and electric potential difference.
- draw the electric field pattern and equipotential line pattern which exist between charged objects.
- determine the magnitude of the potential at a point a known distance from a point charge or an arrangement of point charges.
- state the relationship between electric potential and electric field and determine the potential difference between two points a fixed distance apart in a region where the electric field is uniform.
- determine the kinetic energy in both joules and electric volts of a charged particle which is accelerated through a given potential difference.
- explain what is meant by an electric dipole and determine the magnitude of the electric dipole moment between two point charges.
- given the dimensions, distance between the plates, and the dielectric constant of the material between the plates, determine the magnitude of the capacitance of a parallel plate capacitor.
- given the capacitance, the dielectric constant, and either the potential difference or the charge stored on the plates of a parallel plate capacitor, determine the energy and the energy density stored in the capacitor.

KEY TERMS AND PHRASES

electric potential at point a (V_a) equals the electric potential energy (PE_a) per unit charge (q) placed at that point.

electric potential difference between two points (V_{ab}) is measured by the work required to move a unit of electric charge from point b to point a.

equipotential lines are lines along which each point is at the same potential. On an equipotential surface, each point on the surface is at the same potential.

voltage is a common term used for potential difference. The SI unit of electric potential and potential difference is the volt (V), where 1 V = 1 J/C.

electron volt (eV) is the unit used for the energy gained by a charged particle which is accelerated through a potential difference. 1 eV = 1.6×10^{-19} J.

electric dipole is two equal point charges (q), of opposite sign, separated by a distance ℓ. The SI unit of the dipole moment (p) is the **debye**, where 1 debye = 3.33×10^{-33} C m. In polar molecules, such as water, the molecule is electrically neutral but there is a separation of charge in the molecule. Such molecules have a net dipole moment.

capacitor stores electric charge and consists of two conductors separated by an insulator known as a dielectric. The ability of a capacitor to store electric charge is referred to as **capacitance** (C).

SUMMARY OF MATHEMATICAL FORMULAS

electric potential	$V = PE_a / q$	The electric potential at point a (V_a) equals the electric potential energy (PE_a) per unit charge (q) placed at that point.
electric potential due to a point charge	$V = kq/r$	The electric potential (V) due to a point charge is related to the magnitude of the charge and the distance from the charge (r) to the point in question. If more than one point charge is present, the potential at a particular point is equal to the arithmetic sum of the potential due to each charge at the point in question.

potential difference	$V_{ab} = V_a - V_b = W_{ab}/q$	The electric potential difference between two points (V_{ab}) is measured by the work required to move a unit of electric charge from point b to point a.
	$\Delta PE_{ab} = PE_a - PE_b = q\,V_{ab}$ $V_{ab} = \Delta PE/q$	The potential difference can also be discussed in terms of the change in potential energy of a charge q when it is moved between points a and b.
	$V_{ab} = E\,d\,\cos\theta$	If the charged particle is in an electric field which is uniform, i.e., constant in magnitude and direction, then the potential difference (V_{ab}) equals the product of the electric field strength (E), the distance (d) between points a and b, and the cosine of the angle ($\cos\theta$) between the electric field vector and the displacement vector.
potential gradient	$E_x = -\,\Delta V/\Delta x$	If the electric field is non-uniform, i.e., varies in magnitude and direction between points a and b, then the electric field strength can only be properly defined over an incremental distance. If we consider the x component of the electric field (E_x), then $E_x = -\,\Delta V/\Delta x$, where ΔV is the change in potential over a very short distance Δx. The minus sign indicates that E points in the direction of decreasing V.
electric potential due to an electric dipole	$V = k\,q\,\ell\,\cos\theta/r^2$ or $V = k\,p\,\cos\theta/r^2$	The potential due to two equal point charges (q), of opposite sign, separated by a distance ℓ, depends on the angle θ and the distance r. $q\,\ell$ is the dipole moment (p).

capacitance	$C = Q/V$	Capacitance (C) is the ratio of the charge stored (Q) to the potential difference (V) between the conducting surfaces.
	$C = K \epsilon_o A/d$	For a parallel plate capacitor, the capacitance (C) depends on the dielectric constant (K) of the insulating material between the plates, the permittivity of free space (ϵ_o), the surface area (A) of one side of one plate which is opposed by an equal area of the other plate, and the distance between the plates (d).
energy stored in a charged capacitor	energy = $\frac{1}{2} Q V$ energy = $\frac{1}{2} C V^2$ energy = $\frac{1}{2} Q^2/C$ energy = $\frac{1}{2} \epsilon_o E^2 A d$	The electric energy stored in charging a capacitor depends on the charge stored (Q) and the potential difference (V) across the conducting surfaces. The energy stored by a parallel plate capacitor is related to the electric field existing between the plates. The product of the area (A) and the distance between the plates (d) equals the volume between the conducting surfaces.
energy density	energy density = $\frac{1}{2} \epsilon_o E^2$	The **energy density** is the energy stored per unit volume.

CONCEPT SUMMARY

Electric Potential and Potential Difference

The **electric potential** at point a (V_a) equals the electric potential energy (PE_a) per unit charge (q) placed at that point.

$V_a = PE_a /q$

The **electric potential difference** between two points (V_{ab}) is measured by the work required to move a unit of electric charge from point b to point a.

$V_{ab} = V_a - V_b = W_{ab} /q$

The potential difference can also be discussed in terms of the change in potential energy of a charge (q) when it is moved between points a and b.

$\Delta PE = PE_a - PE_b = q\ V_{ab}$ and therefore $V_{ab} = \Delta PE/q$

Potential difference is often referred to as **voltage**. Both potential and potential difference are scalar quantities which have dimensions of joules/coulomb. The SI unit of electric potential and potential difference is the **volt** (V), where $1\ V = 1\ J/C$.

If the charged particle is in an electric field which is uniform, i.e., constant in magnitude and direction, then the potential difference is related to the electric field as follows:

$V_{ab} = E\ d\ \cos\theta$

E is the electric field strength in N/C, d is the distance between points a and b in meters and θ is the angle between the electric field vector and the displacement vector.

If the electric field is non-uniform, i.e., varies in magnitude and direction between points a and b, then the electric field strength can only be properly defined over an incremental distance. If we consider the x component of the electric field (E_x), then $E_x = -\ \Delta V/\Delta x$, where ΔV is the change in potential over a very short distance Δx. The minus sign indicates that E points in the direction of decreasing V.

EXAMPLE PROBLEM 1. The two horizontal parallel plates of a capacitor are 0.020 m apart and the electric field between the plates is 600 N/C. A deuterium nucleus is initially at rest near the positive plate. Determine the a) magnitude of the electric force acting on the particle, b) work done on the particle as it passes between the plates, c) particle's kinetic energy and speed just before it strikes the negative plate. Note: the mass of the nucleus is 3.34×10^{-27} kg and the charge is 1.6×10^{-19} C.

- - - - - - - - - - - - -

$\Uparrow F_{electric}$

$\oplus$

_____

+ + + + + + + + +

Part a. Step 1.	Solution: (Section 17-1)
The magnitude of the electric field is known. Determine the magnitude of the electric force.	$F = q\ E = (1.6 \times 10^{-19}\ C)(600\ N/C)$ $F = 9.6 \times 10^{-17}\ N$

Part b. Step 1.	$W = F\,d = (9.6 \times 10^{-17}\text{ N})(0.020\text{ m})$
Determine the work done on the particle as it passes between the plates.	$W = 1.9 \times 10^{-18}\text{ J}$ (alternate method) $W = q\,V$ where $V = E\,d$ $W = q\,E\,d = (1.6 \times 10^{-19}\text{ C})(600\text{ N/C})(0.020\text{ m}) = 1.9 \times 10^{-18}\text{ J}$
Part c. Step 1.	$\text{work} = \Delta KE = KE_f - KE_i$
Use the work-energy theorem to determine the particle's final kinetic energy.	$1.9 \times 10^{-18}\text{ J} = KE_f - 0\text{ J}$ $KE_f = 1.9 \times 10^{-18}\text{ J}$
Assume that the particle was initially at rest and determine the particle's final velocity.	$KE_f = \tfrac{1}{2}\,m\,v^2$ $1.9 \times 10^{-18}\text{ J} = \tfrac{1}{2}(3.34 \times 10^{-27}\text{ kg})\,v_f^{\,2}$ $v_f = 3.4 \times 10^4\text{ m/s}$

EXAMPLE PROBLEM 2. The two horizontal parallel plates of a capacitor are 0.0400 m apart and the potential difference between the plates is 100 V. A particle which has an excess charge of +2e, i.e., +3.20 x 10⁻¹⁹ C, is in static equilibrium between the plates. Calculate the a) magnitude and direction of the electric field between the plates and b) mass of the particle.

- - - - - - - - - - - -

$\uparrow F_{\text{electric}}$

$\oplus$

$\downarrow F_{\text{gravitational}}$

+ + + + + + + + + +

Part a. Step 1.	Solution: (Sections 17-1 and 17-2)
Determine the magnitude and direction of the electric field.	The electric field is uniform in the region between the two parallel plates and is directed from the positive plate toward the negative plate. The magnitude of the electric field is given by $V = E\,d\cos\theta$ $100\text{ V} = E\,(0.0400\text{ m})\cos 0°$ $E = (100\text{ V})/(0.0400\text{ m}) = 2500\text{ V/m} = 2500\text{ N/C}$

| Part b. Step 1.

Determine the mass of the particle. | Since the particle is in static equilibrium, the electric force must be equal but oppositely directed from the gravitational force.

$F_{electric} = F_{gravitational}$

$q \, E = m \, g$

$(3.20 \times 10^{-19} \text{ C})(2500 \text{ N/C}) = m \, (9.80 \text{ m/s}^2)$

$m = (3.20 \times 10^{-19} \text{ C})(2500 \text{ N/C})/(9.80 \text{ m/s}^2)$

$m = 8.16 \times 10^{-17} \text{ kg}$ |

Equipotential Lines

Equipotential lines are lines along which each point is at the same potential. On an equipotential surface, each point on the surface is at the same potential. The equipotential line or surface is perpendicular to the direction of the electric field lines at every point. Thus, if the electric field pattern is known, it is possible to determine the pattern of equipotential lines or surfaces and vice versa.

In the following diagrams, the dashed lines represent equipotential lines and the solid lines the electric field lines.

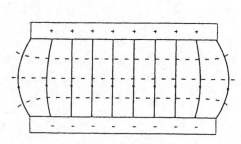

parallel plate conductors

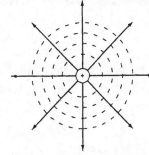

single positive charge

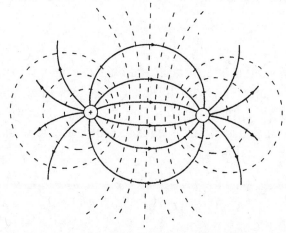

two point charges of opposite sign

TEXTBOOK QUESTION 9. Can two equipotential lines cross? Explain.

ANSWER: The points on equipotential lines (or surfaces) are all at the same potential. If two different equipotential lines were to cross, the point at which they cross would be at two different potentials. Since it is not possible to have two different potentials at the same point, two equipotential lines cannot cross.

Also, equipotential lines are always perpendicular to electric field lines. As discussed in the answer to question 18 in Chapter 16, electric field lines never cross. Therefore, it is not possible for equipotential lines to cross.

TEXTBOOK QUESTION 10. Draw a few equipotential lines in Fig. 16-31b.

ANSWER: The equipotential lines are drawn so that they are perpendicular to the electric field lines at the points where the equipotential lines cross the electric field lines.

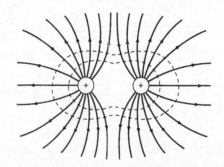

Electron Volt

The energy gained by a charged particle which is accelerated through a potential difference can be expressed in **electron volts** (eV) as well as joules. Higher amounts of energy can be measured in KeV or MeV. $1 \text{ eV} = 1.6 \times 10^{-19}$ J, $1 \text{ KeV} = 10^3$ eV, and $1 \text{ MeV} = 10^6$ eV.

Electric Potential Due to a Point Charge

The electric potential due to a point charge (q) at a distance r from the charge is given by

$V = kq/r$

Note that the zero of potential is arbitrarily taken to be at infinity ($r = \infty$). For a negative charge, the potential at distance r from the charge is less than zero. As r increases, the potential increases toward zero, reaching zero at $r = \infty$. If more than one point charge is present, the potential at a particular point is equal to the arithmetic sum of the potential due to each charge at the point in question.

ANSWER: Electric field is a vector quantity. As shown in the diagram, at the midpoint between the two charges the two electric field vectors, E_1 and E_2, are equal in magnitude but oppositely directed. The magnitude of the resultant electric field is zero.

$$Q_1 + \qquad E_2 \Leftarrow . \Rightarrow E_1 \qquad + Q_2$$

Electric potential is a scalar quantity. Electric potential has magnitude but no direction. At the midpoint between the charges the electric potential due to each charge is positive and equal in magnitude. The resultant electric potential is the arithmetic sum of the two potentials and is greater than zero. The charges are both positive; therefore, the arithmetic sum of the potential of each charge would be greater than zero at every point on the line between the two charges.

ANSWER: At a point where the potential equals zero, the electric field need NOT equal zero. In the diagram shown below, the two point charges are equal in magnitude but one is positive and the other negative. At the midpoint between the two charges, the electric field due to the positive charge is directed away from the positive charge and toward the negative. The electric field due to the negative charge is directed toward the negative charge. The resultant electric field equals the sum of the two vectors and is directed toward the negative charge.

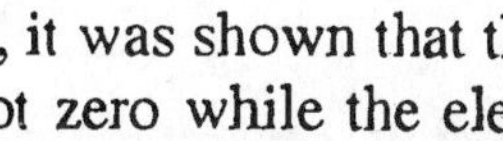

$$Q_1 + \quad \xrightarrow{\quad\quad} \begin{array}{l} E_1 \\ E_2 \end{array} - Q_2$$

Electric potential is a scalar quantity. At the midpoint between the two charges the magnitude of the potential due to the positive charge equals the potential due to the negative charge. However, one potential is positive and the other is negative and the arithmetic sum of the two potentials equals zero.

In the answer to question 5, it was shown that the electric potential at the midpoint between two positive point charges is not zero while the electric field at the midpoint equals zero. The reason is that electric field is a vector quantity while electric potential is a scalar quantity.

Part a. Step 1. Draw a diagram locating the four charges. Determine the distance from each charge to the center of the square.	Solution: (17-5) $r = [(0.0250 \text{ m})^2 + (0.0250 \text{ m})^2]^{1/2}$ $r = 0.0350 \text{ m}$

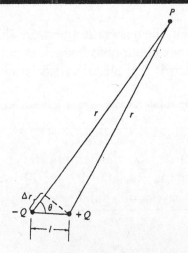

Part a. Step 2. Determine the magnitude of the potential due to one of the charges.	Each charge has the same magnitude and the distance from p is the same. The potential due to one of the charges is given by $V = kq/r$ $V_1 = (9.0 \times 10^9 \text{ Nm}^2/\text{C}^2)(3.00 \times 10^{-6} \text{ C})/(0.0350 \text{ m})$ $V_1 = 7.70 \times 10^5 \text{ volts}$
Part a. Step 3. Determine the total potential of the four charges.	Potential is a scalar quantity. The total potential equals the arithmetic sum of the potential due to the individual charges. $V_{total} = V_1 + V_2 + V_3 + V_4$ $V_{total} = (4) \ k \ q/r = 4 \ V_1$ $V_{total} = 4 \ V_1 = 3.10 \times 10^6 \text{ volts}$
Part b. Step 1. Assume that charges 3 and 4 are negative. Determine V_{total}.	$V_1 = V_2 = 7.70 \times 10^5 \text{ volts}$ $V_3 = V_4 = - 7.70 \times 10^5 \text{ volts}$ $V_{total} = V_1 + V_2 + V_3 + V_4 = 0$

Note: Although the potential at point p is zero, the electric field strength does not equal zero. It is left to the student to show that if charges 1 and 4 are positive and 2 and 3 are negative, then both the potential and the electric field strength equal zero.

Electric Dipole

Two equal point charges (q), of opposite sign, separated by a distance ℓ, are called an **electric dipole**. If $r \gg \ell$, then the potential at point P due to the dipole is given by

$V = k q \ell \cos \theta/r^2$

The product $q\ell$ is the **dipole moment** (p) and the potential can be written as

$V = k p \cos \theta/r^2$

The SI unit of the dipole moment (p) is the **debye**, where 1 debye = 3.33 x 10^{-33} C m. In polar molecules, such as water, the molecule is electrically neutral but there is a separation of charge in the molecule. Such molecules have a net dipole moment.

Capacitance and Dielectrics

A **capacitor** stores electric charge and consists of two conductors separated by an insulator known as a dielectric. The ability of a capacitor to store electric charge is referred to as **capacitance** (C) and is found by the following equation:

$C = Q/V$

Q is the charge stored in coulombs and V is the potential difference between the conducting surfaces in volts.

The SI unit of capacitance is the farad (F), where 1 farad = 1 coulomb/volt (1 F = 1 C/V). Typical capacitors have values which range from 1 picofarad (1 pF) to 1 microfarad (1 μF) where

1 pF = 1 x 10^{-12} F and 1 μF = 1 x 10^{-6} F

The capacitance of a capacitor depends on the physical characteristics of the capacitor as well as the insulating material which separates the conducting surfaces which store the electric charge. For a parallel plate capacitor, the capacitance is given by

$C = K \epsilon_o A/d$ or $C = \epsilon A/d$ where $\epsilon = K \epsilon_o$ and $\epsilon_o = 8.85$ x 10^{-12} C^2/N m^2

K is the dielectric constant of the insulating material between the plates. The constant is dimensionless and depends on the material. For dry air at 20°C, the constant is 1.0006. ϵ_o is the permittivity of free space. A is the surface area of one side of one plate which is opposed by an equal area of the other plate, and d is the distance between the plates. ϵ is the permittivity of the material between the plates.

Energy Stored in a Capacitor

A charged capacitor stores energy. The electric energy stored in charging a capacitor from an uncharged condition to a charge of Q and a potential difference V is given by

energy = ½ Q V = ½ C V^2 = ½ Q^2/C

The energy stored by a parallel plate capacitor is related to the electric field existing between the plates:

$C = \epsilon_o A/d$ and $V = E d$, then energy = ½ C V^2 = ½ ϵ_o E^2 A d

The product of the area (A) and the distance between the plates (d) equals the volume between the conducting surfaces. The **energy density** is the energy stored per unit volume:

energy density = $\frac{1}{2} \epsilon_o E^2$ If a dielectric is present, then ϵ_o is replaced by ϵ.

EXAMPLE PROBLEM 4. A parallel plate capacitor is rated at 500 pF. The potential difference between the plates is 100 volts. Determine the a) charge on each plate and b) energy stored.

Part a. Step 1.	Solution: (Sections 17-7 and 17-9)
Determine the charge on each plate. Note: 1 pf = 10^{-12} F	$C = Q/V$ and $Q = CV$ $Q = (500 \times 10^{-12}$ F$)(100$ volts$) = 5.00 \times 10^{-8}$ C
Part b. Step 1.	energy = $\frac{1}{2} C V^2 = \frac{1}{2} (500 \times 10^{-12}$ F$)(100$ V$)^2$
Determine the energy stored on the capacitor.	energy = 2.50×10^{-6} J

EXAMPLE PROBLEM 5. A 12.0-V battery is connected to a parallel plate capacitor which is rated at 1.00 μF. The plates of the capacitor are separated by 0.00600 m of dry air. Determine the a) charge on each plate and, b) magnitude of the electric field between the plates. c) After the battery is disconnected, a sheet of paraffin 0.00600 m thick and dielectric constant 2.20 is slipped between the plates. Determine the electric field strength and the magnitude of the new capacitance.

Part a. Step 1.	Solution: (Sections 17-7 and 17-8)
Determine the charge stored on each plate.	$Q = C V = (1.00 \times 10^{-6}$ F$)(12.0$ V$) = 1.20 \times 10^{-5}$ C
Part b. Step 1.	$E = V/d = (12.0$ volts$)/(0.00600$ m$)$
Determine the electric field between the plates.	$E = 2.00 \times 10^3$ V/m or 2.00×10^3 N/C
Part c. Step 1. Note: the battery has been disconnected. Determine the new voltage and then determine the the electric field.	Since the battery is disconnected before the dielectric is inserted, the quantity of charge on the plates remains the same. The electric field decreases by a factor equal to the dielectric constant of the paraffin. Since the electric field strength decreases, the potential potential difference between the plates also decreases. $V = V_o/K = (12.0$ volts$)/(2.20) = 5.45$ volts $E = V/d = (5.45$ volts$)/(0.00600$ m$) = 909$ V/m

17-12

Part c. Step 2.	The capacitance increases as a result of the insertion of the dielectric.
Determine the magnitude of the new capacitance.	$C = Q/V = (1.20 \times 10^{-5}\ C)/(5.45\ V) = 2.20 \times 10^{-6}\ F = 2.2\ \mu F$
	or $C = K\ C_o = (2.20)(1.00\ \mu F) = 2.20\ \mu F$

PROBLEM SOLVING SKILLS

For the problems involving a charged particle accelerated through a potential difference:

1. Draw an accurate diagram showing the motion of the particle.
2. Complete a data table.
3. If necessary, review the work-energy theorem, the concept of kinetic energy, Newton's laws of motion, and the kinematics equations.
4. Use the concepts of electric field, electric force, and electric potential along with the concepts of step 3 to solve the problem.

For problems involving the electric potential due to a point charge(s):

1. Remember that potential is a scalar quantity.
2. Solve for the potential due to each charge. The potential due to a negative charge is negative and for a positive charge the potential is positive.
3. The total potential equals the arithmetic sum of the potentials due to the individual point charges.

For problems involving a parallel plate capacitor based on the physical characteristics of the capacitor and the dielectric constant of the material between the plates:

1. Note the value of the dielectric constant.
2. Determine the area of one plate and the distance between the plates.
3. Apply the formula for the capacitance of a parallel plate capacitor.

For problems involving capacitance when a dielectric inserted after the initial conditions are described:

1. If the battery is disconnected before the dieletric is inserted, the charge stored cannot increase. However, the magnitude of both the electric field and potential difference decrease after the dielectric is inserted.
2. If the battery remains connected, the charge stored increases. The magnitude of both the final electric field and the potential difference between the plates equals the value before the dielectric was inserted.
3. In each case, the final capacitance is greater than the original.

SOLUTIONS TO SELECTED TEXTBOOK PROBLEMS

> **TEXTBOOK PROBLEM 12.** What is the speed of a proton whose kinetic energy is 3.2 keV?

Part a. Step 1. Convert keV to joules.	Solution: (Sections 17-1 to 17-4) $(3.2 \text{ keV})[(1 \times 10^3 \text{ eV})/(1 \text{ keV})][(1.6 \times 10^{-19} \text{ J})/(1 \text{ eV})] =$ $5.1 \times 10^{-16} \text{ J}$
Part a. Step 2. Use the formula for the kinetic energy to determine the proton's speed.	$KE_f = \frac{1}{2} m v^2$ $5.1 \times 10^{-16} \text{ J} = \frac{1}{2}(1.67 \times 10^{-27} \text{ kg}) v_f^2$ $v_f = 7.8 \times 10^5 \text{ m/s}$

> **TEXTBOOK PROBLEM 14.** What is the electric potential 15.0 cm from a 4.00-μC point charge?

Part a. Step 1. Use the formula for the electric potential a distance r from a single point charge.	Solution: (Section 17-5) Assume that the potential at infinity is zero. $V = (k\ q)/r = (9.0 \times 10^9 \text{ N m}^2/\text{C}^2)(4.00 \times 10^{-6} \text{ C})/(0.15 \text{ m})$ $V = 2.4 \times 10^5 \text{ volts}$

> **TEXTBOOK PROBLEM 39.** How strong is the electric field between the plates of a 0.80-μF air-gap capacitor if they are 2.0 mm apart and each has a charge of 72 μC?

Part a. Step 1. Determine the potential difference between the plates.	Solution: (Sections 17-7 and 17-8) $C = Q/V$ $0.80\,\mu\text{F} = (72\,\mu\text{C})/V$ $V = (72\,\mu\text{C})/(0.80\,\mu\text{F}) = 90 \text{ volts}$

Part a. Step 2. Determine the electric field between the plates.	$V = E\,d$ 90 volts = E $(2.0 \times 10^{-3}\ m)$ $E = 4.5 \times 10^4\ N/C$

TEXTBOOK PROBLEM 43. What is the capacitance of a pair of circular plates with a radius of 5.0 cm separated by a 3.2 mm of mica?

Part a. Step 1. Complete a data table.	Solution: (Sections 17-7 and 17-8) $r = 5.0\ cm = 5.0 \times 10^{-2}\ m$ $d = 3.2\ mm = 3.2 \times 10^{-3}\ m$ $\epsilon_o = 8.85 \times 10^{-12}\ C^2/N\ m^2$ $K_{mica} = 7.0$
Part a. Step 2. Determine the area of one plate.	$A = \pi\ r^2 = \pi(5.0 \times 10^{-2}\ m)^2$ $A = 7.85 \times 10^{-3}\ m^2$
Part a. Step 3. Determine the capacitance.	$C = (K\ \epsilon_o\ A)/d$ $= [(7.0)(8.85 \times 10^{-12}\ C^2/N\ m^2)(7.85 \times 10^{-3}\ m^2)]/(3.2 \times 10^{-3}\ m)$ $C = 1.5 \times 10^{-10}\ F = 150\ pF$

TEXTBOOK PROBLEM 63. A 3.4-μC and a -2.6-μC charge are placed 1.6 cm apart. At what points along the line joining them is (a) the electric field zero, and (b) the electric potential zero?

Part a. Step 1. Determine the point where the electric field equals zero.	Solution: (Sections 16-7 and 17-5) For convenience, let $Q_1 = 3.4\mu C$ and place Q_1 at x = 0. Let $Q_2 = -2.6\mu C$ and place Q_2 at x = 1.6 cm. At the point in question, the vector sum of the two electric fields equals zero. Since Q_1 is greater than Q_2, the only location that would satisfy this condition is a location on the x-axis to the right of charge Q_2.
Part a. Step 2. Draw a diagram locating each charge.	The distance from Q_2 to Q_3 is 1.6 cm, and r is the distance from Q_2 to the point (p) where the electric field equals zero. Q_1 Q_2 .p $\oplus$ $\ominus$ $E_2 \Leftarrow \oplus \Rightarrow E_1$ $\vdash\leftarrow$ 1.6 cm $\rightarrow \vdash\leftarrow$ r $\rightarrow\vdash$

Part a. Step 3. Determine the point p where the total electric field equals zero.	$E_1 = (k\, Q_1)/(1.6\ cm + r)^2$ and $E_2 = (k\, Q_2)/r^2$ $\vec{E_1} + \vec{E_2} = 0$ $\vec{E_1} = -\vec{E_2}$ $(k\, Q_1)/(1.6\ cm + r)^2 = -(k\, Q_2)/r^2$ k cancels algebraically, and rearranging $(1.6 + r)^2/r^2 = -(3.4\ \mu C)/(-2.6\ \mu C)$ $(1.6 + r)^2/r^2 = 1.31$ and $(1.6 + r)/r = \pm 1.14$ either $(1.6 + r)/r = +1.31$ or $(1.6 + r)/r = -1.3$ $1.6 + r = 1.14\ r$ $1.6 + r = -1.14\ r$ $0.14\ r = 1.6\ cm$ $-2.14\ r = 1.6\ cm$ $r = 11.4\ cm$ $r = -0.75\ cm$ The point p must be located to the right of Q_2, therefore $r = +11.4$ cm and point p is located at $x = 1.6\ cm + 11.4\ cm = 13\ cm$.
Part b. Step 1. Draw a diagram locating each charge as well as the point between the charges where V = 0.	Let p_1 represent the point between the charges where the electric potential equals zero. Let r represent the distance from Q_1 to p_1 while 1.6 cm - r represent the distance from p_1 to Q_2. Q_1 p_1 Q_2 ⊕ · ⊖ ¦← r → ¦← 1.6 cm - r →¦
Part b. Step 2. Mathematically determine the value of r.	Electric potential is a scalar quantity. At the point in question, the arithmetic sum of the two potentials equals zero. $V_T = V_1 + V_2$ $0 = (k\, Q_1)/r + (k\, Q_2)/(1.6 - r)$ $(k\, Q_1)/r = -(k\, Q_2)/(1.6 - r)$ k cancels algebraically, and rearranging $(1.6 - r)/r = -Q_2/Q_1$ but $-Q_1/Q_2 = -(+3.4\ \mu C/-2.6\ \mu C) = 1.31$ $(1.6 - r)/r = 1.31$ and $1.6 - r = 1.31\ r$ $2.31\ r = 1.6$ and $r = 0.69\ cm$ Therefore, the point p_1 is located on the x-axis at $x = 0.69\ cm$.

Part b. Step 3. Draw a diagram locating each charge as well as the second point where V = 0.	Because Q_1 has a greater magnitude than Q_2, the second point (p_2) where V = 0 must be to the right of Q_2. Let s represent the distance from Q_2 to p_2 while 1.6 cm + s represents the distance from Q_1 to p_2. Q_1 Q_2 p_2 ⊕ ⊖ · ¦← 1.6 m →¦← s →¦
Part b. Step 4. Mathematically determine the position to the right of Q_2 where V = 0.	$V_T = V_1 + V_2$ $0 = (k\ Q_1)/(1.6 + s) + (k\ Q_2)/s$ $-(k\ Q_1)/(1.6 + s) = (k\ Q_2)/s$ k cancels algebraically, and rearranging $(1.6 + s)/s = -Q_1/Q_2$ but $-Q_1/Q_2 = -(+3.4\ \mu C/-2.6\ \mu C) = 1.31$ $(1.6 + s)/s = 1.31$ and $1.6 + s = 1.31\ s$ $0.31\ s = 1.6$ $s = 5.16$ cm Therefore, the point p_2 is located on the x-axis at x = 1.6 cm + 5.16 cm = 6.8 cm.

TEXTBOOK PROBLEM 64. A 2600-pF air-gap capacitor is connected to a 9.0-V battery. If a piece of Pyrex glass is placed between the plates, how much charge will then flow from the battery?

Part a. Step 1. Determine the charge stored before the glass is inserted.	Solution: (Sections 17-7 and 17-8) $Q_i = C\ V = (2600 \times 10^{-12}\ F)(9.0\ V)$ $Q_i = 2.34 \times 10^{-8}\ C$
Part a. Step 2. Determine the capacitance with the dielectric inserted. Note: $K_{pyrex} = 5.0$.	$C_f = K\ \epsilon_o\ A/d$ but $\epsilon_o\ A/d = 2600\ pF$ $C_f = (5.0)(2600\ pF) = 13000\ pF$

Part a. Step 3.

Determine the charge stored on the plates after the dielectric is inserted and the increase (ΔQ) in the charge stored.

Since the battery remains connected to the capacitor, the potential difference between the plates stays at 9.0 V. However, because of the dielectric, the amount of charge stored on the plates will increase.

$$Q_f = C \, V = (13000 \times 10^{-12} \text{ F})(9.0 \text{ V})$$

$$Q_f = 1.17 \times 10^{-7} \text{ C}$$

$$\Delta Q = Q_f - Q_i = 1.17 \times 10^{-7} \text{ C} - 2.34 \times 10^{-8} \text{ C}$$

$$\Delta Q = 9.4 \times 10^{-8} \text{ C}$$

CHAPTER 18

ELECTRIC CURRENTS

OBJECTIVES

After studying the material of this chapter, the student should be able to:

- explain how a simple battery can produce an electrical current.
- define current, ampere, emf, voltage, resistance, resistivity, and temperature coefficient of resistance.
- write the symbols used for electromotive force, electric current, resistance, resistivity, temperature coefficient of resistance, and power and state unit associated with each quantity.
- distinguish between a) conventional current and electron current and b) direct current and alternating current.
- know the symbols used to represent a source of emf, resistor, voltmeter, and ammeter and how to interpret a simple circuit diagram.
- given the length, cross-sectional area, resistivity, and temperature coefficient of resistance, determine a wire's resistance at room temperature and at some higher or lower temperature.
- solve simple dc circuit problems using Ohm's law.
- use the equations for electric power to determine the power and energy dissipated in a resistor and calculate the cost of this energy to the consumer.
- distinguish between the rms and peak values for current and voltage and apply these concepts in solving problems involving a simple ac circuit.

KEY TERMS AND PHRASES

source of emf is a device which transforms one form of energy, chemical, mechanical, etc., to electric energy. Examples of sources of emf are a chemical battery and an electric generator.

emf refers to electromotive force. Emf is a measure of the potential difference across the source of voltage.

electric current is the rate of flow of electric charge. The magnitude of the current is measured in **ampere**s (I), where 1 ampere = 1 coulomb/second.

conventional current is the direction of positive charge flow. Both positive and negative ions move through gases and liquids. Only negative charges, i.e., electrons, move through solids and this is referred to as **electron current**. For historical reasons, conventional current is used in referring to the direction of electric charge flow.

electrical resistance refers to the opposition offered by a substance to the flow of electrical current. The resistance of a metal conductor is a property which depends on its dimensions, material, and temperature. The unit of resistance is the ohm (Ω).

Ohm's law states that magnitude of the electric current that flows through a closed circuit depends directly on the voltage between the terminals of the source of emf and inversely to the electrical resistance.

electric power is the rate at which work is done to maintain an electric current in a circuit. The SI unit of power is the watt (W), where 1 W = 1 J/s. The kilowatt is a commonly used unit where 1 kilowatt = 1000 watts.

direct current (dc) refers to a current which flows in one direction only.

alternating current (ac) refers to a current where the direction of current flow through the circuit changes usually at a particular frequency (f). The frequency used in the United States is 60 cycles per second or 60 Hz.

root mean square current (I_{rms}) is the square root of the mean of the square of the current flowing through a circuit.

SUMMARY OF MATHEMATICAL FORMULAS

electric current	$I = Q/t$	Electric current is measured in amperes (I), where 1 ampere = 1 coulomb/second and $I = Q/t$.
electrical resistance	$R = \rho\, L/A$	At a specific temperature, the resistance (R) of a metal wire is related to the length (L), cross-sectional area (A), and a constant of proportionality called the **resistivity** (ρ).
	$\rho_T = \rho_o(1 + \alpha\, T)$	As the temperature of a conductor changes, both the resistivity and resistance of a conductor change. ρ_T and R_T are the values of the resistivity and resistance at temperature T, while ρ_o and R_o are the values of the resistivity and resistance at 20°C. α is the
	$R_T = R_o(1 + \alpha\, T)$	temperature coefficient of resistance which remains constant over a certain range of temperature. The value of α depends on the material and its units are $(C°)^{-1}$.

Ohm's law	$I = V/R$ or $V = I\,R$	The magnitude of the electric current (I) that flows through a closed circuit depends directly on the voltage (V) between the battery terminals and inversely to the circuit resistance (R). The current (I) is measured in amperes, the voltage (V) in volts, and the resistance (R) in ohms.
electric power in a dc circuit	$P = I\,V$	Electric power equals the product of the current I and the potential difference V.
	$P = I^2\,R$	In a circuit of resistance R, the rate at which electrical energy (P) is converted to heat energy is referred to as joule heating. Joule heating is related to the product of the square of the current and the resistance.
	$P = V^2/R$	Since I = V/R, an alternate formula for power can be written in terms of the voltage and the resistance.
alternating current electricity	$V = V_o \sin 2\pi f t$	The emf (V) produced by an ac electric generator is sinusoidal.
	$I = I_o \sin 2\pi f t$	The current produced in a closed circuit connected to an ac generator is sinusoidal. V_o and I_o represent the peak or maximum voltage and current respectively, and t represents the time in seconds.
	$P = I_o^2\,R \sin^2 2\pi f t$	The power delivered to a resistance R at any instant (t).
	$\bar{P} = \tfrac{1}{2} I_o^2\,R$	The average power ($\bar{P}$) delivered to the resistance where I_{rms} is the root mean square current.
	$\bar{P} = (I_{rms})^2\,R$	
	$\bar{P} = I_{rms}\,V_{rms}$	where $I_{rms} = 0.707\,I_o$ and $V_{rms} = 0.707\,V_o$

CONCEPT SUMMARY

The Electric Battery

A **battery** is a source of electric energy. A simple battery contains two dissimilar metals, called **electrodes**, and a solution called the **electrolyte**, in which the electrodes are partially immersed. An example of a simple battery would be one in which zinc and carbon are used as the electrodes, while a dilute acid, such as sulfuric acid (dilute), acts as the electrolyte. The acid dissolves the zinc and causes zinc ions to leave the electrode. Each zinc ion which enters the electrolyte leaves two electrons on the zinc plate. The carbon electrode also dissolves but at a slower rate. The result is a difference in potential between the two electrodes.

The potential difference between the terminals of a battery in which no internal energy losses occur is referred to as the **electromotive force** (emf) and is measured in volts. The symbol electromotive force is ξ. The schematic symbol for a battery is ⊣⊢ .

Electric Current

An electric **current** exists whenever electric charge flows through a region, e.g., a simple light bulb circuit. The magnitude of the current is measured in **ampere**s (I), where 1 ampere = 1 coulomb/second and $I = Q/t$. The direction of **conventional current** is in the direction in which positive charge flows. In gases and liquids both positive and negative ions move. Only negative charges, i.e., electrons, move through solids and this is referred to as **electron current**. For historical reasons, conventional current is used in referring to the direction of electric charge flow.

TEXTBOOK QUESTION 1. What quantity is measured by a battery rating in ampere-hours (A h)?

ANSWER: An ampere-hour is a way of expressing the total charge the battery can supply at its rated voltage. For example, if the battery is rated at 55 A h, then $Q = I t = (55 \text{ C/s})(3600 \text{ s})$ and $Q = 198,000$ C.

Resistivity

When electric charge flows through a circuit it encounters electrical **resistance**. The resistance of a metal conductor is a property which depends on its dimensions, material, and temperature. A resistor is represented by a jagged line, e.g., ⎓⟋⟍⟋⟍⎓ .

At a specific temperature, the resistance (R) of a metal wire of length L and cross-sectional area A is given by

$$R = \rho L/A$$

ρ is a constant of proportionality called the **resistivity**. The unit of resistance is the ohm (Ω) and the unit of resistivity is ohm m. Within a certain range of temperature, the resistivity of a conductor changes according to the following equation:

$$\rho_T = \rho_o(1 + \alpha T)$$

while the resistance changes according to the equation

$$R_T = R_o(1 + \alpha T)$$

18-4

ρ_o and R_o are the values of the resistivity and resistance at 20°C while ρ_T and R_T are the values of the resistivity and resistance at temperature T, where T is measured in °C. α is the **temperature coefficient of resistance**. The value of α depends on the material and its units are $(C°)^{-1}$.

ANSWER: Resistance of a wire depends on the resistivity, length, cross-sectional area, and temperature. The resistivity of aluminum is greater than that of copper. Therefore, if the length, area, and temperature are the same for both, the aluminum wire would have greater resistance. However, an aluminum wire of larger cross-sectional area could be made so that the resistance is the same. Also, it would be possible to keep the aluminum wire at room temperature but raise the temperature of the copper wire until the resistance is the same as that of the aluminum wire.

EXAMPLE PROBLEM 1. a) Determine the electrical resistance of a 20.0 m length of tungsten wire of radius 0.200 mm at 20°C. b) If the temperature of the wire does not change, determine the resistance of the same wire if it is stretched to a length of 60.0 m. The resistivity of tungsten is 5.60×10^{-8} Ω m.

Part a. Step 1.	Solution: (Section 18-4)
Complete a data table.	$A = \pi r^2 = \pi (0.200 \text{ mm})^2 (1.0 \text{ m}/1000 \text{ mm})^2 = 1.26 \times 10^{-7} \text{ m}^2$
	$L = 20.0 \text{ m}, \quad \rho = 5.60 \times 10^{-8} \text{ } \Omega \text{ m}$
Part a. Step 2.	$R = \rho L/A$
Determine the resistance of 20.0 m of wire.	$= (5.60 \times 10^{-8} \text{ } \Omega\text{m})(20.0 \text{ m})/(1.26 \times 10^{-7} \text{ m}^2)$
	$R = 8.89 \text{ } \Omega$
Part b. Step 1. Determine the area of the wire when it is stretched to 60.0 m.	Stretching the wire will affect the wire's length and cross-sectional area. However, the volume (V) of the wire must remain the same. The volume (V) equals the product of the length (L) and cross-sectional area (A).
	$V_f = V_o$
	$A_o L_o = A_f L_f$
	$(1.26 \times 10^{-7} \text{ m}^2)(20.0 \text{ m}) = A_f (60.0 \text{ m})$
	$A_f = 4.20 \times 10^{-8} \text{ m}^2$

Part b. Step 2.	$R_f = \rho \, L/A$
Determine the resistance of 60.0 m of wire.	$= (5.60 \times 10^{-8} \ \Omega m)(60.0 \ m)/(4.20 \times 10^{-8} \ m^2)$
	$R_f = 80.0 \ \Omega$

EXAMPLE PROBLEM 2. a) The resistance of the tungsten wire is 80.0 Ω at 20.0°C. Determine the resistance of the same wire if the temperature is raised to 70.0°C. b) The resistance of a carbon wire is 80.0 Ω at 20.0°C. Determine the resistance of the wire at 70.0°C. Explain why the resistance of each substance changes as it does with the increasing temperature. Note: the temperature coefficient of resistance is 0.0045/C° for tungsten and -0.00050/C° for carbon.

Part a. Step 1.	Solution: (Section 18-4)
Determine the resistance of tungsten at 70.0°C.	$R = R_o \, (1 + \alpha \, T)$
	$= 80.0 \ \Omega \ [1 + (0.0045/C°)(70.0°C - 20.0°C)]$
	$R = 80.0 \ \Omega \ (1 + 0.225) = 80.0 \ \Omega \ (1.23) = 98.0 \ \Omega$
Part a. Step 2. Explain why the resistance of tungsten increases with temperature.	Tungsten is a metal and its resistance increases with temperature. At room temperature the outer electron is free to move throughout the metal. As the temperature increases, the atoms are vibrating more rapidly. The electric field of the individual atoms has a greater probability of interfering with the electrons as they move through the metal, therefore, the resistance increases.
Part b. Step 1. Determine the resistance of carbon at 70.0°C.	$R = R_o \, (1 + \alpha \, T)$
	$= 80.0 \ \Omega \ [1 + -0.00050/C°(70.0°C - 20.0°C)]$
	$= 80.0 \ \Omega \ (1 + -0.025) = 80.0 \ \Omega \ (0.975)$
	$R = 78.0 \ \Omega$
Part b. Step 2. Explain why the resistance of carbon decreases as the temperature increases.	Carbon is a semiconductor and the increase in temperature causes more electrons to break free and become part of the electron current. Because of the increased number of free electrons, the resistance of carbon wire to the flow of current decreases.

Ohm's Law

The magnitude of the electric current that flows through a closed circuit depends directly on the voltage between the battery terminals and inversely to the circuit resistance. The relationship that connects current, voltage, and resistance is known as **Ohm's law** and is written as follows:

$I = V/R$ or $V = IR$

The current (I) is measured in amperes, the voltage (V) in volts, and the resistance (R) in ohms.

The following is the schematic for a simple direct current circuit, e.g., a flashlight circuit.

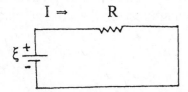

where ξ represents the battery voltage and R represents the electrical resistance of the filament of the light bulb. I is the current. The arrow represents the direction of conventional current. Conventional current is directed away from the positive side of the battery.

TEXTBOOK QUESTION 11. Explain why lightbulbs almost always burn out just as they are turned on and not after they have been on for some time.

ANSWER: Tungsten is a metal and its length increases with temperature. Also, the electrical resistance of a cold tungsten filament is much lower than that of a hot filament. When a tungsten filament light bulb is first turned on, its resistance is low but the current flowing through the bulb is high ($I = V/R$). As a result of the high current and the sudden increase in its length, the rate of conversion of electrical energy ($P = I^2 R$) to heat energy is greatest, the stress on the filament is greatest and the light is more likely to burn out.

Electric Power

Work is required to transfer charge through an electric circuit. The work required depends on the amount of charge transferred through the circuit and the potential difference between the terminals of the battery: $W = QV$. The rate at which work is done to maintain an electric current in a circuit is termed **electric power**. Electric power equals the product of the current I and the potential difference V, i.e., $P = IV$. The SI unit of power is the watt (W), where 1 W = 1 J/s. The kilowatt is a commonly used unit where 1 kilowatt = 1000 watts.

The electric energy produced by the source of emf is dissipated in the circuit in the form of heat. The kilowatt hour (kWh) is commonly used to represent electric energy production and consumption where $1 \text{ kWh} = 3.6 \times 10^6$ J.

In a circuit of resistance R, the rate at which electrical energy is converted to heat energy is given by

$P = I V$ but $V = IR$, then $P = I(IR) = I^2 R$ Note: $I^2 R$ is known as **joule heating**.

Since $I = V/R$, an alternate formula for power can be written as follows:

$P = IV = (V/R)V = V^2/R$

Alternating Current

In a **direct current** (dc) circuit the current flows in one direction only. In an **alternating current** (ac) circuit the direction of current flow through the circuit changes at a particular frequency (f). The frequency used in the United States is 60 cycles per second or 60 Hz.

The emf produced by an ac **electric generator** is **sinusoidal**. The current produced in a closed circuit connected to the generator is also sinusoidal. The equations for the voltage and current are as follows:

$$V = V_o \sin 2 \pi f t \quad \text{and} \quad I = I_o \sin 2 \pi f t$$

V_o and I_o represent the peak or maximum voltage and current respectively, and t represents the time in seconds.

The power delivered to a resistance R at any instant is

$$P = I^2 R = I_o^2 R \sin^2 2 \pi f t$$

The average power delivered to the resistance is

$$\overline{P} = \tfrac{1}{2} I_o^2 R = (I_{rms})^2 R$$

I_{rms} is the **root mean square** current. This current is the square root of the mean of the square of the current. It can be shown that $I_{rms} = 0.707 \, I_o$ and $V_{rms} = 0.707 \, V_o$

A direct current whose values of I and V equal the rms values of I and V of an alternating current produce the same amount of power. In ac circuits, it is usually the rms value that is specified. For example, ordinary ac line voltage is 120 volts. The 120 volts is V_{rms}, while the peak voltage V_o is 170 volts.

TEXTBOOK QUESTION 13. Electric power is transferred over large distances at very high voltages. Explain how the high voltage reduces power loss in the transmission lines.

ANSWER: Power losses in transmission lines are the result of Joule heating $P = I^2 R$. Because of Joule heating, it is necessary to have a low current (I) in the line. Since $P = I V$, this can be accomplished by stepping up the voltage at the power plant. The high voltage is sent at low current through the transmission lines. When it reaches an electrical sub-station in your neighborhood, a step down transformer reduces the voltage and the current increases.

EXAMPLE PROBLEM 3. It is estimated that the typical American child watches 31 hours of television per week. If the average television draws 1.00 ampere of current from a 120-volt line, determine the a) power rating of the television in watts, b) number of kilowatt-hours of electrical energy consumed in one year during the time that the child watches the TV and c) yearly cost to the parents if electricity costs 8.00 cents per kilowatt-hour.

Part a. Step 1. Determine the power rating of the TV.	Solution: (Section 18-6) P = I V = (1.00 ampere)(120 volt) = 120 watts
Part b. Step 1. Determine the total number of kWh used in one year.	W = P t = (120 watts)(1.0 KW/1000 watts)(31 h/1 week)(52 weeks/1 y) W = 193 kWh
Part c. Step 1. Calculate the yearly cost.	cost = (193 kWh)(8.00 cents/1.0 kWh) cost = 1550 cents or \$15.50

EXAMPLE PROBLEM 4. A 10.0-Ω resistor is connected to a 120-volt ac line. Determine the a) average power dissipated in the resistor, b) rms current through the resistor, c) maximum instantaneous current through the resistor, and d) maximum instantaneous power dissipated in the resistor.

Part a. Step 1. Determine the average power dissipated in the resistor.	Solution: (Section 18-8) $\overline{P}$ = I V where I and V refer to the rms values of current and voltage. Since I = V/R, $\overline{P}$ = V^2/R $\overline{P}$ = (120 volts)^2/(10.0 Ω) = 1400 watts
Part b. Step 1. Determine I_{rms}.	I_{rms} = V_{rms} /R = (120 volts)/(10.0 Ω) I_{rms} = 12.0 amp
Part c. Step 1. Determine the maximum instantaneous current that flows through the resistor.	I_{rms} = 0.707 I_o, where I_o is the maximum instantaneous (peak) current that flows through the resistor, therefore, I_o = I_{rms} /0.707 I_o = (12.0 amp)/(0.707) = 17.0 amp
Part d. Step 1. Determine the maximum instantaneous power dissipated in the resistor.	P = I_o^2 R $\sin^2 2\pi$ f t Note: maximum value of $\sin^2 2\pi$ f t = 1.0 P_{max} = I_o^2 R P_{max} = (17.0 amp)^2(10.0 Ω) = 2890 watts

PROBLEM SOLVING SKILLS

For problems involving resistivity and temperature coefficient of resistance:

1. Complete a data table listing information both given and implied. For example, the resistivity, length, cross-sectional area, temperature coefficient of resistance, and the change in temperature.
2. Solve for the resistance and the resistance at some higher temperature.

For problems involving electric power, electric energy, and the cost of electric energy:

1. Complete a data table listing information both given and implied. For example, the current, voltage, resistance, and time the device is used.
2. Determine the power dissipated in watts and kilowatts.
3. Determine the number of KWh of energy used and multiply the number of KWh by the cost per KWh.

For problems involving alternating current:

1. Complete a data table listing information both given and implied. For example, include the rms current and voltage, peak current and voltage, and the resistance.
2. Use the data table to solve for the average and instantaneous power dissipated in the resistor.

SOLUTIONS TO SELECTED TEXTBOOK PROBLEMS

TEXTBOOK PROBLEM 8. A 9.0-V battery is connected to a bulb whose resistance is 1.6 Ω. How many electrons leave the battery per minute?

Part a. Step 1.	Solution: (Sections 18-2 and 18-3)
Determine the current through the bulb.	$I = V/R = (9.0 \text{ V})/(1.6 \ \Omega)$ $I = 5.63 \text{ A} = 5.63 \text{ C/s}$
Part a. Step 2.	$Q = I t$
Determine the number of electrons.	$= (5.63 \text{ C/s})[(6.25 \times 10^{18} \text{ electrons})/(1 \text{ C})][(60 \text{ s})/(1 \text{ min})]$ $Q = 2.1 \times 10^{21} \text{ electrons}$

TEXTBOOK PROBLEM 19. A 100-W light bulb has a resistance of about 12 Ω when cold (20°C) and 140 Ω when on (hot). Estimate the temperature of the filament when hot assuming an average temperature coefficient of resistivity $\alpha = 0.0060(C°)^{-1}$.

Part a. Step 1.	Solution: (Section 18-4)
Determine the change in the filament's temperature.	$R = R_o (1 + \alpha \, \Delta T)$
	$140 \, \Omega = (12 \, \Omega) \, [1 + 0.0060(C°)^{-1} \, \Delta T]$
	$1 + 0.0060(C°)^{-1} \, \Delta T = (140 \, \Omega)/(12 \, \Omega)$
	$\Delta T = (11.7 - 1)/[0.0060(C°)^{-1}]$
	$\Delta T = 1780 \, C°$
Part a. Step 2.	$\Delta T = T_f - T_i$
Estimate the filament's final temperature.	$1780 \, C° = T_f - 20°C$
	$T_f \approx 1800°C$

TEXTBOOK PROBLEM 33. How many kWh of energy does a 550-W toaster use in the morning if it is in operation for a total of 15 minutes? At a cost of 9 cents/kWh, estimate how much this would add to your monthly electric energy bill if you made toast four mornings per week.

Part a. Step 1.	Solution: (Section 18-6)
Determine the number of kWh used each morning.	$W = P \, t$
	$= (550 \text{ watts})(1.0 \text{ kW}/1000 \text{ watts})(15 \text{ min/day})(1 \text{ hour}/60 \text{ min})$
	$W = 0.14 \text{ kWh}.$
Part a. Step 2.	Assume an average month of 30 days.
Determine the number of kWh used each month.	$W = (0.14 \text{ kWh})(4 \text{ mornings/week})(4 \text{ weeks})$
	$W = 2.2 \text{ kWh}$
Part a. Step 3.	$\text{cost} = (2.2 \text{ kWh})[(9 \text{ cents})/(1.0 \text{ kWh})]$
Calculate the monthly cost.	$\text{cost} = 20 \text{ cents}$

TEXTBOOK PROBLEM 41. A small immersion heater can be used to heat a cup of water for coffee or tea. If the heater can heat 120 ml of water from 25°C to 95°C in 8.0 min, (a) approximately how much current does it draw from a 12-V battery and (b) what is its resistance? Assume the manufacturer's claim of 60% efficiency.

Part a. Step 1.	Solution: (Section 18-6)
Determine the amount of heat required to raise the temperature of the system from 25°C to 95°C.	$Q = m_w \, c_w \, \Delta T_w$ $Q = (120 \text{ ml})(1 \text{ g/1 ml})(1.0 \text{ cal/g°C})(95 \text{ °C} - 25 \text{ °C})$ $Q = 8.4 \times 10^3 \text{ cal}$
Part a. Step 2. Convert the energy from calories to joules.	Heat energy is measured in calories while electrical energy is measured in joules, where $4.18 \text{ J} = 1.00$ calories. $(8.4 \times 10^3 \text{ cal})(4.18 \text{ J/cal}) = 3.5 \times 10^4 \text{ J}$
Part a. Step 3. Write a formula for the electrical energy used to heat the water.	Note: the immersion heater is only 60% efficient. (0.60) electrical energy = (60%)(power)(time) = $(0.60) \text{ I V}$ (0.60) electrical energy = $(0.60)[(I)(12 \text{ V})](8.0 \text{ min})(60 \text{ s/1 min})]$
Part a. Step 4. Use the law of conservation of energy to the determine the electric current.	(0.60) electrical energy = heat energy $(0.60)[(I)(12 \text{ V})](8.0 \text{ min})(60 \text{ s/1 min}) = 3.5 \times 10^4 \text{ J}$ $I = 10 \text{ A}$
Part a. Step 5. Determine the resistance of the heating coil.	$R = V/I$ $R = (12 \text{ V})/(10 \text{ A})$ $R = 1.2 \; \Omega$

TEXTBOOK PROBLEM 46. An 1800 W arc welder is connected to a 660 V_{rms} ac line. Calculate (a) the peak voltage and (b) the peak current.

Part a. Step 1. Determine the peak voltage.	Solution: (Section 18-8) $V_{rms} = 0.707 \, V_o$ where V_o represents the peak voltage 660 volts = $0.707 \, V_o$ $V_o = 930$ volts

Part a. Step 2. Determine the rms current.	$P = I_{rms} V_{rms}$ 1800 watt $= I_{rms}$ (660 volts) $I_{rms} = 2.7$ A
Part a. Step 3. Determine the peak current.	$I_{rms} = 0.707\ I_o$ where I_o represents the peak current 2.7 A $= 0.707\ I_o$ $I_o = 3.9$ A

TEXTBOOK PROBLEM 74. A 100-watt, 120-volt lightbulb has a resistance of 12 Ω when cold (20°C) and 140 Ω when on (hot). Calculate its power consumption at (a) the instant it is turned on and (b) after a few moments when it is hot.

Part a. Step 1. Calculate the initial current.	Solution: (Section 18-6) $I = V/R = (120$ V$)/(12\ \Omega) = 10$ A
Part a. Step 2. Determine the initial power.	$P = I^2 R = (10$ A$)^2(12\ \Omega) = 1200$ W
Part b. Step 1. Calculate the current when the bulb is hot.	$I = V/R = (120$ V$)/(140\ \Omega) = 0.86$ A
Part b. Step 2. Determine the power when the bulb reaches operating temperature.	$P = I^2 R = (0.86$ A$)^2(140\ \Omega) = 100$ W Note: the 100 W given in the statement of the problem is the power rating when the bulb reaches its operating temperature.

CHAPTER 19

DC CIRCUITS

OBJECTIVES

After studying the material of this chapter, the student should be able to:

- determine the equivalent resistance of resistors arranged in series or in parallel or the equivalent resistance of a series-parallel combination.
- use Ohm's law and Kirchhoff's rules to determine the current through each resistor and the voltage drop across each resistor in a single loop or multiloop dc circuit.
- distinguish between the emf and the terminal voltage of a battery and calculate the terminal voltage given the emf, internal resistance of the battery, and external resistance in the circuit.
- determine the equivalent capacitance of capacitors arranged in series or in parallel or the equivalent capacitance of a series-parallel combination.
- determine the charge on each capacitor and the voltage drop across each capacitor in a circuit where capacitors are arranged in series, parallel or a series-parallel combination.
- calculate the time constant of an RC circuit. Determine the charge on the capacitor and the potential difference across the capacitor at a particular moment of time and the current through the resistor at a particular moment in time.
- describe the basic operation of a galvanometer and calculate the resistance which must be added to convert a galvanometer into an ammeter or a voltmeter.
- describe how a slide wire potentiometer can be used to determine the emf of a source of emf. Given the emf of a standard cell, use the slide wire potentiometer to calculate the emf of the unknown.
- describe how a Wheatstone bridge circuit can be used to determine the resistance of an unknown resistor. Given three known resistors and a Wheatstone bridge circuit, calculate the resistance of an unknown resistor.

KEY TERMS AND PHRASES

series circuit is an electric circuit with only a single path for electric current to travel. The current through each circuit element is the same.

parallel circuit is an electric circuit with more than one path for electric current to travel. The current is divided among the branches of the circuit. The voltage drop is the same across each branch.

equivalent resistance is the resistance of a single resistor which is equivalent to the total resistance of a network of resistors.

emf refers to electromotive force. emf is a measure of the potential difference across the source of voltage, i.e., a battery or generator.

internal resistance refers to the resistance to electric current inside the voltage source. The internal resistance of the source of emf is always considered to be in a series with the external resistance present in the electric circuit.

terminal voltage is the potential difference available to the circuit outside the source of emf. The terminal voltage equals the difference between the emf and the voltage drop across the internal resistance.

Kirchhoff's first rule or **junction rule** states that the sum of all currents entering any junction point equals the sum of all currents leaving the junction point. This rule is based on the law of conservation of electric charge.

Kirchhoff's second rule or **loop rule** states that the algebraic sum of all the gains and losses of potential around any closed path must equal zero. This law is based on the law of conservation of energy.

RC circuit consists of a resistor and a capacitor connected in series to a dc power source.

time constant (τ) of an RC circuit equals the product of the resistance and the capacitance and is measured in seconds.

galvanometer consists of a moving coil placed in a magnetic field. When an electrical current flows through the galvanometer the interaction of the current in the coil and the magnetic field causes the coil to deflect. The galvanometer is used to detect and measure very low currents, usually in the range of 1 milliampere or less.

ammeter measures the amount of electric current passing a particular point in a circuit. The ammeter is placed in series in the circuit and consists of a galvanometer and a resistor of very low value, called the **shunt resistor**, placed in parallel with the galvanometer.

voltmeter measures the potential difference across a circuit element, e.g., a resistor. A voltmeter consists of a galvanometer of internal resistance r and a resistor of high resistance R placed in series with the galvanometer.

potentiometer circuit is used to give precise measurements of the voltage of the source of emf.

Wheatstone bridge is a device used to give precise measurements of electrical resistance.

SUMMARY OF MATHEMATICAL FORMULAS

resistors arranged in series	$I = I_1 = I_2 = ... = I_n$	The current (I) at every point in a series circuit equals the current leaving the battery.
	$V = V_1 + V_2 + ... + V_n$	The potential difference between the terminals of the battery (V) equals the sum of the potential differences across the resistors.
	$R = R_1 + R_2 + ... + R_n$	The equivalent electrical resistance (R) for a series combination equals the sum of the individual resistors.
resistors arranged in parallel	$I = I_1 + I_2 + ... + I_n$	The battery current (I) equals the sum of the currents in the branches.
	$V = V_1 = V_2 = ... = V_n$	The potential difference across each resistor in the arrangement is the same.
	$1/R = 1/R_1 + 1/R_2 + .. + 1/R_n$	The equivalent resistance (R) of a parallel combination is always less than the smallest of the individual resistors.
terminal voltage of a source of emf	$V = \xi - Ir$	The terminal voltage (V) equals the difference between the emf of the source (ξ) and the drop in potential due to internal resistance (I r).
Kirchhoff's first rule or junction rule	$\Sigma I = 0$	Kirchhoff's first rule or junction rule states that the sum of all currents entering any junction point equals the sum of all currents leaving the junction point. This rule is based on the law of conservation of electric charge.
Kirchhoff's second rule or loop rule	$\Sigma V = 0$	Kirchhoff's second rule or loop rule: The algebraic sum of all the gains and losses of potential around any closed path must equal zero. This law is based on the law of conservation of energy.

capacitors arranged in parallel	$V = V_1 = V_2 = ... = V_n$	The potential difference across each capacitor equals the potential difference (V) of the source of emf.
	$Q = Q_1 + Q_2 + ... + Q_n$	The total charge stored on the capacitor plates (Q) equals the sum of the charges stored on the individual capacitors.
	$C = C_1 + C_2 + ... + C_n$	The equivalent capacitance equals the sum of the individual capacitors.
capacitors arranged in series	$Q = Q_1 = Q_2 = ... = Q_n$	The charge (Q) that leaves the source of emf equals the charge that forms on each capacitor.
	$V = V_1 + V_2 + ... + V_n$	The potential difference across the source of emf (V) equals the sum of the potential differences across the capacitors.
	$1/C = 1/C_1 + 1/C_2 + .. + 1/C_n$	The equivalent capacitance (C) of a parallel combination is always less than the smallest of the individual capacitors.
charging an RC circuit	$I = I_o\, e^{-t/RC}$	The current I in the circuit at time t after the switch is closed depends on the initial current (I_o), the resistance (R), and the capacitance (C).
	$V = \xi(1 - e^{-t/RC})$	The potential difference (V) across the capacitor as it is being charged.
	$Q_t = Q\,(1 - e^{-t/RC})$	The amount of charge (Q_t) accumulated on the capacitor as it is being charged.

discharging an RC circuit	$I = I_o\, e^{-t/R'C}$	the current in an RC discharge circuit as a function of time
	$V = V_o\, e^{-t/R'C}$	the voltage across the resistance (R') in an RC discharge circuit as a function of time
	$Q_t = Q\, e^{-t/R'C}$	the charge (Q_t) on a capacitor in an RC discharge circuit as a function of time

CONCEPT SUMMARY

Resistors in Series and Parallel

A simple dc circuit may contain resistors arranged in series or in parallel or in a series-parallel combination. A simple **series circuit** is shown in the diagram. The current (I) at every point in a series circuit equals the current leaving the battery.

$$I = I_1 = I_2 = I_3 = \ldots = I_n$$

Assuming that the connecting wires offer no resistance to current flow, the potential difference between the terminals of the battery (V) equals the sum of the potential differences across the resistors, i.e.,

$$V = V_1 + V_2 + V_3 + \ldots + V_n$$

The equivalent electrical resistance (R) for this combination is equal to the sum of the individual resistors, i.e.,

$$R = R_1 + R_2 + R_3 + \ldots + R_n$$

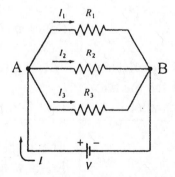

The diagram shown at the right represents a simple **parallel circuit**. The current leaving the battery divides at junction point A and recombines at point B. The battery current (I) equals the sum of the currents in the branches. In general,

$$I = I_1 + I_2 + I_3 + \ldots + I_n$$

The potential difference across each resistor in the arrangement is the same, i.e., $V = V_1 = V_2 = V_3 = \ldots = V_n$. If no other resistance is present, the potential difference across each resistor equals the potential difference across the terminals of the battery.

The equivalent resistance (R) of a parallel combination is always less than the smallest of the individual resistors. The formula for the equivalent resistance is as follows:

$$1/R = 1/R_1 + 1/R_2 + 1/R_3 + \ldots + 1/R_n$$

TEXTBOOK QUESTION 2. Discuss the advantages and disadvantages of Christmas tree lights connected in parallel versus those connected in series.

ANSWER: Lights connected in parallel operate independently from one another. If one light burns out, the remaining lights stay on. If the lights are connected in series, the current through one must pass through all of the other lights. As a result, in a series connection, if one light burns out then they all go out until a replacement bulb is inserted.

TEXTBOOK QUESTION 3. If all you have is a 120-V line, would it be possible to light several 6-V lamps without burning them out? How?

ANSWER: If the lamps are connected in series, the sum of the potential differences across the lamps would be 120 V. If the maximum voltage across each lamp is to be 6 V, then it would be necessary to connect 120 V/6 V = 20 lamps in series.

EXAMPLE PROBLEM 1. Given $R_1 = 2.0\ \Omega$, $R_2 = 6.0\ \Omega$, and $R_3 = 12.0\ \Omega$. Determine the a) equivalent resistance of the following arrangement, b) total current leaving the battery, c) potential drop across each resistor, and d) current through each resistor.

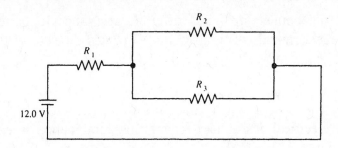

Part a. Step 1.	Solution: (Section 19-1)
Determine the equivalent resistance of the the two resistors in parallel.	$1/R_\| = 1/(6.0\ \Omega) + 1/(12.0\ \Omega) = (2 + 1)/(12.0\ \Omega)$ $R_\| = (12.0\Omega)/3 = 4.0\ \Omega$

Part a. Step 2.

Determine the total resistance.

The parallel combination is in series with the 2.0-Ω resistor. Re-draw the circuit showing the 2.0 Ω and $R_\|$ and solve for R_T.

$R_T = 2.0\ \Omega + 4.0\ \Omega = 6.0\ \Omega$

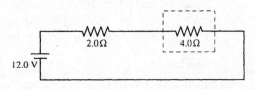

The dotted line drawn about the 4.0-Ω resistor indicates that this resistor is equivalent to the parallel combination.

Part b. Step 1. Determine the current leaving the battery.	Using Ohm's law: $I_T = V/R_T = 12.0 \text{ V}/6.0 \text{ }\Omega = 2.0 \text{ A}$
Part c. Step 1. Determine the drop in potential across each resistor.	The current which leaves the battery passes through the 2.0-Ω resistor. The potential drop across the 2.0-Ω resistor is $V_1 = I_T R_1 = (2.0 \text{ A})(2.0 \text{ }\Omega) = 4.0 \text{ volts}$ The potential drop across the parallel combination can be determined by the product of the total current that enters the combination at the junction point and the equivalent resistance of the parallel combination. $V_\parallel = (2.0 \text{ A})(4.0 \text{ }\Omega) = 8.0 \text{ volts}$
Part d. Step 1. Determine the current in each resistor.	It was shown in part b that the current in the 2.0-Ω resistor is 2.0 A. The potential difference across the 6.0-Ω and 12.0-Ω resistors is 8.0 V. Ohm's law is used to determine each current. $I_2 = 8.0 \text{ V}/6.0 \text{ }\Omega = 1.3 \text{ A}$ $I_3 = 8.0 \text{ V}/12.0 \text{ }\Omega = 0.67 \text{ A}$

TEXTBOOK QUESTION 16. Why is it more dangerous to turn on an electric appliance when you are standing outside in bare feet than when you are inside wearing shoes with thick soles?

ANSWER: The soles of your shoes are made of a material which has high resistance. If the resistance is high then the current flow through your body will be low. The resistance of bare feet to current flow is significantly lower and as a result a much larger current can flow through your body.

Emf and Terminal Voltage

All sources of emf have what is known as **internal resistance** (r) to the flow of electric current. The internal resistance of a fresh battery is usually small but increases with use. Thus the voltage across the terminals of a battery is less than the emf of the battery. The **terminal voltage** (V) is given by the equation $V = \xi - Ir$, where ξ represents the emf of the source of potential in volts, I the current leaving the source of emf in amperes and r the internal resistance in ohms. The internal resistance of the source of emf is always considered to be in a series with the external resistance present in the electric circuit.

Kirchhoff's Rules

Kirchhoff's rules are used in conjunction with Ohm's law in solving problems involving complex circuits:

Kirchhoff's first rule or junction rule: the sum of all currents entering any junction point equals the sum of all currents leaving the junction point. This rule is based on the law of conservation of electric charge.

Kirchhoff's second rule or loop rule: the algebraic sum of all the gains and losses of potential around any closed path must equal zero. This law is based on the law of conservation of energy.

Suggestions for Using Kirchhoff's Laws

Diagram (a) shown below is an example of a complex circuit which can be solved using Kirchhoff's rules.

Step 1. As shown in diagram (b), assign a direction to the current in each independent branch of the circuit. Assume that the current flows away from the positive side of the battery and toward the negative side. Because of the directions of I_1 and I_3 in the diagram, the direction of current I_2 is directed from point A to point B.

Step 2. Place a positive (+) sign on the side of each resistor where the current enters and a negative sign on the side where the current exits. This indicates that a drop in potential occurs as the current passes through the resistor. Place a (+) sign next to the long line of the battery symbol and a (-) sign next to the short line. If you choose the wrong direction for the flow of current in a particular branch, your final answer for the current in that branch will be negative. The negative answer indicates that the current actually flows in the opposite direction.

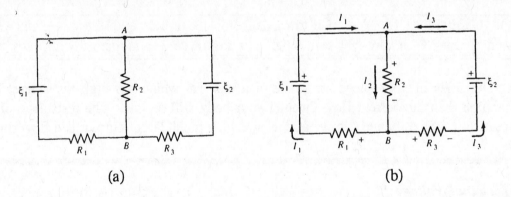

(a)   (b)

Step 2. Select a junction point and apply the junction rule, e.g., at point A in the diagram:

$$I_1 + I_3 = I_2$$

The junction rule may be applied at more than one junction point. In general, apply the junction rule to enough junction points so that each branch current appears in at least one equation.

Step 3. Apply Kirchhoff's loop rule by first taking note whether there is a gain or loss of potential at each resistor and source of emf as you trace the closed loop. Remember that the sum of the gains and losses of potential must add to zero. For example, for the left loop of the sample circuit, start at point B and travel clockwise around the loop. Because the direction chosen for the loop is also the direction assigned for the current, there is a gain in potential across the battery (- to +), but a loss of potential across each resistor (+ to -). For the diagram shown below and using the loop rule, the following equation can be written:

$$- I_1 R_1 + \xi_1 - I_2 R_2 = 0$$

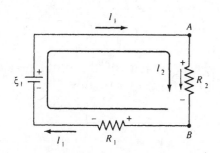

clockwise around loop

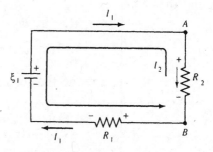

counterclockwise around loop

The direction taken around the loop is arbitrary. As shown in the above right diagram, tracing a counterclockwise path around the circuit starting at B, there is a gain in potential across each resistor (- to +) and a drop in potential across the battery (+ to -). The loop equation is

$$+ I_2 R_2 - \xi_1 + I_1 R_1 = 0$$

Multiplying both sides of the above equation by -1 and algebraically rearranging, it can be shown that the two equations are equivalent.

Be sure to apply the loop rule to enough closed loops so that each branch current appears in at least one loop equation. Solve for each branch current using standard algebraic methods.

EXAMPLE PROBLEM 2. Determine the current in each branch of the complex circuit shown on page 19-8 given ξ_1 = 4.00 V, ξ_2 = 2.00 V, R_1 = 1.00 Ω, R_2 = 2.00 Ω, and R_3 = 3.00 Ω

Part a. Step 1.	Solution: (Sections 19-3 and 19-4)
Apply the junction rule at point A and write an equation for the current.	at point A: $I_1 + I_3 = I_2$ rearranging gives $I_1 - I_2 + I_3 = 0$ (eqn. 1)

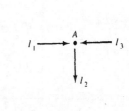

Part a. Step 2.

Note the direction of the current through each resistor. Place + and - signs on the appropriate sides of each resistor. Apply the loop rule. Start at point B and proceed clockwise around the left loop.

Note that at each resistor there will be a potential drop (+ to -) while at the source of emf (ξ_1) there will be a gain.

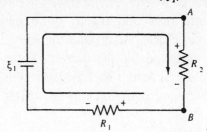

$- I_1 R_1 + \xi_1 - I_2 R_2 = 0$ rearranging and simplifying gives

$+ I_1 R_1 + I_2 R_2 = \xi_1$

Substitute the values given in the statement of the problem and arrange the resulting equation in the form of equation 1.

$(1.00 \ \Omega)I_1 + (2.00 \ \Omega)I_2 = 4.00 \ V$

$1.00 \ I_1 + 2.00 \ I_2 = 4.00 \ A$ (equation 2)

Part a. Step 3.

Again apply the loop rule. Start at point B and proceed clockwise around the right loop.

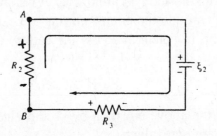

By proceeding clockwise there will be a potential gain (- to +) at each resistor while at the source of emf (ξ_2) there will be a potential drop (+ to -).

$+ I_2 R_2 - \xi_2 + I_3 R_3 = 0$ and rearranging gives

$+ I_2 R_2 + I_3 R_3 = \xi_2$

Substitute the values given in the statement of the problem and arrange as in equation 1.

$(2.00 \ \Omega) \ I_2 + (3.00 \ \Omega) \ I_3 = 2.00 \ V$

$2.00 \ I_2 + 3.00 \ I_3 = 2.00 \ A$ (equation 3)

Part a. Step 4.

Solve the three equations using algebra.

The following method is known as substitution:

$$I_1 \quad - \quad I_2 \quad + \quad I_3 \quad = 0 \qquad \text{(equation 1)}$$

$$1.00 \, I_1 + 2.00 \, I_2 \qquad \qquad = 4.00 \qquad \text{(equation 2)}$$

$$2.00 \, I_2 + \qquad \quad 3.00 \, I_3 = 2.00 \qquad \text{(equation 3)}$$

Subtract equation 2 from equation 1 in order to eliminate I_1.

$$I_1 \quad - \quad I_2 \quad + I_3 \quad = 0 \qquad \text{(equation 1)}$$

$$1.00 \, I_1 + \quad 2.00 \, I_2 \qquad \qquad = 4.00 \quad \text{(equation 2)}$$

$$\overline{\qquad \qquad -3.00 \, I_2 \quad + 1.00 \, I_3 \quad = -4.00 \, \text{(equation 4)}}$$

Multiply both sides of equation 4 by 3 and then subtract equation 3 from equation 4.

$$-9.00 \, I_2 \quad + \quad 3.00 \, I_3 \quad = -12.0 \qquad \text{(equation 4)}$$

$$2.00 \, I_2 \quad + \quad 3.00 \, I_3 \quad = \quad 2.00 \qquad \text{(equation 3)}$$

$$\overline{\quad -11.0 \, I_2 \qquad \qquad \qquad = -14.0}$$

$$I_2 = (-14.0)/(-11.0) = 1.27 \text{ A}$$

Substitute $I_2 = 1.27$ A into equation 3 and determine I_3.

$$2.00 \, I_2 + 3.00 \, I_3 = 2.00 \quad \text{(equation 3)}$$

$$(2.0)(1.27) + 3.00 \, I_3 = 2.00$$

$$I_3 = -0.182 \text{ A}$$

The negative sign indicates that the actual direction of the current is opposite from the direction initially assumed. In this instance, the direction of current I_3 is away from point A.

Substitute the values of I_2 and I_3 into equation 1 to solve for I_1.

$$I_1 - I_2 + I_3 = 0 \quad \text{(equation 1)}$$

$$I_1 - 1.27 \text{ A} + -0.182 \text{ A} = 0$$

$$I_1 = 1.45 \text{ A}$$

Capacitors in Series and Parallel

A circuit with **capacitors in parallel** is shown below. According to Kirchhoff's loop rule, the potential difference (V) of the source of emf: $V = V_1 = V_2 = V_3 = ... = V_n$. The total charge stored on the capacitor plates (Q) equals the amount of charge which left the source of emf: $Q = Q_1 + Q_2 + ... + Q_n$ and since $Q = CV$ then

$$C V = C_1 V_1 + C_2 V_2 + ... + C_n V_n \quad \text{and} \quad C = C_1 + C_2 + ... + C_n$$

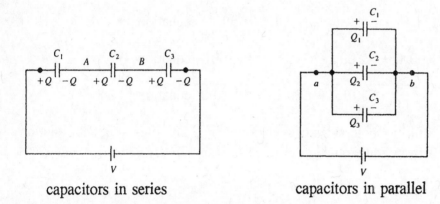

capacitors in series   capacitors in parallel

For **capacitors in series,** the amount of charge (Q) that leaves the source of emf equals the amount of charge that forms on each capacitor: $Q = Q_1 = Q_2 = Q_3 = ... = Q_n$.

From Kirchhoff's loop rule, the potential difference across the source of emf (V) equals the sum of the potential differences across the individual capacitors:

$$V = V_1 + V_2 + V_3 + ... + V_n \qquad \text{and since } V = Q/C \text{ then}$$

$$Q/C = Q_1/C_1 + Q_2/C_2 + ... + Q_n/C_n \quad \text{and} \quad 1/C = 1/C_1 + 1/C_2 + ... + 1/C_n$$

TEXTBOOK QUESTION 15. Suppose that three identical capacitors are connected to a battery. Will they store more energy if they are connected in series or in parallel?

ANSWER: The capacitors store more energy when they are connected in parallel. For example, let each capacitor be rated at 6.00 F and they are connected to a 6.00-volt battery. If the capacitors are connected in **series**:

$$1/C = 1/(6.00 \text{ F}) + 1/(6.00 \text{ F}) + 1/(6.00 \text{ F}) \quad \text{then} \quad 1/C = 3/(6.0 \text{ F}) \text{ and } C = 2.00 \text{ F}$$

The charge across each can now be determined since the charge on each equals the charge (Q) produced by the battery.

$$Q = C V = (2.00 \text{ F})(6.00 \text{ V}) = 12.0 \text{ C}$$

The energy stored on each is the same and can now be determined.

energy = ½ Q_1^2/C_1 = ½(12.0 C)²/(6.00 F) = 12.0 J

The total energy stored by the combination is

energy = ½ Q^2/C = ½(12.0 C)²/(2.0 F) = 36.0 J

The capacitors are identical and if they are connected in **parallel** equal amounts of charge form on each capacitor.

Q_1 = C_1 V = (6.00 F)(6.00 V) = 36.0 C   total charge stored Q = Q_1 + Q_2 + Q_3 = 108 C
the equivalent capacitance of the combination C = C_1 + C_2 + C_3 = 18.0 F

The total energy stored by the combination can now be determined.

energy = ½ Q^2/C = ½ (108 C)²/(18.0 F) = 324 J

EXAMPLE PROBLEM 3. Three capacitors are arranged as shown in the diagram. Determine the a) equivalent capacitance of the combination, b) potential difference across each capacitor, and c) charge on each capacitor.

C_1 = 10 μF

C_2 = 20 μF

C_3 = 30 μF

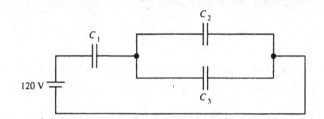

Part a. Step 1.	Solution: (Section 19-6)
Determine the equivalent capacitance of the two capacitors arranged in parallel.	$C_{\|}$ = C_2 + C_3 $C_{\|}$ = 20 μF + 30 μF = 50 μF
Part a. Step 2. Redraw the diagram and solve for the total capacitance.	C_1 and $C_{\|}$ in series; therefore 1/C = 1/C_1 + 1/$C_{\|}$ 1/C = 1/(10μF) + 1/(50μF) 1/C = 6/(50μF) and C = 8.3 μF 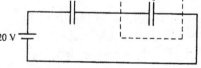
Part b. Step 1. Determine the total charge that leaves the source of emf and the charge that forms on C_1.	Q = C V = (8.3 μF)(120 V) = 1000 μC Q = 1.0 x 10⁻³ C C_1 is in series with the source of emf. The charge that forms on C_1 equals the total charge that left the source of emf, Q_1 = 1000 μC.

Part b. Step 2. Determine the potential difference across C_1.	$V_1 = Q_1/C_1$ $V_1 = 1000\ \mu C/10\ \mu F = 100\ V$
Part b. Step 3. Determine the potential difference across the two capacitors arranged in parallel.	The potential difference across the capacitors in parallel can be determined by applying Kirchhoff's loop rule. Starting at the source of emf and following the loop through C_1 and C_2: $120\ V - V_1 - V_2 = 0$ but $V_1 = 100$ volts. Therefore, $V_2 = 20\ V$ The potential difference is the same across capacitors in parallel; therefore, $V_3 = 20\ V$.
Part c. Step 1. Determine the charge stored on each of the capacitors in the parallel combination.	The charge stored on C_1 was determined to be 1000 C. The charge stored on C_2 and C_3 can be determined as follows: $Q_2 = C_2 V_2 = (20\ \mu F)(20\ V) = 400\ \mu C$ $Q_3 = C_3 V_3 = (30\ \mu F)(20\ V) = 600\ \mu C$ Note that the total charge stored on the parallel combination equals 1000 μC and is equal to the amount stored on C_1.

Circuits Containing Resistor and Capacitor

An **RC circuit** consists of a resistor and a capacitor connected in series to a dc power source. When switch 1 (S_1), shown in the diagram below, is closed, the current will begin to flow from the source of emf and charge will begin to accumulate on the capacitor. Using Kirchhoff's loop rule it can be shown that

$\xi - I\,R - Q/C = 0$

where I R refers to the drop in potential across the resistor and Q/C refers to the drop in potential across the capacitor.

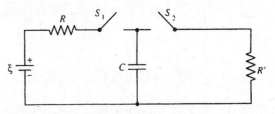

As charge accumulates on the capacitor, the potential difference will increase until it equals ξ, and at that point the current ceases to flow.

By using the methods of calculus, it can be shown that the current I through the circuit at time t after the S_1 is closed is given by

$I = I_o\ e^{-t/RC}$

where I_o is the initial value of the current just after switch 1 is closed. RC is the product of the resistance and the capacitance. RC is referred to as the time constant (τ) of the circuit and is measured in seconds. The potential difference across the capacitor as it is being charged is

$$V = \xi(1 - e^{-t/RC})$$

while the amount of charge (Q_t) accumulated on the capacitor as it is being charged is

$$Q_t = Q (1 - e^{-t/RC})$$ where Q is the charge on the capacitor when it is fully charged.

The variation of current, voltage, and charge is shown graphically below. Note that the time is marked off in terms of the time constant (τ).

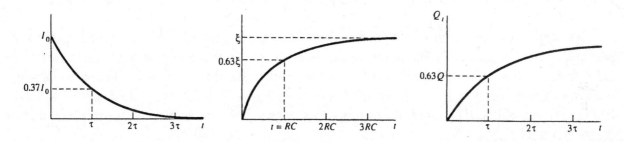

If switch 1 is opened and switch 2 (S_2) is closed then the capacitor will **discharge** through resistor R'. The charge on the capacitor (Q_t) decreases exponentially with time as follows:

$$Q_t = Q\ e^{-t/R'C}$$ where Q is the charge on the capacitor when it is fully charged.

The following diagram shows the variation of charge with time as the capacitor discharges. The time is marked off in time constants.

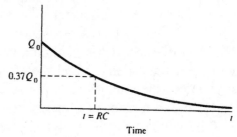

The current in a discharge circuit as a function of time is $I = I_o\ e^{-t/R'C}$, where I_o represents the initial current as the discharge begins.

The voltage across the resistance (R') in a discharge circuit as a function of time is

$$V = V_o\ e^{-t/R'C}$$ where V_o represents the initial voltage across the capacitor as the discharge begins.

Galvanometer, Ammeter, and Voltmeter

A **galvanometer** is a device used to detect and measure very low currents, usually in the range of 1 milliampere or less. It usually consists of a coil of wire suspended by a fine wire and arranged parallel to the magnetic field produced by a permanent magnet. When a current passes through the coil, a torque is produced. This torque causes the coil to twist until a restoring torque provided by the coil suspension balances the torque produced by the interaction of the current in the coil and the magnetic field of the permanent magnet. The torque on the coil and the angle through which the coil deflects is proportional to the current. As a result the galvanometer can be calibrated to determine the amount of current passing through the coil.

Because the coil of a galvanometer consists of wires, the galvanometer presents resistance to the flow of current through it. This internal resistance is usually low.

An **ammeter** is a device designed to measure the amount of electric current passing a particular point in a circuit. The ammeter is placed in series in the circuit and consists of 1) a galvanometer and 2) a resistor of very low value, called the **shunt resistor**, placed in parallel with the galvanometer. Since a potential drop occurs across it, an ammeter is designed to have low overall resistance. Therefore, if it is properly designed, the ammeter will have minimal effect on the external circuit.

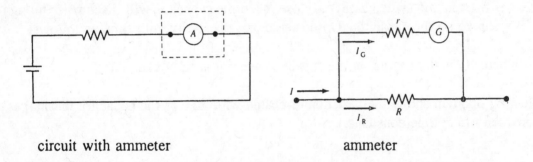

circuit with ammeter   ammeter

A **voltmeter** measures the potential difference across a circuit element, e.g., a resistor. The voltmeter is placed in parallel with the circuit element and therefore the potential difference across the voltmeter is the same as across the circuit element. A voltmeter consists of 1) a galvanometer of internal resistance r and 2) a resistor of high resistance R placed in series with the galvanometer. The resistance of the voltmeter is very large compared to the circuit element and as a result very little current flows through the voltmeter and it has a minimal effect on the circuit.

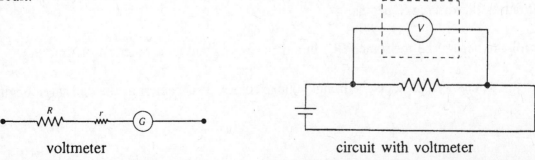

voltmeter   circuit with voltmeter

TEXTBOOK QUESTION 22. Explain why an ideal ammeter would have zero resistance and an ideal voltmeter infinite resistance.

ANSWER: An ammeter is connected in series in the circuit. The total resistance to current flow is the sum of the ammeter resistance and the resistance of the circuit as a whole. If the ammeter has zero resistance then there is no voltage drop across it and it has no effect on the flow of current in the circuit.

The potential difference across resistors connected in parallel is the same. The voltmeter is connected in parallel with the circuit element and the voltage drop across each is the same. If the resistance of the voltmeter is much greater than that of the circuit, the equivalent resistance would equal the resistance of the circuit element. The voltmeter would draw a very small current and would have little effect on the circuit. However, it would record the voltage drop across the circuit element. The ideal voltmeter would have infinite resistance and would not alter the circuit.

The Potentiometer

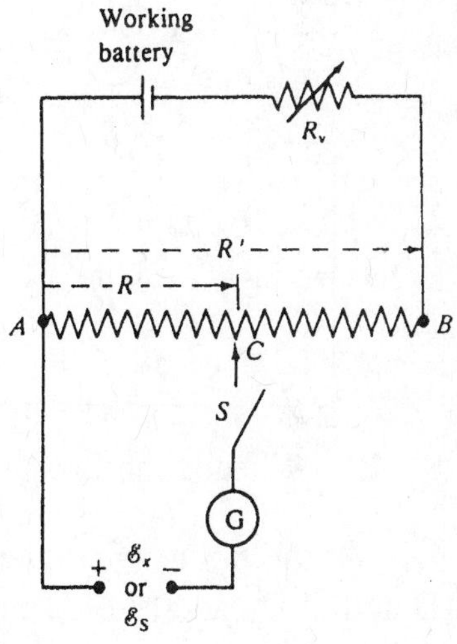

The **slide wire potentiometer** is used to give precise measurements of the source of emf. A typical potentiometer circuit is shown at right. A battery V is connected to a resistor R, the resistance of which can be varied so that a convenient potential difference occurs across the slide wire (AB).

The resistance of the slide wire is proportional to the wire's length, $R = \rho L/A$. Since $V = IR$, then the potential difference between point A and point C along the length of the wire is proportional to the distance between A and C (L_{AC}).

A circuit containing a standard cell of known emf (ξ_s) is then placed in parallel with the slide wire. The position of the slide is then adjusted along the length of the wire until no current flows through the galvanometer (G). The length of the wire between A and C is measured. At this point the potential difference along the wire is equal to ξ_s and therefore $\xi_s \propto L_{AC}$.

The standard cell is then replaced with the unknown and the slide is again adjusted until no current flows. The length $L_{AC'}$ is noted. The emf of the unknown is ξ_x and $\xi_x \propto L_{AC'}$. The emf of the unknown is given by $\xi_x/\xi_s = L_{AC'}/L_{AC}$.

EXAMPLE PROBLEM 4. The emf of the standard cell used in the potentiometer circuit shown above is 1.50 V. Determine the emf of the unknown if the galvanometer current is zero and L_{AC} is 50 cm when the standard cell is in the circuit and 30 cm when the unknown is in the circuit.

Part a. Step 1.	Solution: (Section 19-10)
Determine the emf of the unknown cell.	The emf of the unknown can be determined by using the following equation:
	$\xi_x/\xi_s = L_{AC}/L_{AC}$ and substituting values gives
	$\xi_x/1.50 \text{ V} = 30 \text{ cm}/50 \text{ cm}$
	$\xi_x = 0.90 \text{ V}$

The Wheatstone Bridge

The **Wheatstone bridge** is a device used to give precise measurements of electrical resistance. In the Wheatstone bridge circuit shown below, R_1, R_2, and R_3 are known while R_x has unknown resistance. The variable resistor R_3 is adjusted until no current flows through the galvanometer when the switch (S) is closed.

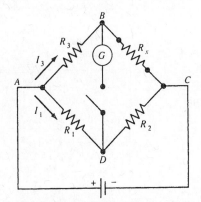

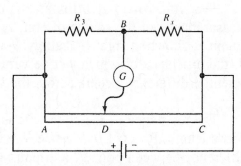

Wheatstone bridge circuit   circuit with slide wire

At the point where the current through the galvanometer (G) is zero, the potential at points B and D is exactly the same, therefore

$V_{AB} = V_{AD}$ and $I_3 R_3 = I_1 R_1$

also $V_{BC} = V_{DC}$ and $I_3 R_x = I_1 R_2$. Therefore,

$R_x/R_3 = R_2/R_1$ and R_x can be determined.

In practice, the slide wire used in the potentiometer circuit is often used since the wire has uniform resistance along its length. The slide is moved until the galvanometer current reaches zero. At this point $R_x/R_3 = L_{CD}/L_{AD}$ and R_x can be determined.

EXAMPLE PROBLEM 5. The Wheatstone bridge circuit shown above is used to measure the resistance of an unknown resistor. If $R_1 = 10 \text{ }\Omega$, $R_2 = 20 \text{ }\Omega$, and the galvanometer current is zero when R_3 is adjusted to 30 ohms, determine the value of R_x.

Part a. Step 1.	Solution: (Section 19-10)
Determine the resistance of the unknown resistor.	The resistance of the unknown can be determined by using the following equation: $R_x/R_3 = R_2/R_1$; therefore, $R_x/30\,\Omega = 20\,\Omega/10\,\Omega$ $R_x = 60\ \Omega$

EXAMPLE PROBLEM 6. The slide wire Wheatstone bridge circuit shown above is used to determine the resistance of an unknown resistor. The galvanometer current is zero when $R_3 = 30\ \Omega$, $L_{DC} = 40$ cm, and $L_{AD} = 60$ cm. Determine the value of R_x.

Part a. Step 1.	Solution: (Section 19-10)
Determine the magnitude of the unknown resistance.	The resistance of the unknown can be determined as follows: $R_x/R_3 = L_{DC}/L_{AD}$ and substituting, $R_x/30\ \Omega = 40\ \text{cm}/60\ \text{cm}$ $R_x = 20\ \Omega$

PROBLEM SOLVING SKILLS

For problems involving equivalent resistance:

1. Determine whether the resistors are arranged in series or in parallel.
2. If necessary, simplify the circuit step by step until it is reduced to a simple series or parallel combination. Use the appropriate formula to determine the equivalent resistance.
3. Use Ohm's law and Kirchhoff's laws to solve for the current through each resistor and the potential difference across each resistor.

For problems involving a complex circuit:

1. Assign a direction to the current in each branch of the circuit. Place a + sign on the side of each resistor where the current enters and a - sign where the current exits.
2. Place a + sign at the positive terminal of each battery and a - sign at the negative terminal.
3. Select a junction point and use Kirchhoff's junction rule to write an equation for the currents entering the point.
4. Apply Kirchhoff's loop rule and write equations based on the gains and losses of potential around selected loops.
5. Use an algebraic technique to solve for the unknown currents. Recall that it is necessary to have as many equations as there are unknowns in order to solve the problem.

For problems involving equivalent capacitance:

1. Determine whether the capacitors are arranged in series or in parallel.
2. If necessary, simplify the circuit step by step until it is reduced to a simple series or parallel combination. Use the appropriate formula to determine the equivalent capacitance.
3. Use $C = Q/V$ and apply Kirchhoff's rules to solve for the charge stored on each capacitor and the potential difference across each capacitor.

For problems involving the construction of an ammeter or voltmeter from a galvanometer and an added resistor.

1. Recall that for an ammeter the shunt resistor is added in parallel while for a voltmeter the additional resistor is added in series.
2. Apply Ohm's law and Kirchhoff's rules and solve for the magnitude of the resistance necessary for the conversion.

SOLUTIONS TO SELECTED TEXTBOOK PROBLEMS

TEXTBOOK PROBLEM 1. Calculate the terminal voltage for a battery with an internal resistance of 0.900 Ω and an emf of 8.50 V when the battery is connected in series with (a) an 81.0-Ω resistor, and (b) an 810-Ω resistor.

Part a. Step 1.	Solution: (Section 19-2)
Determine the equivalent resistance.	The internal resistance (r) is connected in series with the external resistance (R).
	$R_{total} = R + r = 81.0\ \Omega + 0.900\ \Omega = 81.9\ \Omega$
Part a. Step 2.	$I_{total} = V_{total}/R_{total}$
Determine the total current in the circuit.	$= (8.50\ \text{V})/(81.9\ \Omega)$
	$I_{total} = 0.104\ \text{A}$
Part a. Step 3.	$V_{terminal} = \xi - I_{total}\ r$
Determine the terminal voltage.	$= 8.50\ \text{V} - (0.104\ \text{A})(0.900\ \Omega)$
	$V_{terminal} = 8.41\ \text{V}$
Part b. Step 1.	$R_{total} = R + r = 810\ \Omega + 0.900\ \Omega = 811\ \Omega$
Determine the total resistance (R_{total}).	

Part b. Step 2. Determine the total current in the circuit.	$I_{total} = V_{total}/R_{total}$ $\quad = (8.50 \text{ V})/(811 \text{ }\Omega)$ $I_{total} = 0.0104 \text{ A}$
Part b. Step 3. Determine the terminal voltage.	$V_{terminal} = \xi - I_{total}\ r$ $\quad = 8.50 \text{ V} - (0.0104 \text{ A})(0.900 \text{ }\Omega)$ $V_{terminal} = 8.49 \text{ V}$

TEXTBOOK PROBLEM 5. Four 240-Ω light bulbs are connected in series. What is the total resistance of the circuit? What is their resistance if they are connected in parallel?

Part a. Step 1. Determine the equivalent resistance if they are connected in series.	Solution: (Section 19-1) $R = R_1 + R_2 + R_3 + R_4$ $\quad = 240 \text{ }\Omega + 240 \text{ }\Omega + 240 \text{ }\Omega + 240 \text{ }\Omega$ $R = 960 \text{ }\Omega$
Part a. Step 2. Determine the equivalent resistance if they are connected in parallel.	$1/R = 1/R_1 + 1/R_2 + 1/R_3 + 1/R_4$ $1/R = 1/(240 \text{ }\Omega) + 1/(240 \text{ }\Omega) + 1/(240 \text{ }\Omega) + 1/(240 \text{ }\Omega)$ $1/R = 4/(240 \text{ }\Omega)$ $R = (240 \text{ }\Omega)/4 = 60 \text{ }\Omega$

TEXTBOOK PROBLEM 18. A 75-W, 110-V bulb is connected in parallel with a 40-W, 110-V bulb. What is the net resistance?

Part a. Step 1. Determine the total power dissipated by the bulbs.	Solution: (Section 19-1) $P_{total} = 75 \text{ W} + 40 \text{ W}$ $P_{total} = 115 \text{ W}$
Part a. Step 2. Determine the voltage of the power supply.	The bulbs are connected in parallel. Therefore, $V_{total} = V_1 = V_2 = 110 \text{ V}$

Part a. Step 3.	$P_{total} = (V_{total})^2/R_{net}$ and $R_{net} = (V_{total})^2/P_{total}$
Determine the net resistance.	$R_{net} = (110 \text{ V})^2/(115 \text{ W})$
	$R_{net} = 105 \ \Omega$

TEXTBOOK PROBLEM 29. Determine the magnitudes and directions of the currents through each resistor shown in Fig. 19-48. The batteries have emfs of $\xi_1 = 9.0$ V and $\xi_2 = 12.0$ V and the resistors have values of $R_1 = 25 \ \Omega$, $R_2 = 18 \ \Omega$, and $R_3 = 35 \ \Omega$

Part a. Step 1.	Solution: (Sections 19-3 and 19-4)
Choose a direction for the current through each branch.	The choice of direction for the current in each branch is arbitrary. However, a reasonable assumption would be that the current flows away from the batteries as shown in the diagram and toward the left in resistor R_2. 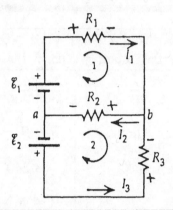

Part a. Step 2.	at point b: $I_1 + I_3 = I_2$ and rearranging gives
Apply the junction rule at point b and write an equation for the current.	$I_1 - I_2 + I_3 = 0$ (equation 1)

| Part a. Step 2.

Apply the loop rule to loop 1 shown in the diagram. Note: loop 1 is the top loop in the diagram. Start at point a and proceed clockwise around the loop. | Note the direction of the current through each resistor. A plus sign is placed on the side of the resistor where the current enters. A minus sign is placed on the side of the resistor where the current exits. + to - indicates that a drop of potential has occurred across the resistor.

Traveling clockwise starting at point a, there is a gain of potential across the source of emf (ξ_1) and a drop of potential across each resistor.

$+ \xi_1 - I_1 R_1 - I_2 R_2 = 0$ rearranging and simplifying gives

$+ I_1 R_1 + I_2 R_2 = \xi_1$

Substitute the values given in the statement of the problem and arrange the resulting equation in the form of equation 1.

$(25 \ \Omega)I_1 + (18 \ \Omega)I_2 = 9.0$ V

$2.5 \ I_1 + 1.8 \ I_2 = 0.90$ A (equation 2) |

19-22

Part a. Step 3.	By proceeding clockwise there will be a gain of potential across each resistor and a drop in potential across the source of emf (ξ_2).

Again apply the loop rule. Start at point *a* and proceed clock-wise around loop 2. Loop 2 is the bottom loop in the diagram.

$+ I_2 R_2 + I_3 R_3 - \xi_2 = 0$ and rearranging gives

$+ I_2 R_2 + I_3 R_3 = \xi_2$

Substitute the values given in the statement of the problem and arrange as in equation 1.

$(18 \ \Omega) I_2 + (35 \ \Omega) I_3 = 12.0 \ V$

$1.8 \ I_2 + 3.5 \ I_3 = 1.2 \ A$ (equation 3)

Part a. Step 4.

Solve the three equations using algebra.

The following method is known as substitution:

$$
\begin{array}{lllll}
I_1 & - \ I_2 & + \ I_3 & = 0 & \text{(equation 1)} \\
2.5 \ I_1 & + 1.8 \ I_2 & & = 0.90 \ A & \text{(equation 2)} \\
& 1.8 \ I_2 & + 3.5 \ I_3 & = 1.2 \ A & \text{(equation 3)}
\end{array}
$$

Multiply equation 3 by -1.0 and add equations 2 and 3.

$$
\begin{array}{llll}
2.5 \ I_1 & + 1.8 \ I_2 & = 0.90 \ A & \text{(equation 2)} \\
& - 1.8 \ I_2 \quad - 3.5 \ I_3 & = - 1.2 \ A & \text{(equation 3)} \\
\hline
2.5 \ I_1 & \quad\quad - 3.5 \ I_3 & = -0.30 \ A & \text{(equations 4)}
\end{array}
$$

Multiply equation 1 by 1.8 and add equations 1 and 2.

$$
\begin{array}{llll}
1.8 \ I_1 & - \ 1.8 \ I_2 \quad + \ 1.8 \ I_3 & = 0 & \text{(equation 1)} \\
2.5 \ I_1 & + 1.8 \ I_2 & = 0.90 \ A & \text{(equation 2)} \\
\hline
4.3 \ I_1 & \quad\quad + 1.8 \ I_3 & = 0.90 \ A & \text{(equation 5)}
\end{array}
$$

Multiply equation 4 by -1.72 and add equations 4 and 5.

$$
\begin{array}{llll}
-4.3 \ I_1 & + \ 6.0 \ I_3 & = 0.52 \ A & \text{(equation 4)} \\
4.3 \ I_1 & + \ 1.8 \ I_3 & = 0.90 \ A & \text{(equation 5)} \\
\hline
& 7.8 \ I_3 & = 1.42 \ A &
\end{array}
$$

$I_3 = (1.42 \ A)/(7.8) = + 0.18 \ A$

The positive sign indicates that the direction of the current is in the direction initially assumed. A negative value would indicate that the direction of current I_2 is opposite from the direction arbitrarily chosen.

Substitute $I_3 = 0.18 \ A$ into equation 3 and determine I_2.

$1.8 \ I_2 \quad + 3.5 \ (0.18 \ A) \quad = \ 1.2 \ A$

$I_2 = 0.32$ A

Substitute the values of I_2 and I_3 into equation 1 to solve for I_1.

$I_1 - I_2 + I_3 = 0$ (equation 1)

$I_1 - 0.32$ A $+ 0.18$ A $= 0$

$I_1 = + 0.14$ A

TEXTBOOK PROBLEM 50. In Fig. 19-56 (same as Fig. 19-20a), the total resistance is 15.0 kΩ, and the battery's emf is 24.0 volts. If the time constant is measured to be 35.0 μs, calculate (a) the total capacitance of the circuit and (b) the time it takes for the voltage across the resistor to reach 16.0 V after the switch is closed.

Part a. Step 1. Determine the total capacitance of the circuit.	Solution: (Section 19-7) $\tau = R\,C$ and $C = \tau/R$ $C = (35.0\ \mu s)/(15000\ \Omega) = 2.3 \times 10^{-9}$ F

Part b. Step 1. Determine the time required for the voltage across the resistor (V_R) to reach 16 volts.	$V_{total} = V_R + V_C$ where $V_C = \xi(1 - e^{-t/RC})$ 24 volts = 16 volts + (24 volts)$(1 - e^{-t/35\mu s})$ 8 volts = (24 volts)$(1 - e^{-t/35\mu s})$ 8 volts = 24 volts - (24 volts)$e^{-t/35\mu s}$ -16 volts = - (24 volts)$e^{-t/35\mu s}$ $0.67 = e^{-t/35\mu s}$ $\ln 0.67 = \ln e^{-t/35\mu s}$ $- 0.40 = -t/35\mu s$ $t = 14\ \mu s = 1.4 \times 10^{-5}$ s

---

TEXTBOOK PROBLEM 55. A galvanometer has an internal resistance of 30 Ω and deflects full scale for a 50-μA current. Describe how to use this galvanometer to make (a) an ammeter to read currents up to 30 A, and (b) a voltmeter to give a full-scale deflection of 250 V.

Part a. Step 1. Draw a diagram locating the resistance which must be added to convert the galvanometer into an ammeter.	**Solution: (Section 19-10)** In order to convert the galvanometer into an ammeter which reads from 0 to 30.0 A, it is necessary to add a shunt resistor (R) in parallel with the galvanometer. 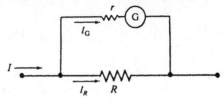
Part a. Step 2. Determine the magnitude of the current which passes through the shunt.	Only 50 μA, i. e., 5.0×10^{-5} A, may pass through the galvanometer when it reads full scale deflection. Therefore, if the ammeter is to measure 30 A when it is fully deflected, the shunt current is $I_{shunt} = 30.0$ A $- 5.0 \times 10^{-5}$ A ≈ 30.0 A
Part a. Step 3. Determine the voltage drop across the galvanometer and across the shunt.	The galvanometer and shunt resistance are in parallel. The potential difference across each will be equal, i.e., $V_{galv} = V_{shunt}$. $V_{galv} = I_{galv}\, r = (5.0 \times 10^{-5}$ A$) (30\ \Omega)$ $V_{galv} = 1.5 \times 10^{-3}$ V but $V_{shunt} = V_{galv} = 1.5 \times 10^{-3}$ V
Part a. Step 4. Determine the shunt resistance.	$V_{shunt} = I_{shunt}\, R_{shunt}$ 1.5×10^{-3} V $= (30$ A$)\, R_{shunt}$ $R_{shunt} = 5.0 \times 10^{-5}\ \Omega$
Part b. Step 1. Draw a diagram locating the resistance which must be added to convert the galvanometer into a voltmeter.	In order to convert the galvanometer into a voltmeter, a resistance R must be placed in series with the galvanometer. 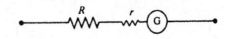

Part b. Step 2.

Determine the magnitude of the resistance which must be added in order to convert the galvanometer into a voltmeter.

The voltage drop across the circuit must be 250 volts when a current of 50 μA, i. e., 5.0×10^{-5} A, passes through the galvanometer.

$$V = I\,(r + R)$$

$$250\ V = (5.0 \times 10^{-5}\ A)(30.0\ \Omega + R)$$

$$30\ \Omega + R = (250\ V)/(5.0 \times 10^{-5}\ A)$$

$$30\ \Omega + R = 5.0 \times 10^{6}\ \Omega$$

$$R = 5.0 \times 10^{6}\ \Omega - 30\ \Omega \approx 5.0 \times 10^{6}\ \Omega \quad \text{or} \quad 5.0\ M\Omega$$

CHAPTER 20

MAGNETISM

OBJECTIVES

After studying the material of this chapter, the student should be able to:

- draw the magnetic field pattern produced by iron filings sprinkled on paper placed over different arrangements of bar magnets.
- determine the magnitude of the magnetic field produced by both a long, straight, current-carrying wire and a current loop. Use the right hand rule to determine the direction of the magnetic field produced by the current.
- explain what is meant by ferromagnetism, and include in the explanation the concept of domains and the Curie temperature.
- state the conventions adopted to represent the direction of a magnetic field, the current in a current-carrying wire, and the direction of motion of a charged particle moving through a magnetic field.
- apply the right-hand rule to determine the direction of the force on either a charged particle traveling through a magnetic field or a current-carrying wire placed in a magnetic field.
- determine the torque on a current loop arranged in a magnetic field and explain galvanometer movement.
- explain how a mass spectrograph can be used to determine the mass of an ion and how it can be used to separate isotopes of the same element.

KEY TERMS AND PHRASES

north pole or "north seeking pole" of a bar magnet tends to align with the Earth's magnetic field and point toward magnetic north. The south pole of a bar magnet tends to point toward magnetic south.

magnetic field surrounds every magnet and is also produced by a charged particles in motion relative to some reference point. The direction of the field at any point is indicated by the north pole of a compass needle placed at that point. The SI unit of magnetic field strength is the tesla (T).

right-hand rule is used to predict the direction of the magnetic field produced by a current-carrying wire. The thumb of the right hand points in the direction of the conventional current in the wire. The fingers encircle the wire in the direction of the magnetic field.

magnetic force acts on a charged particle traveling through a magnetic field. The magnetic force always acts perpendicular to the direction of the magnetic field and the velocity vector. A second right-hand rule is used to predict the direction of the force on the wire.

velocity selector allows only charged particles which have a particular velocity to pass undeflected. The velocity is the same regardless of the magnitude of the charge or the mass of the particle.

mass spectrograph uses charged particles traveling through magnetic fields to determine the relative mass of the particle.

galvanometer movement is the basis of most meters, i.e., ammeters, voltmeters, and ohmeters, as well as electric motors. Galvanometer movement is the result of interaction of the magnetic field of a permanent magnet which is directed perpendicular to a current carried by a loop or coil of wire.

ferromagnetic materials, such as iron, can be permanently magnetized. Each atom has a net magnetic effect and the atoms tend to align their magnetic fields in arrangements known as **domains**. Each domain contributes to the overall magnetic field of the piece of iron. In an ordinary piece of iron or other ferromagnetic material, the magnetic fields produced by the individual domains cancel out so that the object is not a magnet. In a magnet, the domains are larger in one direction than in any other and a net magnetic effect is produced.

Curie temperature refers to the temperature where it is no longer possible to magnetize an object and a permanent magnet loses its magnetic effect.

SUMMARY OF MATHEMATICAL FORMULAS

magnetic field long, straight wire	$B = (\mu_o I)/(2 \pi r)$	The magnitude of the magnetic field strength (B) a perpendicular distance (r) from a long, straight wire carrying a current (I). The SI unit of B is the tesla (T) and $\mu_o = 4\pi \times 10^{-7}$ T/A.
loop of wire	$B = \mu_o I/2r$	The magnitude of the magnetic field strength (B) at the center of a loop of wire of radius (r) which carries a current (I).

force on a current-carrying wire in a magnetic field	$F = I \ell B \sin \theta$	The force (F) on a wire carrying a current depends on the magnitude of the current (I), the length (ℓ), magnetic field strength (B), and the angle (θ) between the directions of the current in the wire and the magnetic field.
force on a charged particle traveling through a magnetic field	$F = q v B \sin \theta$	The force (F) on a charged particle traveling through a magnetic field depends on the magnitude of the charge (q), the velocity of the charge (v), the magnetic field strength (B), and the angle (θ) between the direction of motion of the particle and the direction of the magnetic field.
radius of the circular path followed by an ion in a mass spectrograph	$r = (m E)/(q B B')$	The radius of the path depends on the mass of the particle (m), the electric field strength (E), and the magnetic field strength (B) in the velocity selector, and the magnetic field strength (B') which causes the circular motion.
galvanometer movement	$\tau = N I A B$	If the coil of the galvanometer is pivoted in the center, a torque is produced which depends on the number of turns of wire in the coil (N), the current (I), the area of the coil (A), and the magnetic field strength (B). The quantity N I A is the magnetic moment of the coil.
force between two parallel conducting wire	$F/\ell = (\mu_o I_1 I_2)/(2 \pi L)$	Two parallel, current-carrying conductors produce magnetic fields which result in a force between the conductors. The force per unit length (F/ℓ) depends on the product of the currents (I_1) and (I_2) and is inversely related to the separation (L) between the wires.

CONCEPT SUMMARY

Magnets and Magnetic Fields

Two **bar magnets** exert a force on one another. If two **north poles** (or **south poles**) are brought near, a repulsive force is produced. If a north pole and a south pole are brought near, then a force of attraction results. Thus, "Like poles repel, unlike poles attract."

The concept of a field is applied to magnetism as well as gravity and electricity. A **magnetic field** surrounds every magnet and is also produced by a charged particle in motion relative to some reference point. The presence of the magnetic field about a bar magnet can be seen by placing a piece of paper over the bar magnet and sprinkling the paper with iron filings. The magnetic field produced by certain arrangements of bar magnets are represented in the diagrams shown below.

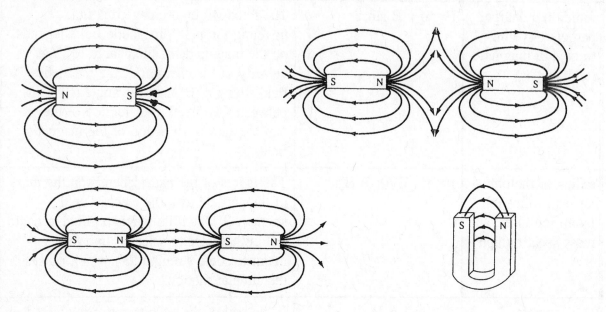

Electric Currents Produce Magnetism

A wire carrying a current (I) produces a magnetic field. The magnitude of the magnetic field strength (B) a perpendicular distance r from a **long, straight wire** is given by

$$B = (\mu_o I)/(2 \pi r) \qquad \text{B is measured in teslas (T) and } \mu_o = 4\pi \times 10^{-7} \text{ T/A,}$$
$$\text{where } \mu_o \text{ is known as the permeability of free space}$$

The direction of the magnetic field produced by a current-carrying wire can be predicted by using the **right-hand rule**. The thumb of the right hand points in the direction of the conventional current in the wire. The fingers encircle the wire in the direction of the magnetic field.

The magnitude of the strength of the magnetic field (B) at the center of a **loop of wire** of radius r which carries a current I is

$$B = \mu_o I/2r$$

As shown in the diagrams at the top of the next page, the direction of the magnetic field at the center of the loop can again be predicted by using the right-hand rule. The thumb is placed tangent to a point on the loop and is directed in the same direction as the current in the loop at that point. The fingers encircle the wire in the same direction as the magnetic field.

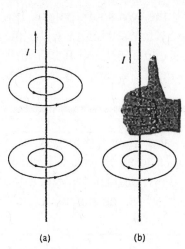
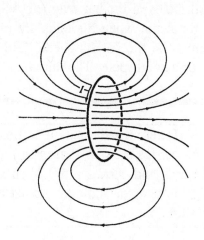
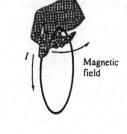

(a)   (b)

magnetic field produced
by a straight wire

magnetic field produced by a
loop of wire

Conventions

Certain **conventions** have been adopted in order to represent the direction of the magnetic field and the current in a wire. A magnetic field directed into the paper is represented by a group of x's, while a magnetic field out of the paper is represented by a group of dots. A current carrying wire which is arranged perpendicular to the page is represented by a circle. If the current is directed into the paper, then an x is placed in the center of the circle. If the current is directed out of the paper, then a dot is placed in the center of the circle.

```
x  x  x  x  x          .  .  .  .  .  .
x  x  x  x  x          .  .  .  .  .  .          ⊙                    ⊗
x  x  x  x  x          .  .  .  .  .  .
```

B field into the paper	B field out of the paper	wire with current out of the paper	wire with current into the paper

---

TEXTBOOK QUESTION 19. Describe electric and/or magnetic fields surround a moving electric charge?

ANSWER: A moving electric charge produces a magnetic field. An electric charge, whether moving or not, has an electric field in the region surrounding the charge. However, the electric charge has mass and must be surrounded by a gravitational field. Therefore, an electric field, a magnetic field, and a gravitational field all surround a moving electric charge.

TEXTBOOK QUESTION 14. Suppose you have three iron rods, two of which are magnetized but the third is not. How would you determine which two are the magnets without using any additional objects?

ANSWER: The two north poles of the bar magnets will repel as will the two south poles. The two rods which exhibit repulsion when the ends are oriented one way but attraction when the orientation is reversed must be the bar magnets. Either end of the unmagnetized iron rod is attracted to a north pole or a south pole. The unmagnetized rod will never be repelled by either end of a bar magnet.

EXAMPLE PROBLEM 1. Determine the magnetic field strength a) at a point 1.0 meters from a long straight wire which carries a 3.0-A current, b) at the center of a loop of wire 0.10 m in radius that contains 100 turns and carries a 5.0-A current. c) Use the right-hand rule and the conventions discussed in the chapter summary to indicate the direction of the magnetic field produced in each of the situations shown.

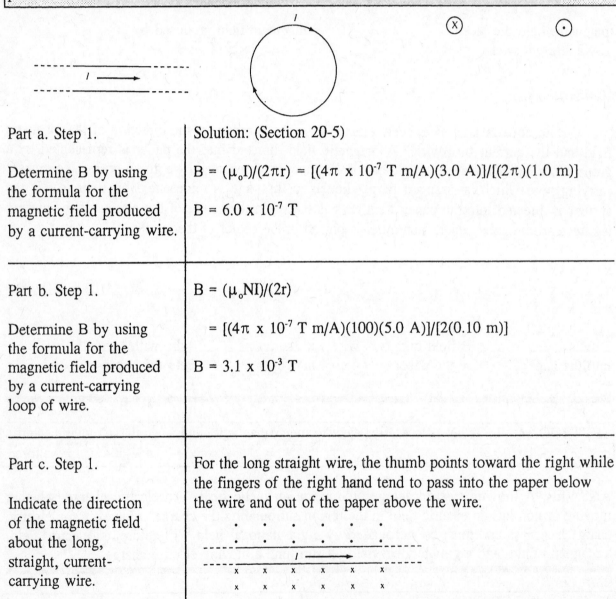

Part a. Step 1.	Solution: (Section 20-5)
Determine B by using the formula for the magnetic field produced by a current-carrying wire.	$B = (\mu_o I)/(2\pi r) = [(4\pi \times 10^{-7}\ T\ m/A)(3.0\ A)]/[(2\pi)(1.0\ m)]$ $B = 6.0 \times 10^{-7}\ T$
Part b. Step 1. Determine B by using the formula for the magnetic field produced by a current-carrying loop of wire.	$B = (\mu_o NI)/(2r)$ $= [(4\pi \times 10^{-7}\ T\ m/A)(100)(5.0\ A)]/[2(0.10\ m)]$ $B = 3.1 \times 10^{-3}\ T$
Part c. Step 1. Indicate the direction of the magnetic field about the long, straight, current-carrying wire.	For the long straight wire, the thumb points toward the right while the fingers of the right hand tend to pass into the paper below the wire and out of the paper above the wire.

Part c. Step 2. Indicate the direction of current-carrying loop of wire.	For the current-carrying loop, the thumb is placed tangent to the loop at arbitrarily selected points. However, it must be placed pointing in the direction of the current at that point. The fingers of the field produced by the right hand tend to pass into the paper inside the loop and out of the paper outside of the loop. 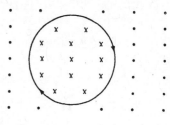
Part c. Step 3. Indicate the direction of the magnetic field produced by a long, straight, current-carrying wire if the current is directed into the paper.	The thumb of the right hand is directed perpendicular to the plane of the paper and into the paper. The fingers encircle the wire in a clockwise direction.
Part c. Step 4. Indicate the direction of the magnetic field produced by a long, straight, current-carrying wire if the current is directed out of the paper.	The thumb of the right hand is directed perpendicular to the plane of the paper and out of the paper. The fingers encircle the wire in a counterclockwise direction.

Force on a Current-Carrying Wire in a Magnetic Field

A current (I) in a wire consists of moving electrical charges and a force (F) may be produced when a current-carrying wire of length ℓ is placed in a magnetic field. The magnitude of the force is given by the equation

$$F = I \ell B \sin \theta$$

where B is the **magnetic field strength** in teslas (T). Other units for magnetic field strength include newtons per ampere meter (N/A m), newtons per coulomb meters per second (N/C m/s), webers per square meter (Wb/m^2), and gauss (G).

$$1 \text{ T} = 1 \text{ N/A m} = 1 \text{ N/C m/s} = 1 \text{ Wb/m}^2 = 10^4 \text{ G}$$

θ is the angle between the directions of the current in the wire and the magnetic field. The force on the wire is zero if θ = 0° (sin 0° = 0) and is a maximum if θ = 90° (sin 90° = 1.0).

As described in the textbook and shown in the diagram, a second **right-hand rule** is used to predict the direction of the force on the wire: "First you orient your right hand so that the outstretched fingers point in the direction of the (conventional) current; from this position when you bend your fingers they should then point in the direction of the magnetic field lines; if they do not, rotate your hand and arm about the wrist until they do, remembering that your straightened fingers must point in the direction of the current. When your hand is oriented in this way, then the extended thumb points in the direction of the force on the wire."

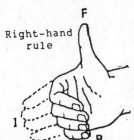

Right-hand rule

EXAMPLE PROBLEM 2. A wire carrying 2.0 A of current is placed in a uniform magnetic field of strength 3.0 T as shown in the diagram. a) Use the right-hand rule to determine the direction of the force on the wire. b) Determine the magnitude of the force on a 0.020 meter section of the wire.

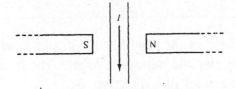

Part a. Step 1.	Solution: (Section 20-3)
Use the right-hand rule to determine the direction the force.	Based on the diagram, the current is directed toward the top of the page. Using the right-hand rule, your fingers point in the direction of the current. Your hand is then arranged so that when you bend your fingers, your fingers bend toward the direction of B which is towards the right side of the page. Your thumb is directed into the paper which means that the force is directed into the page.

Part b. Step 1.
 	
Determine the magnitude of the force.
	The current is at right angles to the direction of the magnetic field, θ = 90° Also, 1.0 T = 1.0 N/A m.
	
	F = I ℓ B sin θ
	
	  = (2.0 A)(0.020 m)(3.0 T)(sin 90°)
	
	  = (2.0 A)(0.020 m)(3.0 N/A m)(1.0)
	
	F = 0.12 N

Force on a Charged Particle Moving in a Magnetic Field

An electrically charged particle (q) moving through a magnetic field (B) at speed v may be acted upon by a force (F). The magnitude of the force (F) on the particle is

F = q v B sin θ

where θ is the angle between the direction of motion of the particle and the direction of the magnetic field. If θ = 0°, then the particle is traveling parallel to the field and no force exists on the particle (sin 0° = 0). If θ = 90°, then sin 90° = 1, the particle is traveling perpendicular to the magnetic field, and the force is a maximum.

The direction of the force can be predicted by again using the right-hand rule: the outstretched fingers of your right hand point along the direction of motion of the positively charged particle (v) and when you bend your fingers they must point along the direction of B; then your thumb points in the direction of the force. If the particle is charged negatively, then the force is directed opposite from the direction of the thumb.

TEXTBOOK QUESTION 10. Three particles, a, b, and c, enter a magnetic field as shown in Fig. 20-46. What can you say about the charge on each particle?

ANSWER: All three particles are shown moving into the magnetic field. Based on the formula F = q v B sin θ, it can be said that particle b has no excess electric charge because it is not deflected. Particles a and c are deflected; therefore, they must carry an excess charge.

Using the right-hand rule: let the outstretched fingers of your right hand point toward the right in the direction of motion of particles a and b. Your bent fingers point in the direction of b and your thumb points in the direction of the force. Using this rule, it can be said that particle a is charged positively while particle c is charged negatively.

EXAMPLE PROBLEM 3. An alpha particle (α), mass 6.68×10^{-27} kg and charge $+3.2 \times 10^{-19}$ C. The particle is traveling at 5.0×10^7 m/s and enters a magnetic field of magnitude 0.20 T. The magnetic field is directed at right angles to the direction of motion of the particle. a) Determine the radius of the circle in which the particle travels. b) Based on the diagram, determine the direction of motion in which the particle travels, i.e., either clockwise or counterclockwise as viewed from above.

```
              .  .  .  .  .  .  .  .
                 .  .  .  .  .  .  .  .
      α ⇒.  .  .  .  .  .  .  .
                 .  .  .  .  .  .  .  .
              .  .  .  .  .  .  .  .
```

Part a. Step 1.	Solution: (Section 20-4)
Determine the radius of the circle in which the particle travels.	The alpha particle is deflected by a magnetic force and travels in a circular path. The magnetic force provides the centripetal acceleration (a_c), thus Net $F = m\ a_c$ but $a_c = v^2/r$ $F_{magnetic} = m\ v^2/r$ $q\ v\ B \sin \theta = m\ v^2/r$ but $\theta = 90°$ and $\sin 90° = 1$ and solving for r $r = (m\ v)/(q\ B)$ $\quad = [(6.68 \times 10^{-27}\ kg)(5.0 \times 10^7\ m/s)]/[(3.2 \times 10^{-19}\ C)(0.20\ N/A\ m)]$ $r = 5.2\ m$
Part b. Step 1. Use the right-hand rule to determine the direction of motion of the particle.	The fingers point toward the right side of the page. Your hand is then arranged so that when you bend your fingers they point out of the paper in the direction of B. Your thumb points toward the bottom on the page. At the point where it enters the magnetic field, the particle is deflected downward. 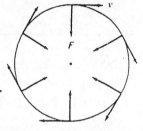 Applying the right-hand rule at several more points indicates that the particle travels in a clockwise circle as viewed from above.

EXAMPLE PROBLEM 4. An electron is accelerated through a potential difference of 100 volts into a magnetic field of 1.0 T. Determine the a) velocity of the particle as it enters the field, b) radius of the circle in which it travels, c) period of its motion. d) Based on the diagram shown below, determine the direction of motion of the electron in the magnetic field, i.e., either clockwise or counterclockwise as viewed from above. Assume that the initial velocity of the electron is zero. Note: $m_e = 9.1 \times 10^{-31}$ kg and $q_e = 1.6 \times 10^{-19}$ C.

x x x x x

x x x x x

x x x x x

↑

⊖

Part a. Step 1. Determine the electron's velocity as it enters the magnetic field.	Solution: (Section 20-4) The electron is accelerated through a potential difference and gains kinetic energy. The work done on the particle is equal to the product of the charge on the particle and the potential difference. $W = q\ V = \frac{1}{2}\ m\ v^2 - \frac{1}{2}\ m\ v_o^2$ but $v_o = 0$ m/s, thus $q\ V = \frac{1}{2}\ m\ v^2$ and rearranging gives $v = (2\ q\ V/m)^{\frac{1}{2}}$ $v = [2\ (1.6 \times 10^{-19}\ C)(100\ V)/(9.1 \times 10^{-31}\ kg)]^{\frac{1}{2}}$ $v = 5.9 \times 10^6$ m/s
Part b. Step 1. Determine the radius of the circle in which the electron travels.	The magnetic force provides the centripetal acceleration and causes the electrons to follow a circular path. $F_{magnetic} = m\ a_{centripetal}$ $q\ v\ B\ \sin \theta = m\ v^2/r$ but $\theta = 90°$ and $\sin 90° = 1$ solving for r gives $r = (m\ v)/(q\ B)$ $= [(9.1 \times 10^{-31}\ kg)(5.9 \times 10^6\ m/s)]/[(1.6 \times 10^{-19}\ C)(1.0\ N/A\ m)]$ $r = 3.4 \times 10^{-5}$ m
Part c. Step 1. Determine the period of the motion. Hint: the period (T) of the motion refers to the time required for the electron to complete one revolution.	The distance traveled in one revolution equals the circumference $(2\ \pi\ r)$ of the circle. The period can be determined by dividing the circumference of the circle by the velocity (v). $T = (2\ \pi\ r)/v$ $= [2\ \pi\ (3.4 \times 10^{-5}\ m)]/(5.9 \times 10^6\ m/s)$ $T = 3.6 \times 10^{-11}$ s

Part d. Step 1.	Based on the diagram, your fingers point toward the top of the page.
Determine the direction of motion of the particle. Note: remember that an electron carries a negative charge.	Your hand is then arranged so that when you bend your fingers they point into the paper in the direction of B. Your thumb points towards the left. However, since the particle carries a negative charge, the force is in the opposite direction. Therefore, the force particle is towards the right. Arbitrarily selecting a few more points along the path shows that the electron will travel in a clockwise circle as viewed from above.

The Mass Spectrograph

The velocity selector (see Chapter 27) allows only charged particles which have a particular velocity to pass undeflected regardless of the magnitude of their charge or their mass. Because of this it can be used along with a second magnetic field (B′) to determine the mass of an ion. As shown in the diagram, the particle passes through the velocity selector into magnetic field (B′) arranged perpendicular to its path. The mass of the particle is given by

$m = (q\ B′r)/v$ and since $v = E/B$, then $m = (q\ B\ B′\ r)/E$

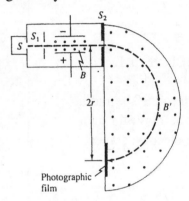

Ions of different elements will not follow the same path because they differ in mass. Even if a pure substance is used, particles having different mass are found. This is because elements contain isotopes, i.e., particles which have the same chemical and physical properties but which have a different number of neutrons in the nucleus. Two isotopes of the same element do not have the same mass; thus the ions will follow paths of different radius.

Galvanometer Movement

Galvanometer movement is the basis of most meters, i.e., ammeters, voltmeters, and ohmeters, as well as electric motors. Galvanometer movement is the result of interaction of the magnetic field of a permanent magnet which is directed perpendicular to a current carried by a loop or coil of wire. Based on the diagram, the force on each vertical segment of length a is

$F = N\ I\ a\ B \sin 90°$

where N refers to the number of turns of wire in the coil. If the coil is pivoted in the center, a torque is produced

which equals $\tau = N\ I\ a\ b\ B$

where b is the distance between the vertical segments. Since (a)(b) equals the area (A) of the coil, then

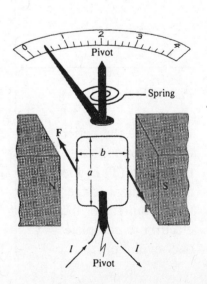

$\tau = N I A B$

and the quantity N I A is the **magnetic moment** of the coil.

Force between Two Parallel Conductors

Two parallel current-carrying conductors produce magnetic fields which result in a force between the conductors. If the currents are in the same direction, the force is of one attraction. A repulsive force results if the currents are in opposite directions. The force per unit length (F/ℓ) exerted by conductor 1 on conductor 2 and vice versa is given by

$F/\ell = (\mu_o I_1 I_2)/(2 \pi L)$ where L is the separation between the wires in meters and I_1 and I_2 represent the magnitude of the current in each conductor

---

EXAMPLE PROBLEM 5. Two long parallel wires carry currents of 2.0 A and 5.0 A, respectively. The wires are 0.20 m apart and the currents flow in the same direction. Determine the a) magnitude and direction of the force on a 0.50-m segment of the wire carrying the 2.0-A current, b) force on a 0.50-m segment of the wire carrying the 5.0-A current.

Part a. Step 1.	Solution: (Section 20-5)
Determine the magnitude of the magnetic field acting on the wire carrying the 2.0-A current.	The magnetic field acting on the wire is due to the magnetic field produced by the wire carrying the 5.0-A current. Therefore, $$B = [(4\pi \times 10^{-7} \text{ T m/A})(5.0 \text{ A})]/[(2\pi)(0.20 \text{ m})] = 5.0 \times 10^{-6} \text{ T}$$
Part a. Step 2. Determine the magnitude of the force acting on the wire carrying the 2.0-A current.	The force on a current-carrying wire is given by the equation $F = I \ell B \sin \theta$. $$F = (2.0 \text{ A})(0.50 \text{ m})(5.0 \times 10^{-6} \text{ N/A m}) \sin 90°$$ $$F = 5.0 \times 10^{-6} \text{ N}$$
Part a. Step 3. Determine the direction of the force acting on the 2.0-A current.	Based on the diagram shown at the top of the next page, and using the right-hand rule, the magnetic field produced by the 5.0-A current encircles the wire such that it is vertically into the page at the position of the 2.0-A current. Using the right-hand rule for the force produced by a magnetic field acting on a current-carrying wire, the fingers point toward the bottom of the page in the direction of the 2.0-A current. The magnetic field is inward, perpendicular to the plane of the paper.

The fingers bend until they point inward. The thumb points to the right. The force is therefore toward the right.

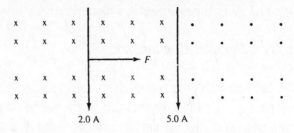

2.0 A   5.0 A

Part b. Step 1. Determine the magnitude of the magnetic field produced by the 2.0-A current.	The magnetic field acting on the 5.0-A current is due to the magnetic field produced by the 2.0-A current. $B = [(4 \pi \times 10^{-7} \text{ T m/A})(2.0 \text{ A})]/[(2\pi)(0.20 \text{ m})]$ $B = 2.0 \times 10^{-6} \text{ N/A m}$
Part b. Step 2. Determine the force acting on the wire carrying the 5.0-A current.	$F = I \ell B \sin \theta \qquad \text{but} \qquad \theta = 90°.$ $F = (5.0 \text{ A})(0.50 \text{ m})(2.0 \times 10^{-6} \text{ N/A m})(\sin 90°)$ $F = 5.0 \times 10^{-6} \text{ N}$
Part b. Step 3. Determine the direction of the force acting on the wire carrying the 5.0-A current.	The magnetic field produced by the 2.0-A current is directed perpendicularly out of the plane of the paper at the position of the 5.0-A current. The fingers point along the direction of the 5.0-A current. The fingers bend to point in the direction of the magnetic field. The thumb points toward the left in the direction of the wire carrying the 2.0-A current. Therefore, the two wires are attracted to one another. Note: the answer to part b can be predicted by using Newton's third law of motion. The magnetic field produced by one wire interacts with the current in the other wire. Each wire is acted upon by a force due to the other wire. The forces should be equal in magnitude but opposite in direction. Alternate solution: As described in section 20-5 of the textbook, the force per unit length between two long, parallel, current-carrying wires is given by $F/\ell = (\mu_o I_1 I_2)/(2 \pi L)$

where L is the separation between the wires and I_1 and I_2 are the currents.

Substituting the data given in the problem:

$F/(0.50 \text{ m}) = [(4\pi \times 10^{-7} \text{ T m/A})(2.0 \text{ A})(5.0 \text{ A})]/[2\pi(0.20 \text{ m})]$

$F = 5.0 \times 10^{-6} \text{ N}$

Ferromagnetism; Domains

If an object is made from a **ferromagnetic material**, such as iron, each atom has a net magnetic effect. The atoms tend to align their magnetic fields in arrangements known as **domains** and each domain contributes to the overall magnetic field of the piece of iron. In an ordinary piece of iron or other ferromagnetic material, the magnetic field produced by the individual domains cancel out so that the object is not a magnet. In a magnet, the domains are larger in one direction than in any other and a net magnetic effect is produced.

If an unmagnetized ferromagnetic object is placed in a magnetic field, the domains which are aligned with the direction of the external magnetic field tend to grow. The increase in the size of the domains comes at the expense of neighboring domains which are not aligned with the external field. The result is that the unmagnetized object is attracted to the external magnetic field and the object is said to exhibit **ferromagnetism.**

When the object is removed from the external field the domains may remain aligned and the object retains a net magnetic effect. The object may lose this net magnetic effect if it is struck, dropped, or heated. Above a certain temperature, known as the **Curie temperature**, it is not possible to magnetize an object and a permanent magnet loses its magnetic effect.

PROBLEM SOLVING SKILLS

For problems involving a current-carrying wire in a magnetic field:

1. Draw an accurate diagram showing the orientation of the wire in the magnetic field. Use the adopted conventions to indicate the direction of the current and the magnetic field.
2. Use the right-hand rule to determine the direction of the force on the wire.
3. Use $F = I \ell B \sin \theta$ to determine the magnitude of the force.

For problems involving a charged particle traveling through a magnetic field:

1. Draw an accurate diagram showing the motion of the particle through the magnetic field. Use the adopted conventions to indicate the direction of the magnetic field and the direction of motion of the particle.
2. Take note of whether the particle is positively charged or negatively charged and then use the

right-hand rule to determine the direction of the force on the particle.

3. If the particle is deflected into circular motion, the magnetic force produces a centripetal acceleration. Use Newton's second law to determine the magnitude of the acceleration.
4. If the speed of the particle as it enters the magnetic field is given or can be calculated, it is possible to determine the radius of the circle in which the particle travels.

For problems involving the magnetic field produced by a current-carrying wire:

1. Draw a diagram showing the orientation of the straight wire or the loop of wire and the direction of the current in the wire.
2. Use the second right-hand rule to determine the direction of the magnetic field. Use the adopted conventions to indicate the direction of the magnetic field.
3. Use the appropriate formula to determine the magnitude of the magnetic field.

SOLUTIONS TO SELECTED TEXTBOOK PROBLEMS

TEXTBOOK PROBLEM 8. Suppose a straight 1.00-mm-diameter copper wire can "just" float horizontally in air because of the force of the Earth's magnetic field $\vec{B}$, which is horizontal, perpendicular to the wire, and of magnitude 5.0×10^{-5} T. What current would the wire carry? Does the answer seem feasible? Explain briefly.

Part a. Step 1.	Solution: (Section 20-3)
Determine the density of copper and the radius of the wire in meters.	From table 10-1, the density of copper $\rho_{Cu} = 8.9 \times 10^3$ kg/m^3 radius = ½ diameter = ½(1.00 mm) = 0.50 mm = 5.0×10^{-4} m
Part a. Step 2. Derive a formula for the wire's mass in terms of its density and volume.	$\rho = m/V$ where $V = \pi r^2 \ell$ $m = \rho V = \rho \pi r^2 \ell$
Part a. Step 3. The magnetic force balances gravity. Use $F = I \ell B \sin \theta$ to determine the magnitude of the current.	The wire is suspended at rest; therefore, net F = 0. $F_{magnetic} - F_{gravity} = 0$ $I \ell B \sin 90° - m g = 0$ $I = (m g)/(\ell B) = [(\rho \pi r^2 \ell) g]/[\ell B] = [(\rho \pi r^2) g]/B$ $I = [(8.9 \times 10^3 \text{ kg/m}^3)(\pi)(5.0 \times 10^{-4} \text{ m})^2(9.8 \text{ m/s}^2)]/(5.0 \times 10^{-5} \text{ T})$ $I = 1370$ A The answer is not feasible. A current of this magnitude would quickly melt the wire.

Part a. Step 1. Express the proton's kinetic energy in joules.	Solution: (Section 20-4) KE = (5.0 MeV)(1.6 x 10^{-13} J)/(1 MeV) = 8.0 x 10^{-13} J
Part a. Step 2. Determine the proton's velocity.	KE = ½ m v^2 8.0 x 10^{-13} J = ½ (1.67 x 10^{-27} kg) v^2 v^2 = 9.6 x 10^{14} m^2/s^2 v = 3.1 x 10^7 m/s
Part a. Step 3. Determine the radius of the proton's path.	The proton is deflected by a magnetic force and travels in a circular path. The magnetic force provides the centripetal acceleration (a_R), therefore, Net F = m a_R but a_R = v^2/r $F_{magnetic}$ = m v^2/r q v B sin θ = m v^2/r but θ = 90° and sin 90° = 1 and solving for r r = (m v)/(q B) = [(1.67 x 10^{-27} kg)(3.1 x 10^7 m/s)]/[(1.6 x 10^{-19} C)(0.20 T)] r = 1.6 m

Part a. Step 1. Determine the strength of the magnetic field produced by the wire.	Solution: (Sections 20-5 and 20-6) The Earth's magnetic field is approximately 5.0 x 10^{-5} T. B_{wire} ≈ (0.01)(5.0 x 10^{-5} T) ≈ 5.0 x 10^{-7} T

Part a. Step 2. Determine the maximum current that the wire can carry.	Assume that the wire is a long straight wire. $B_{wire} = (\mu_o I)/(2\pi r)$ $I = [(B_{wire})(2\pi r)]/\mu_o$ $I \approx [(5.0 \times 10^{-7}\ T)(2\pi)(1.0\ m)]/(4\pi \times 10^{-7}\ T\ m/A)$ $I \approx 2.5\ A$

TEXTBOOK PROBLEM 37. Determine the magnetic field midway between two long straight wires 2.0 cm apart in terms of the current I in one when the other carries 15 A. Assume these currents are (a) in the same direction, and (b) in opposite directions.

Part a. Step 1. Determine the magnitude of the magnetic field produced by each wire at the midpoint between the wires.	Solution: (Sections 20-5 and 20-6) $B_1 = (\mu_o I_1)/(2\pi r) = [(4\pi \times 10^{-7}\ T\ m/A)\ I]/[2\pi(0.010\ m)]$ $B_1 = 2.0 \times 10^{-5}\ I\ T$ $B_2 = (\mu_o I_2)/(2\pi r) = [(4\pi \times 10^{-7}\ T\ m/A)(15\ A)]/[2\pi(0.010\ m)]$ $B_2 = 3.0 \times 10^{-4}\ T$

Part a. Step 2.
 	
Determine the magnitude of the resultant magnetic field when the currents are in the same direction.
	Using the right-hand rule, it can be shown that when the currents are in the same direction, the magnetic fields oppose each other.
	
	$B_{total} = B_1 - B_2 = (2.0 \times 10^{-5}\ I\ T) - 3.0 \times 10^{-4}\ T$
	
	$B_{total} = (2.0 \times 10^{-5}\ I - 3.0 \times 10^{-4})\ T$ $\Uparrow B_2$
	       $\otimes$. $\otimes$
	     I $\Downarrow B_1$ $15\ A$
	The direction of the resultant magnetic field is in the direction of the field that has the greater magnitude.

Part b. Step 1.
 	
Determine the magnitude of the resultant magnetic field when the currents are in the opposite direction.
	Using the right-hand rule, it can be shown that when the currents are in the opposite direction, the magnetic fields are in the same direction.
	
	$B_{total} = B_1 + B_2 = (2.0 \times 10^{-5}\ I\ T) + 3.0 \times 10^{-4}\ T$
	        $B_1 \Uparrow\Uparrow B_2$
	$B_{total} = (2.0 \times 10^{-5}\ I + 3.0 \times 10^{-4})\ T$ $\odot$. $\otimes$
	           I $15\ A$
	Based on the diagram, the direction of the resultant magnetic field is vertically upward.

Part a. Step 1.	Solution: (Sections 20-9 and 20-10)
Determine the velocity of the electron.	The magnetic force provides the centripetal acceleration and causes the electrons to follow a circular path.
	$F_{magnetic} = m \, a_{centripetal}$
	$q \, v \, B \sin \theta = m \, v^2/r$ but $\theta = 90°$ and $\sin 90° = 1$
	solving for v gives
	$v = (q \, B \, r)/m$
	$= [(1.6 \times 10^{-19} \text{ C})(0.566 \text{ T})(0.051 \text{ m})]/(1.67 \times 10^{-27} \text{ kg})$
	$v = 2.77 \times 10^6$ m/s

Part a. Step 2.	In order for the protons to travel in a straight line, the force exerted by the electric field must be equal but opposite to the force exerted by the magnetic field.
Determine the magnitude of the electric field.	
	$F_{electric} = F_{magnetic}$
	$q \, E = q \, v \, B$
	$E = v \, B$
	$E = (2.77 \times 10^6 \text{ m/s})(0.566 \text{ T}) = 1.57 \times 10^6$ N/C

Part a. Step 3.	In order for the electric force to balance the magnetic force, the two fields must be perpendicular to each other. As shown in the following diagram:
Use the right-hand rule to determine the direction of the magnetic field.	

```
 +   +    +   +
    ↑F_magnetic
    ⊕  →
    ↓F_electric


 -   -    -   -
```

Using the right hand rule, the fingers point toward the right side of the page. Your thumb is arranged to point in the direction of the magnetic force. When you bend your fingers they point into the paper. Therefore, the magnetic field is directed perpendicular to the direction of the electric field.

TEXTBOOK PROBLEM 72. A doubly charged helium atom, whose mass is 6.6×10^{-27} kg, is accelerated by a voltage of 2400 V. (a) What will be its radius of curvature in a uniform 0.240-T field? (b) What is its period of revolution?

Part a. Step 1. Determine the ion's velocity as it enters the magnetic field.	Solution: (Section 20-4) The ion is accelerated through a potential difference and gains kinetic energy. The work done on the particle is equal to the product of the charge on the particle and the potential difference. $W = q\,V = KE_f - KE_i$ where $q = 2(1.6 \times 10^{-19} C) = 3.2 \times 10^{-19} C$ $q\,V = \frac{1}{2}\,m\,v^2 - \frac{1}{2}\,m\,v_i^2$ but $v_i = 0$ m/s; therefore, $q\,V = \frac{1}{2}\,m\,v^2$ and rearranging gives $v = (2\,q\,V/m)^{\frac{1}{2}}$ $v = [2(3.2 \times 10^{-19}\ C)(2400\ V)/(6.6 \times 10^{-27}\ kg)]^{\frac{1}{2}}$ $v = 4.8 \times 10^5$ m/s
Part a. Step 2. Determine the radius of the circle in which the electron travels.	The magnetic force provides the centripetal acceleration and causes the electrons to follow a circular path. $F_{magnetic} = m\,a_{centripetal}$ $q\,v\,B\,\sin\theta = m\,v^2/r$ but $\theta = 90°$ and $\sin 90° = 1$ solving for r gives $r = (m\,v)/(q\,B)$ $= [(6.6 \times 10^{-27}\ kg)(4.8 \times 10^5\ m/s)]/[(3.2 \times 10^{-19}\ C)(0.240\ N/A\ m)]$ $r = 4.1 \times 10^{-2}$ m
Part b. Step 1. Determine the period of the motion. Hint: the period (T) of the motion refers to the time required for the electron to complete one revolution.	The distance traveled in one revolution equals the circumference $(2\,\pi\,r)$ of the circle. The period can be determined by dividing the circumference of the circle by the velocity (v). $T = (2\,\pi\,r)/v$ $= [2\,\pi\,(4.1 \times 10^{-2}\ m)]/(4.8 \times 10^5\ m/s)$ $T = 5.4 \times 10^{-7}$ s

CHAPTER 21

ELECTROMAGNETIC INDUCTION AND FARADAY'S LAW

OBJECTIVES

After studying the material of this chapter, the student should be able to:

- determine the magnitude of the magnetic flux through a surface of known area, given the strength of the magnetic field and the angle between the direction of the magnetic field and the surface.
- write a statement of Faraday's law in terms of changing magnetic flux. Use Faraday's law to determine the magnitude of the induced emf in a closed loop due to a change in the magnetic flux through the loop.
- use Faraday's law to determine the magnitude of the induced emf in a straight wire moving through a magnetic field.
- state Lenz' law and use Ohm's law and Lenz's law to determine the magnitude and direction of the induced current.
- explain the basic principle of the electric generator. Determine the magnitude of the maximum value of the induced emf in a loop which is rotating at a constant rate in a uniform magnetic field.
- explain how an eddy current can be produced in a piece of metal. Also, describe situations in which eddy currents are beneficial and situations when they must be eliminated.
- explain how a transformer can be used to step-up or step-down the voltage. Apply the equations which relate number of turns, voltages, and currents in the primary and secondary coils to solve transformer problems.
- explain what is meant by mutual-inductance and self-inductance. List the factors which determine the self-inductance of a solenoid. State the SI unit for inductance.
- write the equations for the average induced emf in a solenoid in which the current is changing at a known rate.
- write the equation for the energy stored in an inductor's magnetic field and also for the energy stored per unit volume, i.e., the energy density.
- write the equation for the voltage across the inductor as a function of time after the inductor is connected to the source of emf in an LR circuit. Graph the current as a function of time after the initial connection is completed. Write the equation for the voltage across the inductor as a function of time if the inductor is disconnected from the source of emf and discharged.

Graph the current as a function of time after the discharge is initiated.
- distinguish between resistance, capacitive reactance, inductive reactance, and impedance in an LR or LRC circuit. Calculate the reactance of a capacitor and/or inductor which is connected to a source of known frequency.
- use a phasor diagram to determine the phase angle and total impedance for an LR, LC, or LRC circuit.
- determine the rms current and power dissipated in an LRC circuit. Determine the voltage drop across each circuit element and the resonant frequency of the circuit.

KEY TERMS AND PHRASES

magnetic flux is the product of the magnetic field strength (flux density), the area of the plane of the loop through which the magnetic field passes, and the cosine of the angle that the magnetic field makes with a line drawn normal to the plane of the loop.

Faraday's law states that a voltage is produced in a conductor when the magnetic flux through the conductor changes. The voltage produced is called the induced emf.

Lenz's law states that an induced emf produces a current whose magnetic field always opposes the change in magnetic flux which caused it.

electric motor is essentially a galvanometer movement that is arranged so that a coil of wire, referred to as the armature, runs continuously. The electric current through the armature interacts with the external field and the resulting torque causes the armature to rotate. Depending on the design of the motor, the motor can be arranged to run on direct current (dc) or alternating current (ac) electricity.

electric generator uses mechanical work to produce electric energy. The armature of the generator is turned by an external torque and an emf is induced as the coil passes through an external magnetic field.

back or **counter emf** is produced as the armature of a motor rotates in the external magnetic field. This induced emf produces a torque which opposes the motion of the armature.

eddy current is an induced current produced by a changing magnetic flux in a piece of metal. The direction of the induced current is such as to oppose the magnetic field which caused the current.

mutual induction occurs when a changing current in one circuit induces a current in another circuit.

self-induction occurs when a changing current in a circuit induces a back emf in the circuit.

transformer is a device used to increase or decrease an ac voltage.

LRC series circuit consists of an inductor, capacitor, and a resistor connected in series with an alternating current source of voltage.

phasor is a rotating vector. A phasor is used to represent either current or voltage in an ac circuit. Vector algebra is used to analyze an LRC circuit.

inductive reactance is a measure of the effect an inductor has on the current through an ac circuit. It is analogous to the effect a resistor has to the flow of current through a dc circuit.

capacitative reactance is a measure of the effect a capacitor has on the current through an ac circuit. It is analogous to the effect a resistor has to the flow of current through a dc circuit.

impedance in an ac circuit is analogous to resistance in a dc circuit. Impedance results from the combination of inductive reactance, capacitative reactance, and resistance.

resonant frequency is the frequency at which an LRC circuit has minimum impedance to current flow. At this frequency current flow is a maximum and the energy transferred to the system is a maximum.

SUMMARY OF MATHEMATICAL FORMULAS

magnetic flux	$\Phi_B = B\ A\ \cos\theta$	Magnetic flux (Φ_B) is the product of the magnetic field strength (B), the area (A) of the plane of the loop through which the magnetic field passes, and the cosine of the angle ($\cos\theta$) that the magnetic field makes with a line drawn normal to the plane of the loop.
Faraday's law	$\xi = -N\ \Delta\Phi_B/\Delta t$	The magnitude of the induced emf (ξ) in a coil depends on the number of turns (N) in the loop and the rate of change of flux ($\Delta\Phi_B/\Delta t$) through the loop.
EMF induced in a moving conductor	$\xi = v\ \ell\ B\ \sin\theta$	The emf (ξ) induced in a conducting rod or wire which moves through a magnetic field depends on the velocity (v) of the rod relative to the magnetic field, the length (ℓ) of the rod, and the angle (θ) between the direction of motion of the rod and the direction of the magnetic field.

emf produced by an electric generator	$\xi = N A B \omega \sin \omega t$ $\xi_o = N A B \omega$	The magnitude of the emf (ξ) produced by a generator rotating continuously with a constant angular velocity (ω) depends on the number of turns in the armature (N), the area of the loop (A), the magnitude of the magnetic field (B), and the angle (ωt) that the face of the loop makes with the direction of the magnetic field at time t. ξ_o represents the peak voltage of the generator.
"ideal" transformer	$V_s/V_p = N_s/N_p$ $I_p V_p = I_s V_s$ $I_s /I_p = N_p/N_s$	The voltage in the secondary coil (V_s) depends on the voltage in the primary coil (V_p) and the number of turns in each coil, N_s and N_p. In an ideal transformer, the power in the primary ($P_{input} = I_p V_p$) equals the power in the secondary ($P_{output} = I_s V_s$) The ratio of the currents is related to the ratio of the turns.
induced voltage due to mutual inductance	$\xi = M \Delta I/\Delta t$	The induced voltage (ξ) in one coil depends on the mutual inductance (M) and is the rate of change of the current in a second coil ($\Delta I/\Delta t$).
induced voltage due to self-inductance	$\xi = - L \Delta I/\Delta t$ $L = \mu_o N^2 A/\ell$	The induced voltage in a single coil depends on the self-inductance (L) in a single coil and the rate of change of current in the coil ($\Delta I/\Delta t$). The self inductance (L) of a long coil, called a solenoid, of length ℓ, cross-sectional area A, and N turns

energy stored in an inductor	energy = ½ L I^2	The energy stored in a coil is related to the self-inductance (L) and the current (I).
	energy = ½ B^2 A ℓ/μ_o	The energy stored in the inductor's magnetic field is related to the magetic field strength (B) and the volume enclosed by the windings (A ℓ).
energy density	energy density = ½ B^2/μ_o	the energy stored per unit volume
current rise in an LR circuit	$I = \xi/R\,(1 - e^{-t/\tau})$	The current (I) in an LR circuit is related to the emf of the source (ξ), the resistance (R), the time constant (τ), and time (t).
	$\tau = L/R$	the inductive time constant of an L circuit
current decay in an LR circuit	$I = \xi/R\,e^{-t/\tau}$	The current (I) in an LR decay circuit depends on the emf of the source (ξ), the resistance (R), the time constant (τ), and time (t).
inductive reactance	$X_L = 2\,\pi\,f\,L = \omega\,L$	The inductive reactance (X_L) is related to both the frequency of the source (f) and the self-inductance of the coil (L).
capacitative reactance	$X_C = 1/2\pi fC = 1/\omega C$	The capacitive reactance (X_C) is inversely proportional to the frequency of the source and the capacitance of the capacitor.
LRC series circuit peak voltage impedance	$V_o = [(V_{Ro})^2 + (V_{Lo} - V_{Co})^2]^{1/2}$ $Z = [(X_L - X_C)^2 + R^2]^{1/2}$	The peak voltage (V_o) across the source is related to the peak voltage across the resistor (V_{Ro}), the inductor (V_{Lo}), and the capacitor (V_{Co}). The impedance (Z) is related to the inductive reactance (X_L), the capacitative reactance (X_C), and the resistance (R).

phase angle	$\tan \phi = (X_L - X_C)/R.$	The phase angle (ϕ) between the voltage and current equals the ratio of the difference between the inductive and capacitative reactance ($X_L - X_C$) and the resistance (R).
power dissipated by the impedance	$\bar{P} = I\ V \cos \phi$ $\bar{P} = I^2\ Z \cos \phi$	Power dissipated in the form of heat is related to the current (I), voltage (V), impedance (Z), and the power factor $\cos \phi$.
power factor	$\cos \phi = R/Z$	$\cos \phi$ is defined as the power factor of the LRC circuit. For a resistor, $\phi = 0°$ and $\cos 0° = 1$. Therefore, all of the power is dissipated in the form of joule heating. For a capacitor or an inductor, $\phi = 90°$ and $\cos 90° = 0$. Therefore, no heat is dissipated by a capacitor (or inductor); all of the energy is stored in the electric (or magnetic) field.
resonant frequency for an LRC circuit	$f_o = 1/[2\pi\ (L\ C)^{1/2}]$	Resonance for an LRC circuit occurs when the impedance is a minimum. At resonance, the energy transferred to the system from the source of emf is a maximum. The resonant frequency is inversely related to the square root of the inductance and the capacitance.

CONCEPT SUMMARY

Faraday's Law of Induction

If a bar magnet is moved toward a coil of wire, an emf will be induced in the wire. The magnitude of the induced emf depends on the magnetic field strength (or **flux density**), the area of the loop, and the time required for the change in **magnetic flux** through the area of the loop to occur. The product of the magnetic flux density (B) and the area of the plane of the loop through which it passes is known as the magnetic flux (Φ_B).

If the magnetic field is given in Wb/m^2 and the area of the plane of the loop in m^2, then the flux has the unit of webers (Wb). The formula for the magnetic flux is

$\Phi_B = B\ A \cos \theta$

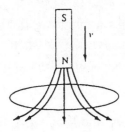 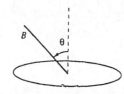

θ is the angle between the direction of B and a line drawn perpendicular to the plane of the loop. If the angle is 0°, the flux passing through the loop is a maximum, cos 0° = 1.0. If the angle is 90°, then no flux passes through the loop, cos 90° = 0.

The magnitude of the induced emf (ξ) in the coil is given by **Faraday's law**:

$$\xi = - N \, \Delta \Phi_B / \Delta t$$

N is the number of turns in the loop and $\Delta \Phi_B / \Delta t$ is the rate of change of flux through the loop.

TEXTBOOK QUESTION 1. What would be the advantage, in Faraday's experiments (Fig. 21-1), of using coils with many turns?

ANSWER: The magnetic field produced in the primary coil which is attached to the battery increases with the number of turns present. Faraday's law states that the voltage (ξ) produced in the secondary coil depends on the rate of change in the magnetic flux ($\Delta \Phi_B / \Delta t$) through the coil changes, where $\Phi_B = B \, A \cos \theta$. Therefore, as the number of turns in the secondary (N) increases, the voltage in the secondary increases, i. e., $\xi = - N \, \Delta \Phi_B / \Delta t$.

Lenz's Law

The minus sign in the Faraday's law formula indicates that the induced emf opposes the change in flux through the loop. This opposition is described by **Lenz's law**, which states that an induced emf produces a current whose magnetic field always opposes the change in magnetic flux which caused it.

An example of Lenz's law would be that of a north pole of a bar magnet being inserted into a coil of wire as shown in the figure below. The x's represent the magnetic field of the bar magnet. According to Lenz's law, an induced current will be produced in the counterclockwise direction through the coil. Using the right-hand rule, it can be determined that this induced current produces a magnetic field (dots in figure) which opposes the magnetic field of the bar magnet.

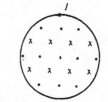

If the bar magnet is held motionless, the flux is no longer changing ($\Delta \Phi_B / \Delta t = 0$) and the induced current disappears. However, if the magnet is withdrawn, an induced current will be produced in the clockwise direction. This current produces a magnetic field which attempts to maintain the magnetic flux through the loop at the same magnitude as when the bar magnet was motionless in the loop. Since the loop contains electrical resistance, this current will quickly be reduced to zero.

It is the relative motion between the magnetic field and the loop which causes the induced emf. Thus, it is possible to induce an emf in a loop by 1) holding the magnetic field constant and changing the area of the loop through which the magnetic flux passes, 2) holding the loop motionless and changing the magnitude of the flux through the loop, or 3) changing the orientation of the loop in the magnetic field by rotating the loop in the field.

TEXTBOOK QUESTION 15. A bar magnet falling inside a vertical metal tube reaches a terminal velocity even if the tube is evacuated so that there is no air resistance. Explain.

ANSWER: The falling magnet causes a change in flux in the region of the tube through which it is passing. According to Faraday's law, this change of flux induces an emf. The induced emf produces a current, which in turn produces a magnetic field which opposes the motion of the falling magnet. The speed of the falling magnet increases until the fields are equal and opposite. The bar magnet stops accelerating and falls at a constant speed which is called the terminal velocity.

EXAMPLE PROBLEM 1. The magnitude of the magnetic field through a 50-turn loop of wire changes from zero to 2.0 Wb/m² in 0.20 s. The radius is 0.40 m and its electrical resistance is 5.0 ohms. a) Determine the magnitude of the average induced emf and the current in the loop. b) Determine the direction of the induced current in the loop if the loop is in the plane of the paper and the magnetic field is directed out of the paper.

Part a. Step 1.	Solution: (Sections 21-1 to 21-4)
Determine the area of of the loop of wire.	$A = \pi r^2 = \pi(0.40 \text{ m})^2 = 0.50 \text{ m}^2$
	Note: The surface area vector is directed out of the plane of the page.
Part a. Step 2.	The magnetic field is directed perpendicular to the coil. Using the convention discussed in the textbook, the angle θ is 0°.
Determine the change in magnetic flux through the loop.	$\Delta \Phi = \Phi_f - \Phi_i$
	$\quad = B_f A \cos \theta - B_i A \cos \theta = (B_f - B_i) A \cos \theta$
	$\Delta \Phi = (2.0 \text{ Wb/m}^2 - 0 \text{ Wb/m}^2)(0.50 \text{ m}^2) \cos 0° = 1.0 \text{ Wb}$

Part a. Step 3. Use Faraday's law to determine the magnitude of the induced emf and Ohm's law to determine the induced current.	$\xi = -N \Delta\Phi/\Delta t = -(50)(1.0 \text{ Wb}/0.20 \text{ s})$ $\xi = -250 \text{ V}$ $I = \xi/R = -250 \text{ V}/5.0 \text{ }\Omega = -50 \text{ A}$

Part b. Step 1.

Draw a diagram showing the loop and the external magnetic field.

. external magnetic field

Part b. Step 2.

Use Lenz's law to predict the direction of the induced current.

Based on Lenz's law, the direction of the induced current is such that it must oppose the external magnetic field that caused it. In the diagram shown below, the external magnetic field is directed out of the paper, the current in the loop must produce a magnetic field which is directed into the paper. The right-hand rule indicates that a clockwise current in the loop will produce a magnetic field that is directed into the paper.

. external magnetic field

x magnetic field produced by the induced current

EXAMPLE PROBLEM 2. After the external magnetic field in loop described in the problem 1 reaches a steady value of 2.0 Wb/m^2, the loop is rotated through 90° in 0.50 s. Determine the magnitude of the induced emf in the loop.

Part a. Step 1.

Determine the change in flux through the loop.

Solution: (Sections 21-1 to 21-4)

The area of the loop and the strength of the external magnetic field are not changing. However, the angle between the loop and the magnetic field changes from 0° to 90°. An emf will be induced in the loop because as the loop rotates it cuts through lines of magnetic flux. The change in flux can be determined as follows:

$$\Delta\Phi = \Phi_f - \Phi_i$$

$$= B A \cos 90° - B A \cos 0° = B A (\cos 90° - \cos 0°)$$

	$= (2.0\ \text{Wb/m}^2)(0.50\ \text{m}^2)(0 - 1.0)$
	$\Delta\Phi = -1.0\ \text{Wb}$
Part a. Step 2.	$\xi = -N\ \Delta\Phi/\Delta t$
Use Faraday's law to determine the magnitude of the induced emf.	$= -(50)(-1.0\ \text{Wb}/0.50\ \text{s})$
	$\xi = +100\ \text{V}$

EMF Induced in a Moving Conductor

An emf is induced in a conducting rod or wire which moves through a magnetic field. This can be predicted by the right-hand rule. In the diagram shown below, a conducting rod is moved along a U-shaped conductor in a uniform magnetic field that points into the paper. The positive charges in the rod are moving with the rod as the rod moves through the magnetic field. Using the right-hand rule, the outstretched fingers point in the direction that the rod is moving. When the fingers bend they must point in the direction of the magnetic field, i.e., into the paper. The thumb points toward point A, indicating that the positive charges tend to move away from B and toward A.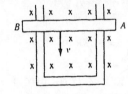

A difference in potential exists between the ends of the rod. This difference in potential is the induced emf. The induced emf in the wire is given by the formula

$\xi = v\ \ell\ B \sin\theta$

where v is the velocity of the rod relative to the magnetic field, ℓ is the length in meters of the segment of the rod, and θ is the angle between the direction of motion of the rod and the direction of the magnetic field.

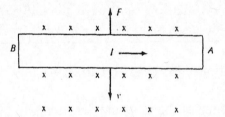

Since the rod is part of a closed circuit, a current will flow through the circuit. As shown in the diagram, the current flows from B toward point A and then clockwise through the circuit. The magnitude of this induced current is determined from Ohm's law, $\hat{I} = \xi/R$.

The direction of the induced current is such that a force is produced that opposes the external force that causes the rod to move through the magnetic field. This also can be predicted by using the right-hand rule. The outstretched fingers point in the direction of the induced current. When the fingers bend to point in the direction of the magnetic field, the thumb of the right hand points toward the top of the page. The force F, which is produced by the interaction of the induced current and the magnetic field, opposes the motion of the wire. This is another example of Lenz's law.

If the rod travels at constant speed and friction between the rod and the U-shaped conductor is negligible, the induced force is equal in magnitude but opposite in direction to the external force moving the rod through the field.

EXAMPLE PROBLEM 3. A metal rod, 1.0 m in length, is rotated at 10.0 rev/s in a circle about one end. As it rotates, it cuts perpendicularly through a magnetic field of strength 5.0 Wb/m² as shown in the diagram. Determine the magnitude of the average induced emf in the rod.

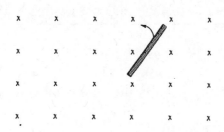

Part a. Step 1.	Solution: (Sections 21-1 to 21-4)
Determine the area "swept out" by the rod during one revolution.	$A = \pi r^2 = \pi (1.0 \text{ m})^2$ $A = 3.14 \text{ m}^2$
Part a. Step 2.	$\Delta \Phi = B \, \Delta A$
Determine the change in flux through the rod during one revolution.	$= (5.0 \text{ Wb/m}^2)(3.14 \text{ m}^2 - 0 \text{ m}^2)$ $\Delta \Phi = 16 \text{ Wb}$
Part a. Step 3. Determine the time required for the wire to complete one revolution.	The time required for the rod to complete one revolution equals the period of the motion. $T = 1/f = 1/(10.0 \text{ rev/s})$ $T = 0.10 \text{ s}$

Part a. Step 4.	Note: N = 1 since there is only one wire.
Use Faraday's law to determine the magnitude of the induced emf.	$\xi = - N \, \Delta\Phi/\Delta t = - (1)(16 \text{ Wb})/(0.10 \text{ s})$
	$\xi = - 160 \text{ wb/s} = - 160 \text{ volts}$

EXAMPLE PROBLEM 4. The metal rails of the U-shaped conductor in the diagram shown on page 21-10 are 1.00 m apart and the magnetic field is 0.100 Wb/m². The magnetic field is directed perpendicular to the paper. The electrical resistance of the closed circuit is 2.00 ohm and is assumed to be constant. Determine the a) magnitude and direction of the induced current in the circuit when the rod is moving at 4.00 m/s, b) magnitude and direction of the external force required to keep the rod moving at a constant speed, and c) power required to keep the rod moving at constant speed as compared to the rate of Joule heating in the circuit.

Part a. Step 1.	Solution: (Sections 21-1 to 21-4)
Determine the magnitude of the induced emf.	Since every point in the wire is moving at the same speed, use
	$\xi = v \, \ell \, B \sin \theta$
	$= (4.00 \text{ m/s})(1.00 \text{ m})(0.100 \text{ Wb/m}^2) \sin 90°$
	$\xi = 0.400 \text{ Wb/s} = 0.400 \text{ volts}$
Part a. Step 2.	$I = \xi/R$
Determine the magnitude of the induced current.	$I = (0.400 \text{ V})/(2.00 \text{ }\Omega) = 0.200 \text{ A}$
Part a. Step 3. Use Lenz's law to determine the direction of the induced current.	Lenz's law predicts that the induced current must oppose the change that causes it. The current opposes the change by interacting with the external magnetic field to produce a force which opposes the rod's motion. Using the right-hand rule, the current can be shown to be flowing from B to A and therefore in a clockwise direction through the U-shaped conductor.
Part b. Step 1. Determine the magnitude of the external force.	Note: 1 Wb/m² = 1 T = 1 N/A m $F = I \, \ell \, B \sin \theta$ $= (0.200 \text{ A})(1.00 \text{ m})(0.100 \text{ N/A m}) \sin 90°$ $F = 0.0200 \text{ N}$

Part c. Step 1. Determine the mechanical power required to keep the rod moving at constant speed.	The external force is moving the wire through the magnetic field, the angle between the force and the direction of motion is 0°. The wire travels at constant speed, therefore $P = F v \cos 0°$ $P = (0.0200 \text{ N})(4.00 \text{ m/s})(\cos 0°) = 0.0800$ watts
Part c. Step 2. Determine the rate of Joule heating.	$P = I^2 R = (0.200 \text{ A})^2(2.00 \ \Omega) = 0.0800$ watts The power supplied by the external force is dissipated in the form of heat in the circuit.

EXAMPLE PROBLEM 5. A 0.100-m long wire segment oriented in the east-west direction is attached by a thin, flexible wire to a galvanometer. The wire segment has a mass of 0.0100 kg and the electrical resistance in the circuit is 1.00 Ω. The wire is dropped from rest into a magnetic field of strength 0.200 T which is directed toward the north. Determine the a) maximum velocity achieved by the wire and b) direction of the induced current in the circuit. Assume that air resistance is negligible.

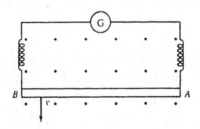

Part a. Step 1. Locate the forces acting on the wire. What is the net force on the wire when it reaches its maximum velocity?	Solution: (Sections 21-1 to 21-4) Gravity causes the wire to accelerate downward. As it accelerates, the induced current gradually increases and the force produced by the interaction of the induced current increases until it is equal but opposite to the wire's weight. The net force on the wire when it reaches maximum speed is zero.
Part a. Step 2. Determine the magnitude of the current flowing through the wire.	$F_{gravity} = F_{magnetic}$ $mg = I \ell B \sin \theta$ but $\theta = 90°$ and $\sin 90° = 1.0$ $I = (mg)/(\ell B) = [(0.0100 \text{ kg})(9.8 \text{ m/s}^2)]/[(0.100 \text{ m})(0.200 \text{ N/A m})]$ $I = 4.90$ A

Part a. Step 3. Use Ohm's law to determine the magnitude of the induced emf.	$\xi = I R$ $\quad = (4.90 \text{ A})(1.00 \ \Omega)$ $\xi = 4.90$ volts
Part a. Step 4. Determine the maximum velocity of the wire.	$\xi = v \ \ell \ B \sin \theta$ $4.90 \text{ volts} = v \ (0.100 \text{ m})((0.200 \text{ N/A m})(\sin 90^\circ)$ $v = 250$ m/s
Part b. Step 1. Use the right-hand rule to determine the direction of the induced current.	Since the force produced by the induced current is directed upward, the thumb points upward. The magnetic field is directed outward, therefore, when you bend your fingers they must point outward. When your hand is properly oriented you will note that the straightened fingers point toward B. Thus, the current flows from A toward B and clockwise around the closed path.

Electric Motor and Generator

The **electric motor** is essentially a galvanometer movement that is arranged so that a coil of wire, referred to as the **armature,** runs continuously. The electric current through the armature interacts with the external field and the resulting torque causes the armature to rotate. Depending on the design of the motor, the motor can be arranged to run on direct current (dc) or alternating current (ac) electricity.

The **electric generator** uses mechanical work to produce electric energy. The armature is turned by an external torque and an emf is induced as the coil passes through an external magnetic field.

If the loop rotates continuously with a constant angular velocity (ω), then the magnitude of the induced emf is given by

$\xi = N A B \omega \sin \omega t$

where ωt is the angle that the face of the loop makes with the direction of the magnetic field at time t and A is the area of the loop.

The output emf of the generator is sinusoidally alternating. However, depending on the design, the generator can be made to produce either dc or ac current.

TEXTBOOK QUESTION 9. Explain why, exactly, the lights may dim briefly when a refrigerator motor starts up. When an electric heater is turned on, the lights may stay dimmed as long as the heater is on. Explain the difference.

ANSWER: When a motor starts up, the current through the armature is large because the induced counter emf is small and the electrical resistance to current flow is small. Electrical outlets on the same circuit as the motor will have a voltage drop and any house lights on the same circuit with the motor will dim. As the armature speeds up, it produces an increasingly large counter emf which reduces the flow of current through the motor. The voltage at the electrical outlets rises and the lights return to their former brightness.

When an electric heater is turned on, it also draws a large current. However, there is no induced counter emf to reduce the electric current to the heater. The heater has a large resistance and the product of the electrical current times the resistance results in a large voltage drop in the heater. The result is that the voltage available to other electrical outlets remains reduced.

EXAMPLE PROBLEM 6. An ac generator consists of a 400-turn flat rectangular coil 0.0300 m long and 0.0400 m wide. The coil rotates at 2400 rpm in a magnetic field of 0.500 T. Determine the a) angular velocity of the coil and b) peak output voltage.

Part a. Step 1.	Solution: (Section 21-5)
Determine the angular velocity of the coil.	The angular velocity ω is related to the frequency (f) by the formula
	$\omega = 2\pi f = 2\pi (2400 \text{ rev/min})(1 \text{ min}/60 \text{ sec})$
	$\omega = 250$ rad/s
Part b. Step 1.	The peak voltage is given by $\xi_o = N A B \omega$
Determine the peak output voltage.	where $A = (0.0300 \text{ m})(0.0400 \text{ m}) = 1.20 \times 10^{-3} \text{ m}^2$
	$\xi_o = (400)(1.20 \times 10^{-3} \text{ m}^2)(0.500 \text{ T})(250 \text{ rad/s})$
	$\xi_o = 60.0$ V

Counter EMF and Torque

As the armature of a motor rotates in the external magnetic field, an induced emf is produced. This induced emf, which is called a **back** or **counter emf**, produces a torque which opposes the motion of the armature (Lenz's law).

21-15

The magnitude of the back emf is proportional to the speed of the armature of the motor. When a motor is turned on, the magnitude of the back emf increases until a balance point is reached and the armature rotates at a constant speed. If the motor speed increases, the back emf increases until a balance point is again achieved.

The above also applies to a generator. The external torque causing the armature to rotate is opposed by a counter torque produced by the counter emf. If this did not occur, it would violate the law of conservation of energy, since it would then be possible to start the armature rotating and produce electrical energy without expending mechanical energy.

Eddy Currents

An induced current, called an **eddy current**, is produced by a changing magnetic flux in a piece of metal. The direction of the induced current is such as to oppose the magnetic field which caused the current (Lenz's law). Eddy currents can be produced by moving the metal through a magnetic field or by allowing a changing magnetic field to pass through a stationary piece of metal.

Eddy currents can be beneficial, e.g., electromagnetic damping in certain analytical balances and the braking system of some electric transit cars. However, eddy currents are often undesirable. For example, the coils of wire which make up the armature of a motor or generator as well as the primary and secondary of a transformer are often wound on an iron core. Eddy currents produced in the iron core dissipate energy in the form of **"joule heating"** ($P = I \xi = I^2 R$). To avoid this problem the core is laminated, which means that it is made of very thin sheets insulated from one another. This insulation causes a large electrical resistance along the path of the eddy current and the magnitude of the eddy current is small. The result is negligible energy losses due to joule heating.

Transformers

A **transformer** is a device used to increase or decrease an ac voltage. It consists of two coils of wire, known as the **primary** and **secondary coils**. The primary is connected to a source of emf and the secondary to a device usually referred to as the load (R_L). A schematic of a simple transformer is shown at right. When the current changes in the primary, the magnetic field it produces changes. As the changing field produced by the primary passes through the secondary, an induced emf is produced. If the secondary is part of a closed circuit, an induced current will pass through it. The advantage of a transformer is that it is possible to produce a higher or lower voltage in the secondary as compared to the primary. This is accomplished by having different numbers of turns of wire in the two coils. If the number of turns of wire in the primary is greater than the secondary, then the voltage in the secondary is lower than in the primary and the transformer is a **step-down** transformer. If the number of turns in the primary is less than the secondary, then the voltage in the secondary is higher than in the primary and the transformer is a **step-up** transformer.

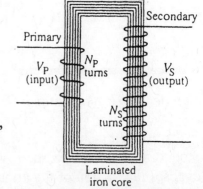

V_s and V_p are the voltages in the secondary and primary coils and N_s and N_p are the number of turns in each coil. The following equation reflects the theoretical limit for an "ideal" transformer:

$$V_s / V_p = N_s / N_p$$

Based on the law of conservation of energy, the power output in the secondary cannot be greater than the power input in the primary. In the "ideal" transformer they would be equal; thus

$$P_{input} = P_{output}$$

$$I_p V_p = I_s V_s \quad \text{and} \quad I_s / I_p = V_p / V_s$$

Power losses in transmission lines are due to joule heating, $P = I^2 R$. Because of joule heating, there is considerable advantage to transmitting power over long distances at high voltage but low current. Thus transformers can step the voltage up at the source and step the voltage down at the output. The ratio of the currents is related to the ratio of the turns as follows:

$$I_s / I_p = N_p / N_s$$

EXAMPLE PROBLEM 7. The primary winding of a transformer contains 100 turns and draws 6.00-A from a 120-V source of emf. The secondary contains 300 turns. Determine the emf and current in the secondary. Assume that the transformer is 100% efficient.

Part a. Step 1.	Solution: (Section 21-7)
Complete a data table.	V_p = 120 volts V_s = ?
	N_p = 100 turns N_s = 300 turns
	I_p = 6.00 I_s = ?
Part a. Step 2.	$V_s/V_p = N_s/N_p$
Determine the emf in the secondary.	$V_s/120$ V = 300/100
	V_s = 360 V
Part b. Step 1.	The transformer is 100% efficient, the power output in the secondary equals the power input in the primary.
Determine the current in the secondary.	$P_{input} = P_{output}$
	$I_p V_p = I_s V_s$
	(6.00 A)(120 V) = I_s (360 V)
	I_s = 2.00 A

Mutual Inductance

If two coils are placed near one another, a changing current in coil 1 creates a changing magnetic field which passes through coil 2. According to Faraday's law, an induced emf is produced in coil 2. The induced emf in coil 2 is given by

$$\xi_2 = M\ \Delta I_1/\Delta t$$

where M is a proportionality constant called the **mutual inductance** and $\Delta I_1/\Delta t$ is the rate of change of the current in the first coil. If the current is changing in coil 2 rather than coil 1; then the induced emf occurs in coil 1 and

$$\xi_1 = M\ \Delta I_2/\Delta t$$

The value of M is the same whether the emf occurs in coil 1 or coil 2. The unit of mutual inductance is the **Henry (H)** where 1 H = 1 ohm second. The value of M depends on whether or not iron is present, the size of the coils, number of turns, and the distance between the coils.

Self-Inductance

Self- inductance appears in a single coil when the current in the coil changes. The changing current produces a changing magnetic field in the coil, and this induces a back emf in the coil. The back emf tends to retard the flow of current if the current in the coil is increasing and induces an emf in the same direction as the current flow if the current is decreasing. The average induced emf is

$$\xi = -\ L\ \Delta I/\Delta t$$

where $\Delta I/\Delta t$ is the rate of change of current in the coil and L is a proportionality constant measured in henries (H).

The self-inductance (L) of a long coil, called a solenoid, of length ℓ, and cross-sectional area A which contains N turns is $L = \mu_o\ N^2\ A/\ell$.

EXAMPLE PROBLEM 8. The current in a 200-mH coil changes from -30.0 mA to +30.0 mA in 3.00 ms. Determine the magnitude of the induced emf.

Part a. Step 1.	Solution: (Section 21-9)
Use the formula for the induced emf in a coil of self-inductance L.	The induced emf in a coil in which the current is changing is given by $$\xi = -\ L\ \Delta I/\Delta t$$

but $I = I_f - I_i = 30.0 \text{ mA} - -30.0 \text{ mA} = +60.0 \text{ mA}$

$\xi = -[(200 \times 10^{-3} \text{ H})(+60.0 \times 10^{-3} \text{ A})]/(3.00 \times 10^{-3} \text{ s})$

$\xi = 4.00 \text{ V}$

EXAMPLE PROBLEM 9. A 2000-turn, air-filled coil is 2.00 m long, 0.100 m in diameter and carries a current of 1.00 A. Determine the a) self-inductance of the coil and b) energy stored in the coil's magnetic field.

Part a. Step 1.	Solution: (Sections 21-9 and 21-10)
Determine the cross-sectional area of the coil.	$A = \pi \, d^2/4 = \pi \, (0.100 \text{ m})^2/4 = 7.85 \times 10^{-3} \text{ m}^2$
Part a. Step 2.	$L = \mu_o \, N^2 \, A/\ell$
Determine the self-inductance of the coil.	$= (4\pi \times 10^{-7} \text{ T m/A})(2000)^2(7.85 \times 10^{-3} \text{ m}^2)/(2.00 \text{ m})$
	$L = 0.0197 \text{ H} = 19.7 \text{ mH}$
Part b. Step 1.	$B = \mu_o \, N \, I/\ell$
Determine B if the magnitude of the field produced by a long coil is given by $B = \mu_o \, N \, I/\ell$.	$= (4\pi \times 10^{-7} \text{ T m/A})(2000)(1.00 \text{ A})/(2.00 \text{ m})$
	$B = 1.26 \times 10^{-3} \text{ T}$
Part b. Step 2.	energy $= \tfrac{1}{2}(B^2/\mu_o) \, A \, \ell$
Determine the energy stored in the magnetic field.	$= \tfrac{1}{2}[(1.26 \times 10^{-3} \text{ T})^2/(4\pi \times 10^{-7} \text{ T m/A})](7.85 \times 10^{-3} \text{ m}^2)(2.00 \text{ m})$
	energy $= 9.86 \times 10^{-3} \text{ J}$

Energy Stored in a Magnetic Field

The energy stored in a coil of inductance L, carrying a current I, is given by the formula energy $= \tfrac{1}{2} L \, I^2$. The energy stored in the inductor's magnetic field is given by the formula energy $= \tfrac{1}{2} B^2 \, A \, \ell/\mu_o$. The volume enclosed by the windings of the coil equals the product of A and ℓ. The energy stored per unit volume, or **energy density**, is given by the formula energy density $= \tfrac{1}{2} B^2/\mu_o$.

LR Circuit

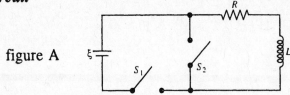

figure A

Figure A is a schematic drawing of an **LR circuit**. If switch 1 (S_1) is closed, the source of emf ξ is connected to the inductor L which has resistance R. The current through the circuit increases according to the formula

$$I = \xi/R \; (1 - e^{-t/\tau})$$

t is the time in seconds since the switch was closed, and τ is the inductive time constant of the circuit, $\tau = L/R$. It can be shown that when $t = \tau$, the current in the circuit has reached 63% of its maximum value, i.e., $I = 0.63 \; I_{max}$. Figure B, shown below, shows the growth of current in the circuit as a function of time.

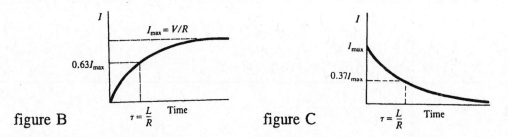

figure B   figure C

If switch 1 is disconnected after a steady current I is reached, the battery is then removed from the circuit and no current flows. If switch 2 is now connected, the energy stored in the inductor L causes a current to flow as follows:

$$I = \xi/R \; e^{-t/\tau}$$

This decay current is represented graphically in figure C. It can be shown that for a decay current, the current reaches 37% of its initial value ($0.37 \; I_{max}$) after 1 time constant ($t = \tau$).

AC Circuits and Impedance

Resistor

An ac source connected to a resistor produces a current (I) and a voltage drop (V) across the resistor given by the following equations:

$$I = I_o \cos 2 \pi f t \quad \text{and} \quad V = V_o \cos 2 \pi f t$$

where I_o and V_o are the peak values, f is frequency of the source in hertz, and t is the time in seconds.

The instantaneous voltage and current are said to be in phase since both are zero at the same moment of time and both reach their maximum values in either direction at the same time. The graphs shown below represent the voltage and current through the resistor as a function of time.

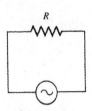

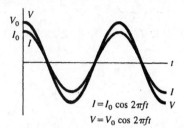

The relationship between V and I follows Ohm's law I = V/R and the average power dissipated is given by

$$P = \overline{I V} = (I_{rms})^2 R = (V_{rms})^2/R \quad \text{where rms refers to root-mean-square.}$$

and $I_{rms} = 0.707 I_o$, $V_{rms} = 0.707 V_o$

Inductor

The current and voltage in an inductor connected to an ac source are given by

$$I = I_o \cos 2\pi f t \quad \text{and} \quad V = -V_o \sin 2\pi f t$$

The current "lags" behind the voltage and is out of phase by 90°, i.e., the current reaches its maximum and minimum values ¼ cycle after the voltage.

An inductor produces a back emf $V = L \, \Delta I/\Delta t$ and therefore resists the flow of current through it. Energy from the source is momentarily stored in the magnetic field and as the field decreases, the energy is transferred back to the source. Thus no power is dissipated in the inductor.

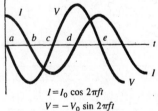

The magnitude of the current is related to the applied voltage at a given frequency by the equation

$$V_L = I X_L \quad \text{where } X_L = 2\pi f L$$

X_L is the **inductive reactance** of the inductor and has units of ohms. The inductive reactance is related to both the frequency of the source and the magnitude of the self-inductance of the coil.

The current and the voltage are not in phase, i.e., the peak current and peak voltage in the inductor are not reached at the same time. The equation $V = I\ X_L$ is valid on the average but not at a particular instant of time.

Capacitor

The current and voltage in a capacitor connected to an ac source are given by

$I = I_o \cos 2\pi ft$ and $V = V_o \sin 2\pi ft$

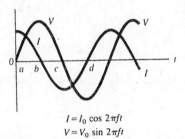

$I = I_0 \cos 2\pi ft$
$V = V_0 \sin 2\pi ft$

The current leads the voltage and is out of phase by 90°, i.e., the current reaches maximum and minimum values ¼ cycle before the voltage.

The average power dissipated in the capacitor is zero. Energy from the source is stored in the electric field between the plates of the capacitor. As the electric field decreases the energy is transferred back to the source.

The magnitude of the current is related to the applied voltage at a given frequency by the equation

$V = I\ X_C$ where $X_C = 1/2\pi fC$ where X_C is the **capacitive reactance** and has units of ohms

Capacitive reactance is inversely proportional to the frequency of the source and the capacitance of the capacitor. As in the case of the inductor, the current and voltage are not in phase, and while the equation $V = I\ X_C$ is valid on the average, it is not valid at a particular instant of time.

EXAMPLE PROBLEM 10. a) Determine the reactance of a 70.0-mH inductor connected to a 60.0-Hz ac line. b) The inductor is replaced with a 10.0-μF capacitor. Determine the reactance.

Part a. Step 1.	Solution: (Section 21-12)
Determine the inductive reactance.	$X_L = 2\pi f L$
	$= 2\pi (60.0 \text{ Hz})(70.0 \times 10^{-3} \text{ H})$
	$X_L = 26.4\ \Omega$

21-22

Part b. Step 1.	$X_C = 1/(2 \pi f C)$
Determine the reactance of the capacitor.	$X_C = 1/[2\pi (60.0 \text{ Hz})(10.0 \times 10^{-6} \text{ F})]$
	$X_C = 265 \ \Omega$

LRC Series Circuit

An LRC circuit (see diagram A) can be analyzed by using a phasor diagram. In a phasor diagram, vector-like arrows are drawn in an xy coordinate system to represent V_{Ro}, V_{Lo}, and V_{Co}. As shown in diagram B, the magnitude of V_{Ro} is represented by an arrow drawn along the +x axis, V_{Lo} along the +y axis, and V_{Co} along the -y axis.

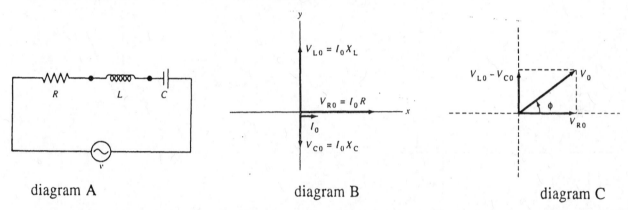

| diagram A | diagram B | diagram C |

Assuming that V_{Lo} is greater than V_{Co}, then diagram B reduces to diagram C. The resultant of $(V_{Lo} - V_{Co})$ and V_{Ro} is V_o. The peak voltage from the source (V_o) is out of phase with V_{Ro} and the current by the angle ϕ, where ϕ is the phase angle. Using the Pythagorean theorem, it can be shown that

$$V = [(V_{Ro})^2 + (V_{Lo} - V_{Co})^2]^{\frac{1}{2}}$$

The **impedance** Z of an LRC circuit can be determined in a like manner.

$$Z = [(X_L - X_C)^2 + R^2]^{\frac{1}{2}}$$ where Z is measured in ohms

The phase angle between the voltage and current can be determined from $\cos \phi = R/Z$ or $\tan \phi = (X_L - X_C)/R$. The power dissipated by the impedance is given by

$$P = I_{rms} V_{rms} \cos \phi$$

where $\cos \phi$ is the power factor. If the circuit contains only resistance, $\phi = 0°$, $\cos 0° = 1$, and $P = I_{rms} V_{rms}$.

If the circuit contains a pure capacitor and/or inductor, $\phi = 90°$ and $\cos 90° = 0$, $P = 0$, and no power is dissipated.

21-23

EXAMPLE PROBLEM 11. An LRC circuit has a 100-mH inductor with 10.0 Ω of resistance connected in series with a 150-μF capacitor and a 60.0-Hz, 120-volt ac source. Determine the a) total impedance, b) phase angle, c) rms current, and d) power dissipated in the circuit.

Part a. Step 1.	Solution: (Section 21-13)
Determine the inductive reactance, capacitive reactance, and total impedance.	$X_L = 2 \pi f L$

$X_L = 2 \pi (60.0 \text{ Hz})(100 \times 10^{-3} \text{ H}) = 37.7 \ \Omega$

$X_C = 1/(2\pi f C) = 1/[2\pi (60.0 \text{ Hz})(150 \times 10^{-6} \text{ F})]$

$X_C = 17.7 \ \Omega$

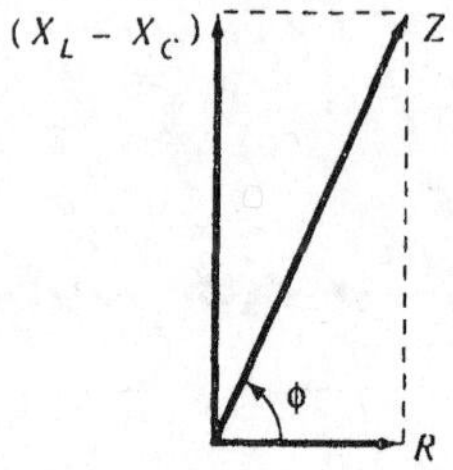

$Z = [(X_L - X_C)^2 + R^2]^{\frac{1}{2}}$

$\quad = [(37.7 \ \Omega - 17.7 \ \Omega)^2 + (10.0 \ \Omega)^2]^{\frac{1}{2}}$

$Z = 22.4 \ \Omega$

Part b. Step 1.	$\tan \phi = (X_L - X_C)/R$
Determine the phase angle.	$\tan \phi = (37.7 \ \Omega - 17.7 \ \Omega)/(10.0 \ \Omega) = 2.0$

$\phi = 63.4°$

Part c. Step 1.	$I_{rms} = V_{rms}/Z$
Determine the rms current.	$\quad = (120 \text{ V})/(22.4 \ \Omega)$

$I_{rms} = 5.36 \text{ A}$

Part d. Step 1.	$P = I \, V \cos \phi$
Determine the power dissipated in the circuit.	$\quad = (5.36 \text{ A})(120 \text{ V}) \cos 63.4°$

$P = 288 \text{ watts}$

or $P = (I_{rms})^2 R = (5.36 \text{ A})^2 (10.0 \ \Omega)$

$P = 287 \text{ watts}$

Resonance in AC Circuits

Since $Z = [(X_L - X_C)^2 + R^2]^{1/2}$, the impedance in an ac circuit is a minimum when $X_L = X_C$. At this point $I_{rms} = V_{rms}/R$ and I_{rms} is a maximum. Also, the energy transferred to the system is a maximum and a condition known as **resonance** occurs.

For example, a resonant circuit in a radio is used to tune in a particular station. A range of frequencies reach the circuit; however, a significant current flows only at or near the resonant frequency. At resonance $X_L = X_C$; therefore,

$2\pi f_0 L = 1/(2\pi f_0 C)$ and rearranging gives $f_0 = 1/[2\pi (L C)^{1/2}]$

where f_0 is the **resonant frequency** of the circuit in hertz.

TEXTBOOK QUESTION 24. Describe how to make the impedance in an LRC circuit a minimum.

ANSWER: For an LRC circuit $Z = [R^2 + (X_L - X_C)^2]^{1/2}$. The impedance is a minimum when $X_L = X_C$. At this point $Z = R$, the impedance is a minimum, and a condition known as resonance occurs. The resonant frequency (f_0) is inversely related to both L and C, i.e.,

$f_0 = 1/[2\pi (L C)^{1/2}]$.

EXAMPLE PROBLEM 12. An LRC circuit has a 1.50-mH inductor with 100 Ω of resistance connected in series with a 100-μF capacitor and a 120-volt ac source. Determine the a) resonant frequency and b) peak current for the circuit.

Part a. Step 1.

Derive a formula for the resonant frequency.

Solution: (Section 21-14)

The impedance reaches its minimum value at resonance. $X_L = X_C$, and $Z = R$.

$X_L = X_C$

$2\pi f_0 L = 1/(2\pi f_0 C)$ Rearranging gives

$f_0^2 = 1/(4\pi^2 L C)$

$= 1/[4 \pi^2 (1.50 \times 10^{-3} \text{ H})(100 \times 10^{-6} \text{ F})]$

$f_0^2 = 1.69 \times 10^5 \text{ Hz}^2$ and $f_0 = 411$ Hz

21-25

Part b. Step 1. Determine the peak voltage.	V_{rms} = 120 volts The peak voltage $V_o = (2)^{1/2} V_{rms}$. Therefore, $V_o = (2)^{1/2} (120 \text{ V}) = 170 \text{ V}$
Part b. Step 2. Determine the peak current.	The peak current occurs at the resonant frequency when the impedance to current flow is at a minimum. At this frequency, $Z = R$, thus $I = V_o/R = 170 \text{ V}/100 \text{ } \Omega = 1.70 \text{ A}$

PROBLEM SOLVING SKILLS

For a straight wire moving at speed v through a uniform magnetic field:

1. Use $\xi = v \ell B \sin \theta$ to determine the magnitude of the induced emf.
2. If the wire is part of a closed circuit, use Ohm's law to determine the current through the wire and the equation $F = I \ell B \sin \theta$ to determine the magnitude of the force which opposes the mechanical force acting on the wire.
3. The power dissipated in the form of Joule heating is given by $P = I^2 R$.

For problems involving a changing magnetic flux through a loop of wire:

1. Determine the rate of change of flux $(\Delta \Phi / \Delta t)$.
2. Apply Faraday's law to determine the magnitude of the induced emf.
3. Apply Ohm's law to determine the magnitude of the induced current and Lenz's law to determine its direction.

For "ideal" transformer problems:

1. Complete a data table listing the voltage, current, and number of turns of wire in both the primary and secondary coils.
2. Use the mathematical equation which relates the voltage and number of turns in the primary to the voltage and number of turns in the secondary.
3. An ideal transformer is 100% efficient. The power in the primary equals the power in the secondary. Use this concept to solve for the current in the secondary.

For problems involving ac circuits and impedance:

1. Use the formulas which relate the frequency to reactance to determine the inductive reactance and capacitive reactance.
2. Draw phasor diagrams and determine the voltage drop across each circuit element as well as the impedance in the circuit.
3. Use a phasor diagram to determine the phase angle. Determine the power factor.
4. If requested, determine the rms current and the average power dissipated.
5. If requested, determine the resonant frequency.

SOLUTIONS TO SELECTED TEXTBOOK PROBLEMS

TEXTBOOK PROBLEM 1. The magnetic flux through a coil of wire containing two loops changes from -50 Wb to +38 Wb in 0.42 s. What is the emf induced in the coil?

Part a. Step 1.	Solution: (Sections 21-1 to 21-4)
Determine the change in flux through the loop.	$\Delta \Phi = \Phi_f - \Phi_i$
	$\Delta \Phi = +38$ Wb - -50Wb = +88 Wb

Part a. Step 2.	$\xi = - N \Delta \Phi / \Delta t$
Use Faraday's law to determine the magnitude of the induced emf.	$= - (2)(+88$ Wb$)/0.42$ s
	$\xi = - 420$ V

TEXTBOOK PROBLEM 14. The moving rod in Fig. 21-12 is 13.2 cm long and generates an emf of 120 mV while moving in a 0.90-T magnetic field. (a) What is its speed? (b) What is the electric field in the rod?

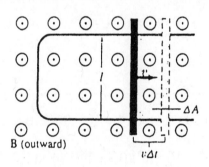

B (outward)

Part a. Step 1.	Solution: Section 21-3
Determine the emf in the rod.	Based on the diagram, the angle between the direction of motion of the rod and the magnetic field is 90°.
	$\xi = B \ell v \sin \theta$
	120×10^{-3} volts $= (0.90$ T$)(0.132$ m$)(v) \sin 90°$
	$v = (100 \times 10^{-3}$ volts$)/[(0.90$ T$)(0.132$ m$)]$
	$v = 1.01$ m/s

Part b. Step 1. Determine the electric field in the rod.	Based on Lenz's law, the induced current must oppose the motion of the rod. Using the right-hand rule, it can be shown that the induced current is clockwise around the circuit. $E = \xi/\ell = (120 \times 10^{-3}$ volts$)/(0.132$ m$) = 0.909$ V/m Based on the diagram, the direction of the electric field is toward the bottom of the rod.

TEXTBOOK PROBLEM 24. A 450-loop circular armature coil with a diameter of 8.0 cm rotates at 120 rev/s in a uniform magnetic field of strength 0.55 T. (a) What is the rms voltage output of the generator? (b) What would you do to the rotation frequency in order to double the rms voltage output?

Part a. Step 1. Determine the angular velocity of the coil in radians per second.	Solution: (Section 21-5) The angular velocity ω is related to the frequency (f) by the formula $\omega = 2\pi f = 2\pi (120$ rev/s$)$ $\omega = 754$ rad/s
Part a. Step 2. Determine the area of the loop.	$A = \pi d^2/4 = \pi (0.080$ m$)^2/4$ $A = 5.0 \times 10^{-3}$ m^2
Part a. Step 3. Determine the peak output voltage.	The peak voltage is given by $\xi_o = N A B \omega$ $\xi_o = (450)(5.0 \times 10^{-3}$ m$^2)(0.55$ T$)(754$ rad/s$)$ $\xi_o = 930$ V
Part a. Step 4. Determine the rms voltage.	$\xi_{rms} = \xi_o/(2)^{\frac{1}{2}} = (0.707)(930$ V$)$ $\xi_{rms} = 660$ V
Part b. Step 1. Determine the rotation frequency needed to double the rms voltage.	The rms voltage is related to the angular frequency (ω). However, the angular frequency (ω) is directly proportional to the rotation frequency (f). Therefore, in order to double the rms voltage it is necessary to double the rotation frequency (f). $f = 2(120$ rev/s$) = 240$ rev/s

Part a. Step 1.	Solution: (Section 21-7)
Complete a data table.	N_p = 330 turns　　N_s = 1340 turns　　V_p = 120 volts I_s = 15.0 A　　V_s = ?　　I_p = ?
Part a. Step 2. Determine the output voltage.	$V_s/V_p = N_s/N_p$ $V_s/120$ V = (1340 turns)/(330 turns) V_s = 490 V
Part b. Step 1. Determine the input current.	The transformer is 100% efficient, the power output in the secondary equals the power input in the primary. $P_{input} = P_{output}$ $I_p V_p = I_s V_s$ I_p (120 V) = (15.0 A)(490 V) I_p = 60.9 A

Part a. Step 1.	Solution: (Sections 21-13)
Determine the inductive reactance.	$X_L = 2\pi f L = 2\pi$ (50 Hz)(0.045 H) X_L = 14.1 Ω
Part a. Step 2. Determine the impedance in the circuit.	The impedance is determined by using a phasor diagram. $Z = (X_L^2 + R^2)^{1/2}$ $Z = [(14.1 \ \Omega)^2 + (30000 \ \Omega)^2]^{1/2} \approx 30000 \ \Omega$ or 30 kΩ

Part b. Step 1.	$X_L = 2\pi\, f\, L = 2\pi\, (3.0 \times 10^4 \text{ Hz})(0.045 \text{ H})$
Determine the inductive reactance.	$X_L = 8500\ \Omega$

Part b. Step 2.	The impedance is determined by using a phasor diagram.
Determine the impedance in the circuit.	$Z = (X_L{}^2 + R^2)^{\frac{1}{2}}$
	$\quad = [(8500\ \Omega)^2 + (30000\ \Omega)^2]^{\frac{1}{2}}$
	$Z = 31000\ \Omega$ or 31 kΩ

TEXTBOOK PROBLEM 73. A square loop 24.0 cm on a side has a resistance of 5.20 Ω. It is initially in a 0.665-T magnetic field, with its plane perpendicular to $\overrightarrow{\mathbf{B}}$, but it is removed from the field in 40.0 ms. Calculate the electric energy dissipated in this process.

Part a. Step 1.	Solution: (Sections 21-1 to 21-4)
Determine the area of of the loop of wire.	Area = length x width
	$A = (0.240 \text{ m})(0.240 \text{ m}) = 0.0576 \text{ m}^2$

Part a. Step 2.	The magnetic field is directed perpendicular to the coil. Using the convention discussed in the textbook, the angle θ is 0°.
Determine the change in magnetic flux through the loop.	$\Delta\Phi = \Phi_f - \Phi_i$
	$\quad = B_f\, A \cos\theta - B_i\, A \cos\theta = (B_f - B_i)\, A \cos\theta$
	$\quad = (0 \text{ T} - 0.665 \text{ T})(0.0576 \text{ m}^2) \cos 0°$
	$\Delta\Phi = -0.0383 \text{ T m}^2$

Part a. Step 3.	$\xi = -\, N\, \Delta\Phi/\Delta t = -\, (1)(-0.0383 \text{ T m}^2)/[(40.0 \text{ ms})(1 \text{ s}/1000 \text{ ms})]$
Use Faraday's law to determine the magnitude of the induced emf and Ohm's law to determine the average current.	$\xi = 0.958 \text{ V}$
	$I = \xi/R = (0.958 \text{ V})/(5.20\ \Omega) = 0.184 \text{ A}$

Part a. Step 4.

Determine the energy
dissipated in the form
of heat.

energy dissipated = power x time

energy = P t but P = I ξ

   = (0.184 A)(0.958 V)[(40.0 ms)(1.0 s)/(1000 ms)]

energy = 7.06 x 10^{-3} J

CHAPTER 22

ELECTROMAGNETIC WAVES

OBJECTIVES

After studying the material of this chapter, the student should be able to:

- give a non-mathematical summary of Maxwell's equations.
- describe how electromagnetic waves are produced.
- draw a diagram representing the field strengths of an electromagnetic wave produced by a sinusoidally varying source of emf.
- calculate the velocity of electromagnetic waves in a vacuum if both the permittivity and permeability of free space are given.
- state the names given to the different segments of the electromagnetic spectrum.
- state the approximate range of wavelengths associated with each segment of the electromagnetic spectrum.
- state the equation which relates the speed of an electromagnetic to the frequency and wavelength and use this equation in problem solving.
- determine the peak magnitude of both the electric and magnetic field strength if the energy density of the electromagnetic wave is given.
- solve problems related to the time average value of the Poynting vector at a particular point and calculate the peak values of both the electric and magnetic fields at this point.

KEY TERMS AND PHRASES

Maxwell's equations consist of four basic equations that describe all electric and magnetic phenomena.

electromagnetic waves are transverse waves which have both an electric and a magnetic component. Maxwell, based on his equations, concluded that a changing magnetic field (B) will produce a changing electric field (E) which is at right angles to the magnetic field and that changing electric field will produce a changing magnetic field. The net result of the interaction of the changing E and B fields is the production of an electromagnetic wave which propagates away from the wave source at the speed of light.

electromagnetic spectrum includes radio waves, microwaves, infrared radiation, visible light, x rays, and gamma rays. They differ in frequency (f) and wavelength (λ) but all travel at the speed of light.

intensity refers to the energy transported per unit time per unit area. The unit of intensity is joules per second per square meter (J/s m^2).

SUMMARY OF MATHEMATICAL FORMULAS

speed of electromagnetic waves in a vacuum	$v = (\epsilon_o \mu_o)^{-\frac{1}{2}}$	The speed of EM waves in a vacuum is inversely related to the square root of the permittivity of free space (ϵ_o), where $\epsilon_o = 8.85 \times 10^{-12}$ C^2/N m^2 and the permeability of free space (μ_o), where $\mu_o = 4 \pi \times 10^{-7}$ T m/A.
speed of electro-magnetic waves	$v = f \lambda$	The speed (v) of electromagnetic waves equals the product of the frequency (f) and the wavelength (λ). The speed of light (v) in a vacuum equals 3.00×10^8 m/s.
energy density	$u = \frac{1}{2} \epsilon_o E^2 + \frac{1}{2} B^2/\mu_o$ $u = (\epsilon_o/\mu_o)^{\frac{1}{2}} E B$	The total energy stored per unit volume (u) at any particular instant. E and B represent the instantaneous magnitudes of the electric and magnetic fields at a particular point.
energy in an electromagnetic wave	$\bar{I} = E_o B_o/2\mu_o$	The average energy transported per unit time per unit area perpendicular to the wave direction is referred to as the intensity (*I*). The unit of *I* is joules per second per square meter (J/s m^2). E$_o$ represents the maximum values of the electric field vector while B$_o$ represents the maximum value of the magnetic field vector.

CONCEPT SUMMARY

Maxwell's Equations

All electric and magnetic phenomena can be described in four basic equations known as **Maxwell's equations**. The following is a summary of each equation:

1) Gauss' law for electricity is a generalized form of Coulomb's law and relates electric charge to the electric field that the charge produces.
2) Gauss' law of magnetism describes magnetic fields and predicts that magnetic fields are continuous, i.e., they have no beginning or ending point. The equation predicts that there are no magnetic monopoles.
3) Faraday's law of induction describes the production of an electric field by a changing magnetic field.
4) Maxwell's extension of Ampere's law describes the magnetic field produced by a changing electric field or by an electric current.

Production of Electromagnetic Waves

Maxwell concluded that a changing magnetic field (B) will produce a changing electric field (E) and the changing electric field will produce a changing magnetic field. The net result of the interaction of the changing E and B fields is the production of a wave which has both an electric and a magnetic component and travels through empty space. This wave is referred to as an electromagnetic wave (EM).

Electromagnetic waves are produced by accelerated electric charges. For example, accelerated electrons in atoms give off visible light while oscillating electric charges in an antenna are undergoing acceleration and produce radio waves. The following diagram represents the field strengths of an electromagnetic wave produced by a sinusoidally varying source of emf.

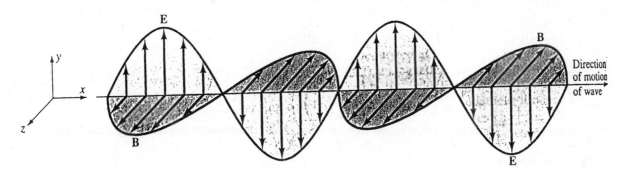

From the figure it should be noted that electromagnetic waves are transverse waves with E and B at right angles to one another and both are perpendicular to the direction of travel.

The speed of EM waves in a vacuum is given by $v = (\epsilon_o \mu_o)^{-\frac{1}{2}}$ where

ϵ_o is the permittivity of free space, $\epsilon_o = 8.85 \times 10^{-12} \ C^2/N \ m^2$ and

μ_o is the permeability of free space, $\mu_o = 4 \pi \times 10^{-7} \ T \ m/A$

TEXTBOOK QUESTION 1. The electric field in an EM wave traveling north oscillates in an east-west plane. Describe the direction of the magnetic field vector in this wave.

ANSWER: The plane of the magnetic field vector is perpendicular to both the direction of travel of the wave and the plane of the electric field vector. Therefore, the magnetic field vector is in a vertical plane which is perpendicular to the surface of the Earth.

EXAMPLE PROBLEM 1. A radar pulse travels from the Earth to the Moon and back in 2.60 seconds. Calculate the distance from the Earth to the Moon in a) meters and b) miles.

Part a. Step 1. Calculate the distance from the Earth to the Moon in meters.	Solution: (Sections 22-3 and 22-4) A radar pulse is an electromagnetic wave which travels at 3.00×10^8 m/s. The time required for the pulse to travel from the Earth to the Moon is equal to $(\frac{1}{2})(2.60 \text{ s}) = 1.30$ s. Since it travels in a straight line at a constant speed, the distance traveled can be determined as follows: $d = c\,t = (3.00 \times 10^8 \text{ m/s})(1.30 \text{ s})$ $d = 3.90 \times 10^8$ m
Part b. Step 1. Calculate the distance from the Earth to the Moon in miles.	1.000 miles = 1609 m Therefore, $d = 3.90 \times 10^8$ m x (1.000 mile/1609 m) $d = 2.42 \times 10^5$ miles Note: the Moon follows an elliptical path in its orbit about the Earth; therefore, the value obtained in this problem would be the distance at a certain point in the Moon's orbit.

Electromagnetic Spectrum

Radio waves and visible light are only a small part of what is known as the electromagnetic spectrum (EM). The spectrum includes radio waves, microwaves, infrared radiation, visible light, x rays, and gamma rays. They differ in frequency (f) and wavelength (λ). The velocity, frequency, and wavelength are related by the following equation which represents the speed of all EM waves in free space.

$v = f\,\lambda$ where the speed (v) of electromagnetic waves in a vacuum = 3.00×10^8 m/s

The following table gives the approximate range of wavelengths for each portion of the EM spectrum:

radio waves 10^4 m to 1 m   ultraviolet 10^{-8} m to 10^{-9} m

microwaves 1 m to 10^{-4} m	X-rays 10^{-9} m to 10^{-11} m
infrared 10^{-4} m to 10^{-7} m	gamma rays less than 10^{-12} m
visible light 4 x 10^{-7} m to 7 x 10^{-7} m	

TEXTBOOK QUESTION 5. Are the wavelengths of radio and television signals longer or shorter than those detectable by the human eye?

ANSWER: Based on Figure 22-8, radio and television signals have longer wavelengths than visible light.

TEXTBOOK QUESTION 7. In the electromagnetic spectrum, what type of EM wave would have a wavelength of 10^3 km? 1 km? 1 m? 1 cm? 1 mm? 1 μm?

ANSWER: Based on Figure 22-8, 10^3 km is in the sub-radio wave region of the EM spectrum (or very long radio wave region), 1 km is in the radio wave region, 1 m is in the TV signal and microwave regions, 1 cm is in the microwave and satellite TV signal regions, 1 mm is in the microwave and infrared regions, and 1 μm is in the infrared region.

EXAMPLE PROBLEM 2. Determine the frequency of red light of wavelength 7.00 x 10^{-7} meters.

Part a. Step 1.	Solution: (Sections 22-3 and 22-4)
Use c = f λ to calculate the frequency of red light.	The frequency, wavelength, and velocity of light are related by the equation c = f λ, where c refers to the speed of light in a vacuum. $$f = c/\lambda = (3.00 \times 10^8 \text{ m/s})/(7.00 \times 10^{-7} \text{ m})$$
	$$f = 4.29 \times 10^{14} \text{ Hz}$$

Energy Density of an EM Wave

Electromagnetic waves transport energy from one region of space to another. This energy is associated with the electric and magnetic fields of the wave. The energy per unit volume (**energy density**) stored in the electric field is given by

$u = \frac{1}{2} \epsilon_o E^2$ where u is the energy density and has units of joules per cubic meter (J/m^3)

The energy stored per unit volume in the magnetic field is given by $u = \frac{1}{2} B^2/\mu_o$.

The total energy stored per unit volume at any particular instant is given by

$$u = \tfrac{1}{2} \, \epsilon_o \, E^2 + \tfrac{1}{2} \, B^2/\mu_o$$

E and B represent the instantaneous magnitudes of the electric and magnetic fields at a particular point. The total energy density can be stated in terms of the electric field only as $u = \epsilon_o \, E^2$ or the magnetic field only as $u = B^2/\mu_o$ or in terms of both E and B as

$$u = (\epsilon_o/\mu_o)^{\frac{1}{2}} \, E \, B$$

EXAMPLE PROBLEM 3. The energy density of an EM wave is $8.00 \times 10^{-5} \, J/m^3$. Determine the peak magnitude of the a) electric field strength and b) magnetic field strength.

Part a. Step 1. Determine the peak magnitude of the electric field strength.	Solution: (Section 22-5) The energy density is related to the electric field strength by the equation $u = \epsilon_o \, E^2$ but $\quad \overline{E^2} = E_o^2/2 \quad$ Therefore, $u = \tfrac{1}{2} \, \epsilon_o \, E_o^2 \quad$ Rearranging gives $E_o = (2u/\epsilon_o)^{\frac{1}{2}}$ $E_o = [(2)(8.00 \times 10^{-5} \, J/m^3)/(8.85 \times 10^{-12} \, C^2/N \, m^2)]^{\frac{1}{2}}$ $E_o = 4.25 \times 10^3 \, N/C$
Part b. Step 1. Determine the peak magnitude of the magnetic field strength.	The energy density can be expressed in terms of the magnetic field strength as follows: $u = B^2/\mu_o \quad$ but $\quad \overline{B^2} = B_o^2/2 \quad$ Therefore, $u = \tfrac{1}{2} \, B_o^2/\mu_o$ $8.00 \times 10^{-5} \, J/m^3 = \tfrac{1}{2} \, B_o^2/(4\pi \times 10^{-7} \, N \, s^2/C^2)$ and solving for B_o: $B_o = 1.42 \times 10^{-5} \, T$ Alternate solution: It can be shown that

$c = E_o/B_o$ Therefore,

$B_o = E_o/c = (4.25 \times 10^3 \text{ N/c})/(3.00 \times 10^8 \text{ m/s})$

$B_o = 1.42 \times 10^{-5} \text{ T}$

Energy in Electromagnetic Waves

The energy transported per unit time per unit area is the intensity I. The magnitude of the intensity is given by

$$I = \epsilon_o c E^2 \quad \text{or} \quad I = E B/\mu_o$$

The unit of intensity is joules per second per square meter (J/s m^2). The average intensity transported per unit area per unit time is given by

$$\overline{I} = \tfrac{1}{2} \epsilon_o c E^2 = \tfrac{1}{2} c/\mu_o B_o^2 = E_o B_o/2\mu_o$$

where E_o and B_o represent the maximum values of E and B.

EXAMPLE PROBLEM 4. A source of EM waves radiates uniformly in all directions. The amplitude of the electric field 1.00×10^4 m from the source is 200 V/m. Determine the time average value of a) the intensity I and b) time average power radiated by the source.

Part a. Step 1.	Solution: (Section 22-5)
Determine the time average value of the intensity.	The time average value of the intensity is given by $$\overline{I} = \tfrac{1}{2} \epsilon_o c E_o^2$$ E_o is the peak value (amplitude) of the electric field; thus $$\overline{I} = \tfrac{1}{2}(8.85 \times 10^{-12} \text{ C}^2/\text{Nm}^2)(3.00 \times 10^8 \text{m/s})(200 \text{ V/m})^2$$ $$\overline{I} = 53.1 \text{ J/s m}^2$$
Part b. Step 1. Determine the time average power radiated by the source.	Because the EM wave radiates uniformly in all directions, it is possible to assume that all of the energy passes through an imaginary sphere of surface area $4\pi r^2$. The radius of this sphere is taken to be 1.00×10^4 m with the source at the center. The time average power $\overline{(P)}$ radiated by the source is related to the intensity as follows:

$$\overline{P} = \overline{I} \ A = \overline{I} \ (4\pi \ r^2)$$

$$= (53.1 \text{ J/s m}^2)(4\pi)(1.00 \times 10^4 \text{ m})^2$$

$$\overline{P} = 6.67 \times 10^{10} \text{ J/s} = 6.67 \times 10^{10} \text{ watt}$$

PROBLEM SOLVING SKILLS

For problems involving the velocity, frequency, or wavelength of a periodic electromagnetic wave in a vacuum:

1. Express the wavelength of the wave in meters and the frequency of the wave in hertz.
2. Use the equation $c = f \lambda$ to solve the problem where $c = 3.0 \times 10^8$ m/s.

For problems related to the energy in electromagnetic waves:

1. Note the instantaneous magnitude of the electric and/or magnetic field vector.
2. Solve for the energy density stored in the electric field using the equation $u = \frac{1}{2} \epsilon_o E^2$. Solve for the energy density stored in the magnetic field using the equation $u = \frac{1}{2} B^2/\mu_o$.

For problems related to the time average value of the intensity:

1. Determine the time average value of the intensity by using the equation

$$\overline{I} = E_o B_o/2\mu_o$$

For problems related to the time average power a specified distance from a source of electromagnetic waves which radiates uniformly in all directions:

1. Determine the surface area of an imaginary sphere through which the energy radiates.

2. Determine the time average power using the equation $\overline{P} = \overline{I} \ A$.

SOLUTIONS TO SELECTED TEXTBOOK PROBLEMS

TEXTBOOK PROBLEM 8. An EM wave has a wavelength 650 nm. What is its frequency, and how would we classify it?

Part a. Step 1.	Solution: (Sections 22-3 and 22-4)
Use $c = f \lambda$ to calculate the frequency of the light.	The frequency, wavelength, and velocity of light are related by the equation $c = f \lambda$, where c refers to the speed of light in a vacuum.
	$f = c/\lambda = (3.00 \times 10^8 \text{ m/s})/[(650 \text{ nm})(1.00 \times 10^{-9} \text{ m/1 nm})]$
	$f = 4.62 \times 10^{14}$ Hz

Part a. Step 2.	The visible portion of the electromagnetic spectrum extends from approximately 400 nm to 700 nm. Therefore, this wavelength is in the orange-red portion of the visible spectrum.
Determine the portion of the EM spectrum that contains this wavelength.	

TEXTBOOK PROBLEM 13. How long would it take a message sent as a radio wave from Earth to reach Mars, (a) when nearest Earth, (b) when farthest from Earth? [Hint: see Table 5-2].

Part a. Step 1.	Solution: (Sections 22-3 and 22-4)
Use Table 5-2, p. 125 in the textbook to determine the closest distance between Earth Mars in meters.	Based on information provided in the textbook, the mean (average) distance from the Sun to the Earth is 149.6 x 10^6 km, while the mean distance from the Sun to Mars is 227.9 x 10^6 km. Assuming that Earth and Mars are on the same side of the sun in their orbit, the distance (d) the message would travel is
	$d = 227.9 \times 10^6$ km $- 149.6 \times 10^6$ km $= 78.3 \times 10^6$ km
	$d = 7.83 \times 10^{10}$ m
Part a. Step 2.	Radio waves travel at the speed of light, i. e. 3.0 x 10^8 m/s. The time required to travel this distance can be determined as follows:
Determine the time for the message to travel this distance.	$t = d/c = (7.83 \times 10^{10}$ m$)/(3.00 \times 10^8$ m/s$) = 261$ s or 4.35 minutes
Part b. Step 1.	The furthest distance occurs when Earth and Mars are on the opposite sides of the Sun. The distance (d) the message would travel is
Determine the furthest distance between Earth and Mars.	$d = 227.9 \times 10^6$ km $+ 149.6 \times 10^6$ km $= 377.5 \times 10^6$ km
	$d = 3.775 \times 10^{11}$ m
Part b. Step 2.	The time required to travel this distance can be determined as follows:
Determine the time for the message to travel this distance.	$t = d/c = (3.775 \times 10^{11}$ m$)/(3.00 \times 10^8$ m/s$) = 1260$ s or 21.0 minutes

Part a. Step 1.	Solution: (Section 22-3 and 22-4)
Calculate the round-trip distance the light travels.	$(35 \text{ km} + 35 \text{ km})[(1000\text{m})/(1 \text{ km})] = 7.00 \times 10^4 \text{ m}$

Part a. Step 2.	time = distance/speed of light
Determine the time for light to travel 7.0×10^4 m.	$t = (7.00 \times 10^4 \text{ m})/(3.00 \times 10^8 \text{ m/s})$
	$t = 2.33 \times 10^{-4} \text{ s}$

Part a. Step 3.	In order for the light from the distant mirror to reflect into the observer's eye, the apparatus must rotate at least ⅛ revolution during the time the light is traveling the 70 km.
Determine the minimum angular distance the apparatus must rotate for the light to be reflected into the observer's eye.	(⅛ revolution)$[(2 \pi \text{ rad})/1 \text{ rev})] = 0.785 \text{ rad}$

Part a. Step 4.	$\omega = \Delta\theta/\Delta t$
Determine the minimum angular velocity of the apparatus in radians per second and revolutions per minute.	$= (0.785 \text{ rad})/(2.33 \times 10^{-4} \text{ s}) = 3370 \text{ rad/s}$
	$= (3370 \text{ rad/s})[(1 \text{ rev})/(2 \pi \text{ rad})][(60 \text{ s})/(1 \text{ min})$
	$\omega = 3.22 \times 10^4 \text{ rev/min}$

Part a. Step 1.	Solution: (Sections 22-3 and 22-4)
Determine the time for the radio wave to travel 3000 km, i.e., 3.0×10^6 m.	Radio waves travel at the speed of light, i.e., 3.0×10^8 m/s. The time required to travel 3.0×10^6 m can be determined as follows:
	$t = d/c = (3.00 \times 10^6 \text{ m})/(3.00 \times 10^8 \text{ m/s}) = 0.010 \text{ s}$

Part a. Step 2.	At 20°C, the speed of sound is 343 m/s.
Determine the time for the sound wave to travel 50 m.	$t = d/v = (50.0 \text{ m})/(343 \text{ m/s}) = 0.146$ s

Part a. Step 3.	difference in time = 0.146 s - 0.010 s = 0.136 s
Determine the difference in the time.	The person listening on the radio will hear the singer before the person in the audience.

TEXTBOOK PROBLEM 33. A certain FM radio tuning circuit has a fixed capacitor C = 840 pF. Tuning is done by a variable inductance. What range of values must the inductance have to tune stations from 88 MHz to 108 MHz?

Part a. Step 1.	Solution: (Section 22-7)
Derive a formula for resonant frequency.	Resonance in ac circuits was discussed in section 21-14. At resonance, the inductive reactance (X_L) equals the capacitive reactance (X_C) and the impedance is at a minimum.
	$X_L = X_C$
	$2\pi f_o L = 1/(2\pi f_o C)$
	$f_o^2 = 1/(4\pi^2 L C)$ and rearranging gives
	$L = 1/(4\pi^2 f_o^2 C)$

Part a. Step 2.	$L = 1/(4\pi^2 f_o^2 C)$
Determine the inductance to tune an 88-MHz station.	$= 1/[4\pi^2 (88 \times 10^6 \text{ Hz})^2 (840 \times 10^{-12} \text{ F})]$
	$L = 3.90 \times 10^{-9}$ H = 3.90 nH

Part a. Step 3.	$L = 1/(4\pi^2 f_o^2 C)$
Determine the inductance to tune a 108-MHz station.	$= 1/[4\pi^2 (108 \times 10^6 \text{ Hz})^2 (840 \times 10^{-12} \text{ F})]$
	$L = 2.59 \times 10^{-9}$ H = 2.59 nH

Part a. Step 4.	The required inductance ranges from 2.59 nH to 3.90 nH
State the range of required inductance.	

TEXTBOOK PROBLEM 42. What are E_o and B_o 2.00 m from a 95.0-W light source? Assume the bulb emits radiation of a single frequency uniformly in all directions.

Part a. Step 1.	Solution: (Section 22-5)
Determine the energy radiated per unit area per unit time.	$I = P/A$ where $A = 4\pi r^2$
	$\quad = (95.0 \text{ J/s})/[4 \pi (2.00 \text{ m})^2]$
	$I = 1.89 \text{ J/m}^2 \cdot \text{s}$

Part a. Step 2.	$I = \frac{1}{2} \epsilon_o c E_o^2$
Determine the peak value (E_o) of the electric field vector.	$1.89 \text{ J/s m}^2 = \frac{1}{2}(8.85 \times 10^{-12} \text{ C}^2/\text{Nm}^2)(3.00 \times 10^8 \text{ m/s}) E_o^2$
	$E_o^2 = 1.42 \times 10^3 \text{ N}^2/\text{C}^2$
	$E_o = (E_o^2)^{\frac{1}{2}} = (1.42 \times 10^3 \text{ N}^2/\text{C}^2)^{\frac{1}{2}}$
	$E_o = 37.7 \text{ N/C}$

Part a. Step 3.	$B_o = E_o/c$
Determine the peak value (B_0) of the magnetic field vector.	$\quad = (37.7 \text{ N/C})/(3.0 \times 10^8 \text{ m/s})$
	$B_o = 1.26 \times 10^{-7} \text{ T}$

CHAPTER 23

LIGHT: GEOMETRIC OPTICS

OBJECTIVES

After studying the material of this chapter, the student should be able to:

- distinguish between mirror reflection and diffuse reflection.
- draw a ray diagram and locate the position of the image produced by an object placed a specified distance from a plane mirror. State the characteristics of the image.
- distinguish between a convex and a concave mirror. Draw rays parallel to the principal axis and locate the position of the principal focal point of each type of spherical mirror.
- draw ray diagrams and locate the position of the image produced by an object placed a specified distance from a concave or convex mirror. State the characteristics of the image.
- use the mirror equations and the sign conventions to determine the position, magnification, and size of the image produced by an object placed a specified distance from a spherical mirror.
- state Snell's law and use this law to predict the path of a light ray as it travels from one medium into another. Explain what is meant by the index of refraction of a medium.
- explain what is meant by total internal reflection. Use Snell's law to determine the critical angle as light travels from a medium of higher index of refraction into a medium of lower index of refraction.
- distinguish between a convex and a concave lens. Draw rays parallel to the principal axis and locate the position of the principal focal points for each type of thin lens.
- draw ray diagrams and locate the position of the image produced by an object placed a specified distance from either type of thin lens. State the characteristics of the image.
- use the thin lens equations and the sign conventions to determine the position, magnification, and size of the image produced by an object placed a specified distance from a concave or convex lens.

KEY TERMS AND PHRASES

ray of light is a single beam of light which travels in a straight line until it strikes an obstacle. If the ray strikes a highly reflective surface, then **specular** (or mirror) reflection occurs. If the ray strikes a rough surface, then **diffuse** reflection results.

laws of specular reflection 1) The incident ray, reflected ray, and the normal to the surface are all in the same plane and 2) the angle of incidence equals the angle of reflection.

virtual image refers to an image that only "appears" to be behind the surface of a mirror (or lens). The image is not actually located at its apparent position and cannot be formed on a screen.

real image of an object refers to an image that can be formed on a screen placed at the position where rays from a point on the object pass through a common point.

spherical mirrors are curved mirrors which are sections of a sphere. The surface of a convex mirror curves toward the observer while the surface of a concave mirror curves away from the observer.

focal point of a spherical mirror is the point where rays parallel and very close to the principal axis all pass through (or appear to come from) after reflecting from the mirror.

focal point of a lens is the point where rays parallel and very close to the principal axis all pass through (or appear to come from) after refraction by the lens.

focal length of a mirror (or lens) is the distance from the center of the mirror (or lens) to the focal point.

linear magnification refers to the ratio of the height of the image to the height of the object.

refraction occurs when light changes direction as it passes from one transparent substance into another. The light changes direction at the interface between the two substances.

Snell's law describes the relationship between the incident ray and the refracted ray during refraction.

index of refraction equals the ratio of the speed of light in a vacuum to the speed of light in a transparent substance. The indices of refraction, along with the angle of incidence, determine the angle of bending of the light at the interface between two transparent substances.

total internal reflection occurs when light traveling through a medium of higher index of refraction is completely reflected at the interface with a medium of lower index of refraction.

critical angle is the minimum angle of incidence at which total internal reflection occurs.

convex lens causes light to converge as it passes through the lens. The surfaces of a convex lens curve outward toward the observer.

concave lens causes light to diverge as it passes through the lens. The surfaces of a concave lens curve away from the observer.

SUMMARY OF MATHEMATICAL FORMULAS

second law of specular reflection	$\theta_i = \theta_r$	The angle of incidence (θ_i) equals the angle of reflection (θ_r).
mirror and lens equations	$1/d_o + 1/d_i = 1/f$	Equation relates the distance from the center of the mirror (lens) to the object (d_o) and the image (d_i) to the focal length (f) of the mirror (lens).
	$f = r/2$	The focal length (f) equals one-half the radius of curvature (r) of the mirror. The focal point is real for a concave mirror but virtual for a convex mirror.
	$m = h_i/h_o = -d_i/d_o$	The linear magnification (m) refers to the ratio of the size of the image (h_i) to the size of the object (h_o). The magnification equals the ratio of the image distance (d_i) to the object distance (d_o).
Snell's law	$n_1 \sin \theta_1 = n_2 \sin \theta_2$	The law of refraction states the relationship between the angle of incidence (θ_1) and the angle of refraction (θ_2). n_1 is the index of refraction of the incident medium while n_2 is the index of refraction of the medium into which the light passes.
critical angle for total internal reflection	$\sin \theta_c = n_2/n_1$	At the critical angle (θ_c) and at all angles greater than this angle, light is totally reflected back into the medium of the incident ray.

CONCEPT SUMMARY

Laws of Reflection

A **ray** of light is a single beam of light which travels in a straight line until it strikes an obstacle. As shown in the diagram at the top of the next page, if the ray strikes a highly reflective surface, then **specular** (or mirror) reflection occurs. If the ray strikes a rough surface, then **diffuse** reflection results. In specular reflection, it is found that 1) the **incident ray (i)**, **reflected ray (r)**, and the **normal (N)** to the surface are all in the same plane and 2) the **angle of incidence** (θ_i) equals the **angle of reflection** (θ_r).

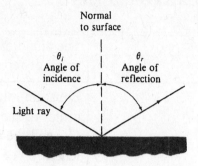

Plane Mirrors

Using the laws of specular reflection it is possible to show that the image formed by an object placed in front of a plane (flat) mirror has the following characteristics: it is erect or upright, **virtual**, and the same size as the object. Also, the apparent distance from the mirror to the image is equal to the actual distance from the mirror to the object.

The word **virtual** refers to the fact that the image of the object only "appears" to be behind the surface of the mirror but the light does not pass through the mirror and the image is not actually located at its apparent position. It is not possible to form a virtual image on a piece of paper or photographic film placed at the image position.

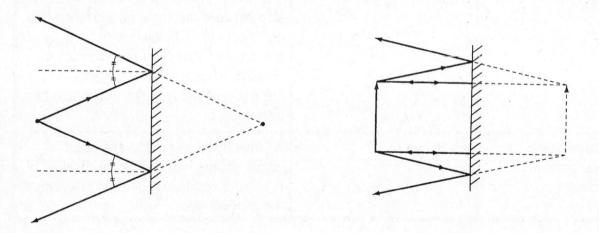

In order to locate the position of the image of any point on an object placed in front of a mirror, it is necessary to draw at least two rays of light which leave that point and reflect from the mirror. The figures shown above show the use of ray diagrams in locating the image produced by 1) a point and 2) an arrow placed in front of a plane mirror.

Spherical Mirrors

Spherical mirrors are curved mirrors which are sections of a sphere. Convex and concave mirrors are two types of spherical mirrors.

The **principal focus** or **focal point** of a spherical mirror is the point where rays parallel and very close to the **principal axis** all pass through (or appear to come from) after reflecting from the mirror. As shown in the following diagrams, the focal point (F) of the mirror is located halfway between the center of curvature and the center of the mirror. The focal length (f) equals one-half the radius of curvature of the mirror, i.e., f = r/2. The focal point is real for a concave mirror but virtual for a convex mirror.

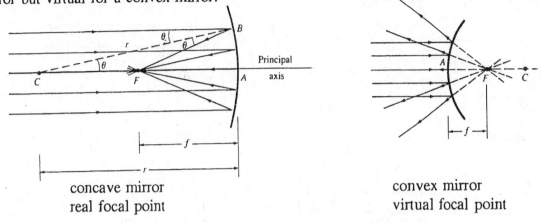

concave mirror
real focal point

convex mirror
virtual focal point

Images Formed by Spherical Mirrors

The characteristics of the images formed by spherical mirrors depend on the distance from the mirror to the object (d_o), the focal length of the mirror (f), and whether the mirror is concave or convex. The following summarizes the possibilities:

type of mirror	object distance as compared to the focal length	characteristics of the image
concave	$d_o > 2f$	real, inverted, diminished
	$d_o = 2f$	real, inverted, same size
	$f < d_o < 2f$	real, inverted, magnified
	$d_o = f$	no image formed
	$d_o < f$	virtual, erect, magnified

Convex mirrors produce only virtual, erect, and diminished images.

The images formed by spherical mirrors may be located by using two rays. 1) A ray from the top of the object reflects from the center of the mirror. The principal axis is the normal to the mirror at this point. The angles of incidence and reflection are easily measured so that the reflected ray may be drawn. 2) A ray from the top of the object, which is parallel to the principal axis, will reflect through the focal point of the concave mirror and appear to be coming from the focal point of a convex mirror.

The image of the top of the object is located at the point where the rays cross in the case of a real image and appear to cross after being traced behind the mirror in the case of a virtual image. The following diagrams represent each of the possibilities discussed in the table on the previous page. Note: d_i is the distance from the image to the mirror.

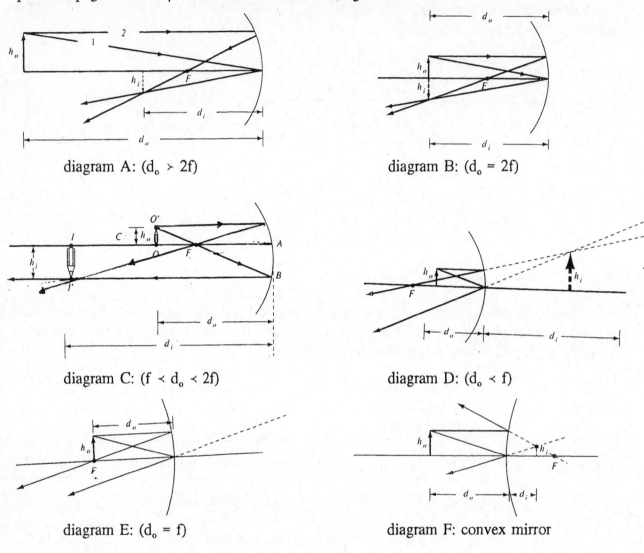

diagram A: ($d_o > 2f$)

diagram B: ($d_o = 2f$)

diagram C: ($f < d_o < 2f$)

diagram D: ($d_o < f$)

diagram E: ($d_o = f$)

diagram F: convex mirror

Mirror Equation

The mirror equation for both concave and convex spherical mirrors is as follows:

$1/d_o + 1/d_i = 1/f$ where $f = r/2$ and r is the radius of curvature of the mirror.

Magnification and Size of Image

The linear magnification (m) refers to the ratio of the size of the image (h_i) to the size of the object (h_o). The magnification produced by a curved mirror is given by

$m = h_i/h_o = - d_i/d_o$ and $h_i = m \, h_o$

Sign Conventions

The following sign conventions are given in the text to be used with the mirror equations: When the object, image, or focal point is on the reflecting side of the mirror (on the left side in all of the drawings), the corresponding distance is considered positive. If any of these points are behind the mirror (on the right), the corresponding distance is considered to be negative. The object height and the image height are considered positive if they lie above the principal axis and negative if they lie below the principal axis. A negative sign is inserted in the magnification equation so that for an upright image the magnification is positive and for an inverted image it is negative.

TEXTBOOK QUESTION 2. Archimedes is said to have burned the whole Roman fleet in the harbor of Syracuse by focusing the rays of the Sun with a huge spherical mirror. Is this reasonable?

ANSWER: In theory, a very large spherical mirror ground to a very high degree of precision with a very long focal length could be built. However, it is doubtful that the technology was available 2300 years ago to construct and mount such a mirror.

In 1973, scientists and engineers used highly polished shields held by a number of soldiers to set fire to the sail of a modern ship. The soldiers stood along the shore (or pier) and reflected the Sun's light to a point on the sail of the ship.

TEXTBOOK QUESTION 7. What is the focal length of a plane mirror? What is the magnification of a plane mirror?

ANSWER: The focal length of a curved mirror equals one-half the radius of curvature of the mirror ($f = r/2$). When the radius of curvature is very large, i.e., approaches infinity, a curved mirror approximates a plane mirror. Therefore, the focal length of a plane mirror is infinite. For a plane mirror $d_i = -d_o$, and since $m = -d_i/d_o$, then $m = +1$.

The linear magnification of the mirror is $+1$. Therefore, the image is the same size as the object and the $+$ sign indicates that the image is upright. The characteristics of the image are that it is virtual, upright, and the same size as the object.

EXAMPLE PROBLEM 1. An object 5.0 cm high is placed 60 cm from a concave mirror of focal length 20 cm. a) Draw a ray diagram and locate the position of the image formed. Draw in the image. b) Mathematically determine the image distance from the mirror, magnification, and height of the image. c) State the characteristics of the image.

Part a. Step 1. Choose an appropriate scale factor and draw an accurate ray diagram.	Solution: (Section 23-3) 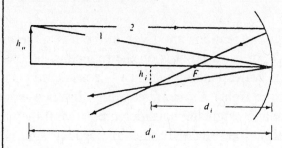

Part b. Step 1. Mathematically determine the image distance from the mirror, magnification, and height of the image.	$1/d_o + 1/d_i = 1/f$ $1/60$ cm $+ 1/d_i = 1/20$ cm $1/d_i = 1/20$ cm $- 1/60 = 2/60$ cm $d_i = 30$ cm $m = -d_i/d_o = -30$ cm$/60$ cm $= -0.50$ $h_i = m\,h_o = (-0.50)(5.0$ cm$) = -2.5$ cm

Note: Recall that the negative sign for both the magnification and image height indicates that the object is real and inverted.

Part c. Step 1. State the characteristics of the image.	The image is real, inverted, and diminished.

EXAMPLE PROBLEM 2. An object 1.0 cm high is placed 22 cm from a concave mirror of focal length 15 cm. a) Draw a ray diagram and locate the position of the image formed. Draw in the image. b) Mathematically determine the image distance from the mirror, magnification, and height of the image. c) State the characteristics of the image.

Part a. Step 1. Choose an appropriate scale factor and draw an accurate ray diagram.	Solution: (Section 23-3) 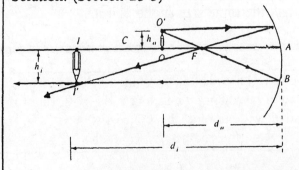

Part b. Step 1.	$1/d_o + 1/d_i = 1/f$
Mathematically determine the image distance from the mirror, magnification, and height of image.	$1/22$ cm $+ 1/d_i = 1/15$ cm
	$1/d_i = 1/15$ cm $- 1/22$ cm
	$1/d_i = (22$ cm $- 15$ cm$)/(15$ cm$)(22$ cm$) = (7$ cm$)/(330$ cm$^2)$
	$d_i = (330$ cm$^2)/(7$ cm$) = 47$ cm
	$m = -d_i/d_o = -(47$ cm$)/(22$ cm$) = -2.1$
	$h_i = m\, h_o = (-2.1)(1.0$ cm$) = -2.1$ cm
Part c. Step 1. Characteristics of the image.	The image is real, inverted, and magnified.

EXAMPLE PROBLEM 3. An object 1.50 cm high is placed 10 cm from a concave mirror of focal length 20 cm. a) Draw a ray diagram and locate the position of the image formed. Draw in the image. b) Mathematically determine the image distance from the mirror, magnification, and height of the image. c) State the characteristics of the image.

Part a. Step 1. Choose an appropriate scale factor and draw an accurate ray diagram.	Solution: (Section 23-3) 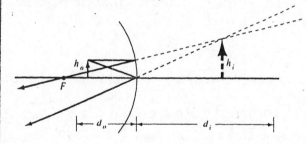
Part b. Step 1. Mathematically determine the image distance from the mirror, magnification, and height of image.	$1/d_o + 1/d_i = 1/f$
	$1/10$ cm $+ 1/d_i = 1/20$ cm
	$1/d_i = 1/20$ cm $- 1/10$ cm and solving gives
	$d_i = -20$ cm
	$m = -d_i/d_o = -(-20$ cm$)/(10$ cm$) = +2.0$
	$h_i = m\, h_o = (+2.0)(1.5$ cm$) = +3.0$ cm

Part c. Step 1. State the characteristics of the image.	The positive sign for both the magnification and the image height indicates that the image is virtual and erect. The image is erect and magnified.

EXAMPLE PROBLEM 4. An object 5.0 cm high is placed 20 cm from a convex mirror of focal
length 20 cm. a) Draw a ray diagram and locate the position of the image formed. Draw in the
image. b) Mathematically determine the image distance from the mirror, magnification, and
height of the image. c) State the characteristics of the image.

Part a. Step 1. Choose an appropriate scale factor and draw an accurate ray diagram.	Solution: (Section 23-4) 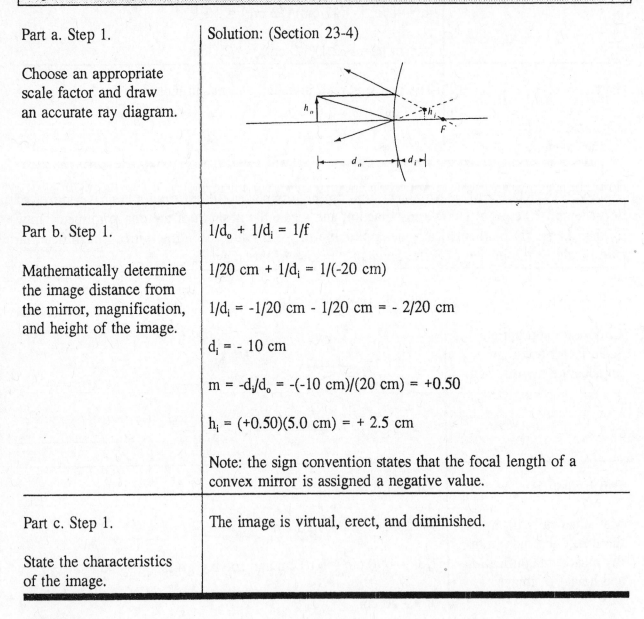
Part b. Step 1. Mathematically determine the image distance from the mirror, magnification, and height of the image.	$1/d_o + 1/d_i = 1/f$ $1/20 \text{ cm} + 1/d_i = 1/(-20 \text{ cm})$ $1/d_i = -1/20 \text{ cm} - 1/20 \text{ cm} = -2/20 \text{ cm}$ $d_i = -10 \text{ cm}$ $m = -d_i/d_o = -(-10 \text{ cm})/(20 \text{ cm}) = +0.50$ $h_i = (+0.50)(5.0 \text{ cm}) = +2.5 \text{ cm}$ Note: the sign convention states that the focal length of a convex mirror is assigned a negative value.
Part c. Step 1. State the characteristics of the image.	The image is virtual, erect, and diminished.

Refraction

As light passes from one medium into another it changes direction at the interface between
the two media. This change of direction is known as **refraction**. The relationship between the

angle of incidence and the **angle of refraction** is given by **Snell's law**:

$$n_1 \sin \theta_1 = n_2 \sin \theta_2$$

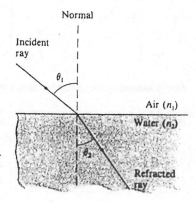

θ_1 is the angle of incidence and is measured from the normal to the surface to the line of the incident ray.

θ_2 is the angle of refraction and is measured from the normal to the surface to the line of the refracted ray.

n_1 is the index of refraction of the medium in which the light is initially traveling. The index of refraction is a dimensionless number which varies from a low of 1.00 for a vacuum to 2.42 for diamond.

n_2 is the index of refraction of the medium into which the light passes.

TEXTBOOK QUESTION 9. What is the angle of refraction when a light ray meets the boundary between two materials perpendicularly?

ANSWER: When a light ray is perpendicular to the boundary between two materials the angle of incidence is $0°$. According to Snell's law :

$$n_1 \sin \theta_1 = n_2 \sin \theta_2 \quad \text{but} \quad \theta_1 = 0° \quad \text{and} \quad \sin \theta_1 = 0$$

$$n_1 (0) = n_2 \sin \theta_2 \quad \text{and} \quad \sin \theta_2 = 0 \quad \text{and} \quad \theta_2 = 0°$$

     In this special case, the angle of refraction is zero and the light ray is not bent as it passes from one medium to the other.

EXAMPLE PROBLEM 5. A ray of light strikes the surface of a flat glass plate at an incident angle of $30°$. The index of refraction of air and glass are 1.00 and 1.50, respectively. Determine the a) angle of reflection and b) angle of refraction.

Part a. Step 1.	Solution: (Section 23-5)
Draw an accurate diagram showing the incident ray, reflected ray, and the approximate path of the refracted ray.	The reflected ray follows the laws of specular reflection, thus the angle of reflection is $30°$.

Part b. Step 1.	$n_1 \sin \theta_1 = n_2 \sin \theta_2$	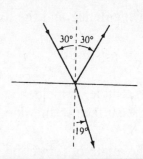
Use Snell's law to determine the angle of refraction.	$(1.00) \sin 30° = (1.50) \sin \theta_2$	
	$\sin \theta_2 = (1.00)(0.50)/(1.50) = 0.33$	
	$\theta_2 = 19°$	

EXAMPLE PROBLEM 6. Water fills the space between two parallel glass plates as shown in the figure below. If the angle of incidence of the ray of light entering the top glass plate is 60°, determine the angle of refraction of the light in the a) top glass plate, b) water. c) Draw the approximate path taken by the ray of light as it passes through each medium.

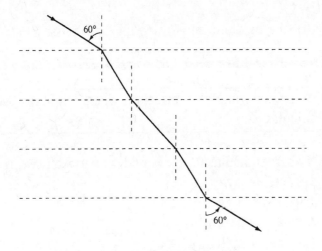

Part a. Step 1.	Solution: (Section 23-5)
Determine the angle of refraction in the top glass plate.	$n_1 \sin \theta_1 = n_2 \sin \theta_2$
	$(1.00) \sin 60° = (1.50) \sin \theta_2$
	$\sin \theta_2 = (1.00)(0.87)/(1.50) = 0.58$
	$\theta_2 = 36°$
Part b. Step 1. Determine the angle of incidence as the light strikes the glass-water interface. Next, determine the angle of refraction in the water.	Because the interface between the media are parallel, the 36° angle of refraction of the light as it passes through the top glass plate is equal to the angle of incidence as the light strikes the interface between the glass and the water. $n_2 \sin \theta_2 = n_3 \sin \theta_3$ $(1.50) \sin 36° = (1.33) \sin \theta_3$

$$\sin \theta_3 = (1.50)(0.58)/(1.33) = 0.654$$

$$\theta_3 = 41°$$

The approximate diagram of the path taken by the ray is shown on the previous page. Applying geometry and Snell's law, it is possible to show that the angle of refraction in the lower glass plate is 36° and the angle in the air is 60°.

Total Internal Reflection

As light passes from a medium of higher index of refraction into a medium of lower index of refraction an angle of incidence is reached at which the angle of refraction is 90°. This angle is known as the **critical angle** (θ_c) and at all angles greater than this angle, the light is totally reflected back into the medium of the incident ray. The equation for the critical angle can be determined by using Snell's law.

$$n_1 \sin \theta_c = n_2 \sin 90° \quad \text{where} \quad \sin \theta_c = n_2/n_1$$

EXAMPLE PROBLEM 7. Determine the critical angle for light passing from a) diamond into air, b) glass into water. The index of refraction for each substance is as follows: diamond 2.42, air 1.00, glass 1.50, water 1.33.

Part a. Step 1.	Solution: (Section 23-6)
Use Snell's law to determine the critical angle for an air-diamond interface.	At the critical angle, $\theta_{air} = 90°$. Using Snell's law, $$n_{diamond} \sin \theta_{diamond} = n_{air} \sin \theta_{air}$$ $$(2.42) \sin \theta_c = (1.00) \sin 90°$$ $$\sin \theta_c = (1.00)(1.00)/(2.42) = 0.41$$ $$\theta_c = 24°$$
Part b. Step 1. Use Snell's law to determine the critical angle angle for a water-glass interface.	$$n_{glass} \sin \theta_{glass} = n_{water} \sin \theta_{water}$$ $$(1.50) \sin \theta_c = (1.33) \sin 90°$$ $$\sin \theta_c = (1.33)(1.00)/(1.50) = 0.89$$ $$\theta_c = 63°$$

> Note: total internal reflection occurs only when light is traveling from a medium of higher index of refraction into a medium of lower index of refraction.

Thin Lenses

Rays of light parallel to the principal axis of a **convex lens** converge at the focal point (F) of the lens after refraction. A convex lens, which is also known as a **converging lens**, has a real focal point.

Rays of light parallel to the principal axis of a **concave lens** diverge after passing through the lens. If the refracted rays are traced on a straight line back through the lens, they "appear" to converge at a focal point. A concave lens, which is also known as a **diverging lens**, has a virtual focal point.

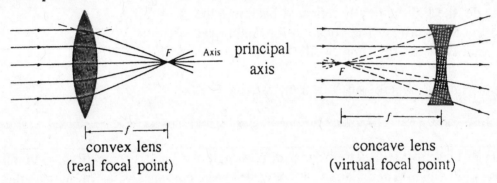

convex lens
(real focal point)

concave lens
(virtual focal point)

Images Formed by a Thin Lens

As in the case of images formed by spherical mirrors, the characteristics of images formed by thin lenses depend on the object distance (d_o), focal length of the lens (f), and whether or not the lens is convex or concave. The following table summarizes the

type of lens	object distance as compared to the focal length	characteristic of image
convex	$d_o > 2f$	real, inverted, diminished
	$d_o = 2f$	real, inverted, same size
	$f < d_o < 2f$	real, inverted, magnified
	$d_o = f$	no image formed
	$d_o < f$	virtual, erect, magnified

Concave lenses produce only virtual, erect, and diminished images.

The images formed by thin lenses may be located by using the following two rays. 1) A ray from the top of the object which strikes the center of the lens. At this point the two sides of the lens are parallel. The ray emerges from the lens slightly displaced but traveling parallel to its original direction. 2) A ray from the top of the object which is parallel to the principal axis of a convex lens passes through the focal point of the side of the lens opposite the object. If the lens is concave, then the ray appears to have come from the virtual focal point located on the same side of the lens as the object.

For a real image, the image of any point on the object is located where the two rays from the point intersect after refraction. For a virtual image it is necessary to trace the path of the refracted rays back through the lens. The image is located at the point where the rays "appear" to cross.

The following diagrams represent each of the situations described in the table.

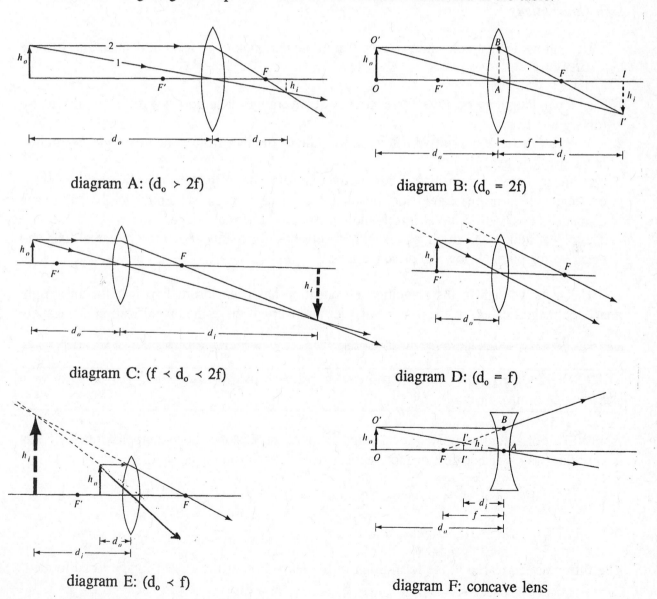

diagram A: $(d_o > 2f)$

diagram B: $(d_o = 2f)$

diagram C: $(f < d_o < 2f)$

diagram D: $(d_o = f)$

diagram E: $(d_o < f)$

diagram F: concave lens

Lens Equation

The lens equation for both convex and concave lenses is

$$1/d_o + 1/d_i = 1/f$$

where d_o is the object distance from the lens, d_i is the image distance from the lens, and f is the focal length of the lens.

Magnification and Size

The linear magnification (m) of a lens is given by

$$m = h_i/h_o = -d_i/d_o \quad \text{and} \quad h_i = m\, h_o \quad \text{where} \quad h_i \text{ is the image height and } h_o \text{ is the object height.}$$

Sign Conventions

The following sign conventions are given in the textbook to be used in connection with the lens equations:

1. The focal length is positive for a convex (converging) lens and negative for a concave (diverging) lens.
2. The object distance is positive if it is on the side of the lens from which the light is coming; otherwise it is negative.
3. The image distance is positive if it is on the opposite side of the lens from where the light is coming; if it is on the same side, the image distance is negative. Equivalently, the image distance is positive for a real image and negative for a virtual image.
4. The object height and the image height are positive for points above the principal axis and negative for points below the principal axis.

The negative sign in the magnification equation has been inserted so that for an upright image the magnification (m) is positive and for an inverted image the magnification is negative.

TEXTBOOK QUESTION 18. Where must the film be placed if a camera lens is to make a sharp image of an object very far away?

ANSWER: For an ordinary fixed focal length camera, any distance beyond approximately 6 feet is very far away. Thus 6 feet or more approximates infinity in equation 23-8.

$$1/d_o + 1/d_i = 1/f \quad \text{but} \quad d_o = \infty \quad \text{and} \quad 1/d_o = 1/\infty = 0$$

$$0 + 1/d_i = 1/f, \quad 1/d_i = 1/f, \quad \text{and} \quad d_i \approx f$$

The film must be placed at the focal point of the lens in order for a clear image to be formed.

EXAMPLE PROBLEM 7. An object 5.0 cm high is placed 40 cm from a convex lens of focal length 20 cm. a) Draw a ray diagram and locate the position of the image formed. Draw in the image. b) Mathematically determine the image distance from the lens, magnification, and height of the image. c) State the characteristics of the image.

Part a. Step 1.	Solution: (Section 23-7 and 23-8)
Choose an appropriate scale factor and draw an accurate ray diagram.	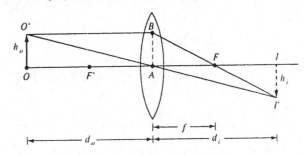

Part b. Step 1.	
Mathematically determine the image distance from the mirror, magnification, and height of the image.	$1/d_o + 1/d_i = 1/f$
	$1/(40 \text{ cm}) + 1/d_i = 1/(20 \text{ cm})$
	$1/d_i = 1/(20 \text{ cm}) - 1/(40 \text{ cm}) = (2 - 1)/(40 \text{ cm})$
	$1/d_i = 1/(40 \text{ cm})$ and $d_i = 40 \text{ cm}$
	$m = -d_i/d_o = -(40 \text{ cm})/(40 \text{ cm}) = -1.0$
	$h_i = m \, h_o = (-1.0)(5.0 \text{ cm}) = -5.0 \text{ cm}$

Part c. Step 1.	
State the characteristics of the image.	The image is real, inverted, and the same size as the object.

EXAMPLE PROBLEM 8. An object 5.0 cm high is placed 30 cm from a convex lens of focal length 20 cm. a) Draw a ray diagram and locate the position of the image formed. Draw in the image. b) Mathematically determine the image distance from the lens, magnification, and height of the image. c) State the characteristics of the image.

Part a. Step 1.	Solution: (Sections 23-7 and 23-8)
Choose an appropriate scale factor and draw an accurate diagram.	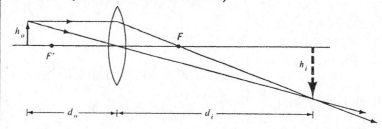

Part b. Step 1.	$1/d_o + 1/d_i = 1/f$
Mathematically determine the image distance from the mirror, magnification, and height of the image.	$1/(30 \text{ cm}) + 1/d_i = 1/(20 \text{ cm})$
	$1/d_i = 1/(20 \text{ cm}) - 1/(30 \text{ cm})$
	$1/d_i = (3 - 2)/(60 \text{ cm}) = 1/(60 \text{ cm})$
	$d_i = 60 \text{ cm}$
	$m = -d_i/d_o = -(60 \text{ cm})/(30 \text{ cm}) = -2.0$
	$h_i = m\, h_o = (-2.0)(5.0 \text{ cm}) = -10.0 \text{ cm}$
Part c. Step 1. State the characteristics of the image.	The image is real, inverted, and magnified.

EXAMPLE PROBLEM 9. An object 3.0 cm high is placed 6.7 cm from a convex lens of focal length 10 cm. a) Draw a ray diagram and locate the position of the image formed. Draw in the image. b) Mathematically determine the image distance from the lens, magnification, and height of the image. c) State the characteristics of the image.

Part a. Step 1.	Solution: (Sections 23-7 and 23-8)
Choose an appropriate scale factor and draw an accurate ray diagram.	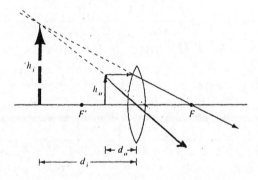

Part b. Step 1.	$1/d_o + 1/d_i = 1/f$
Mathematically determine the image distance from the lens, magnification, and height of image.	$1/6.7 \text{ cm} + 1/d_i = 1/10 \text{ cm}$
	$d_i = -20 \text{ cm}$
	$m = -d_i/d_o = -(-20 \text{ cm})/(6.7 \text{ cm}) = +3.0$
	$h_i = m\, h_o = (+3.0)(3.0 \text{ cm}) = +9.0 \text{ cm}$

| Part c. Step 1.

State the characteristics of the image. | The positive sign for both the magnification and the image height indicates that the image is virtual and erect. The image is erect and magnified. |

EXAMPLE PROBLEM 10. An object 5.0 cm high is placed 40 cm from a concave lens of focal length 20 cm. a) Draw a ray diagram and locate the position of the image formed. Draw in the image. b) Mathematically determine the image distance from the lens, magnification, and height of the image. c) State the characteristics of the image.

Part a. Step 1. Choose an appropriate scale factor and draw an accurate ray diagram.	Solution: (Sections 23-7 and 23-8) 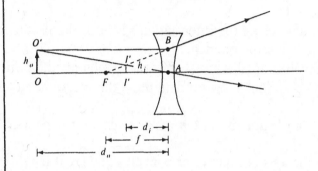
Part b. Step 1. Mathematically determine the image distance from the mirror, magnification, and height of the image.	$1/d_o + 1/d_i = 1/f$ $1/40 \text{ cm} + 1/d_i = 1/(-20 \text{ cm})$ $1/d_i = -1/20 \text{ cm} - 1/40 \text{ cm} = -3/40 \text{ cm}$ $d_i = -13.3 \text{ cm}$ $m = -d_i/d_o = -(-13.3 \text{ cm})/(40 \text{ cm}) = 0.33$ $h_i = m \, h_o = (0.33)(5.0 \text{ cm}) = 1.7 \text{ cm}$
Part c. Step 1. State the characteristics of the image.	The image is virtual, erect, and diminished.

PROBLEM SOLVING SKILLS

For problems involving image formation by a convex or a concave mirror:

1. Identify whether the mirror is concave or convex. The focal length is positive for a concave mirror and negative for a convex mirror.

2. Choose an appropriate scale factor to represent the focal length, object distance, and height of the object.
3. Use the two suggested rays to draw an accurate ray diagram. Draw in the image at the point where the two rays cross after reflection from the mirror.
4. Use the mirror equations and sign conventions to mathematically determine the image distance, image height, and magnification.
5. State the characteristics of the image: real or virtual; erect or inverted; magnified, diminished, or same size as the object.

For problems involving refraction of light as it passes from one medium to another:

1. Draw an accurate diagram locating the incident ray and normal to the surface. Determine the angle of incidence.
2. Complete a data table using the information given in the problem.
3. Use Snell's law to solve the problem.
4. If the problem involves total internal reflection, then at the critical angle, the angle of refraction is 90°. Use Snell's law to determine the magnitude of the critical angle.

For problems involving image formation by a concave or a convex lens:

1. Identify whether the lens is concave or convex. The focal length is positive for a convex lens and negative for a concave lens.
2. Lens problems involve refraction of light while mirror problems involve reflection. Apply the steps listed above for spherical mirror problems to solve problems involving lenses.

SOLUTIONS TO SELECTED TEXTBOOK PROBLEMS

TEXTBOOK PROBLEM 9. If you look at yourself in a shiny Christmas tree ball with a diameter of 9.0 cm when your face is 30.0 cm away from it, where is your image? Is it real or virtual? Is it upright or inverted?

Part a. Step 1.	Solution: (Section 23-3)
Choose an appropriate scale factor and draw an accurate ray diagram. Note: Assume that the ornament acts like a convex mirror.	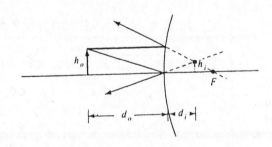

Part a. Step 2.

Determine the focal length of the ball.

radius = ½ diameter = ½ (- 9.0 cm) = - 4.5 cm

f = ½ r = ½ (- 4.5 cm) = - 2.25 cm

	Note: the sign convention states that the focal length of a convex mirror is assigned a negative value.
Part a. Step 3. Mathematically determine the image distance from the mirror, and magnification of the image.	$1/d_o + 1/d_i = 1/f$ $1/(30.0 \text{ cm}) + 1/d_i = 1/(-2.25 \text{ cm})$ $1/d_i = -1/(2.25 \text{ cm}) - 1/(30.0 \text{ cm}) = -0.478 \text{ cm}$ $d_i = -2.09 \text{ cm}$ $m = -d_i/d_o = -(-2.09 \text{ cm})/(30 \text{ cm}) = +0.070$
Part a. Step 4. State the characteristics of the image.	The image is virtual, erect, and diminished.

TEXTBOOK PROBLEM 11. A dentist wants a small mirror that, when 2.20 cm from a tooth, will produce a 4.5x upright image. What kind of mirror must be used and what must its radius of curvature be?

Part a. Step 1. State the characteristics of the image.	Solution: (Section 23-3) The image is virtual, erect, and magnified.
Part a. Step 2. Determine the type of mirror which produces the characteristics needed. Draw an approximate ray diagram locating the image.	The only type of mirror which produces a virtual image with a magnification greater than one is a concave mirror.
Part a. Step 3. Mathematically determine the image distance from the mirror.	$m = -d_i/d_o$ $+4.5 = -d_i/(2.20 \text{ cm})$ $d_i = -9.90 \text{ cm}$
Part a. Step 4. Determine the focal length of the mirror.	$1/d_o + 1/d_i = 1/f$ $1/(2.20 \text{ cm}) + 1/(-9.90 \text{ cm}) = 1/f$

23-21

	$1/f = (0.354)/cm$
	$f = 2.83$ cm
Part a. Step 5.	$r = 2 f$
Determine the mirror's radius of curvature.	$r = 2(2.83$ cm$) = 5.66$ cm

---

TEXTBOOK PROBLEM 31. In searching the bottom of a pool at night, a watchman shines a narrow beam of light from his flashlight, 1.3 m above the water level, onto the surface of the water at a point 2.7 m from the pool (Fig 23-50). Where does the spot of light hit the bottom of the pool, measured from the wall beneath his foot, if the pool is 2.1 m deep?

Part a. Step 1.	Solution: (Section 23-5)
Choose an appropriate scale factor and draw an accurate ray diagram.	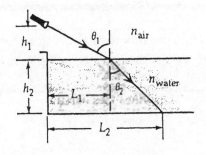
Part a. Step 2.	Based on the diagram,
Determine the angle (θ_1) the incident ray makes with the vertical.	$\tan \theta_1 = L_1/h_1) = (2.7$ m$)/(1.3$ m$) = 2.1$
	$\theta_1 = \tan^{-1} 2.1 = 64.3°$
Part a. Step 3.	$n_1 \sin \theta_1 = n_2 \sin \theta_2$
Use Snell's law to determine the angle of refraction in the water.	$(1.00) \sin 64.3° = (1.33) \sin \theta_2$
	$\theta_2 = 42.6°$
Part a. Step 4.	$\tan \theta_2 = (L_2 - L_1)/h_2 = (L_2 - L_1)/(2.1$ m$)$
Determine the horizontal distance ($L_2 - L_1$) from the point where the ray strikes the water to the point where it strikes the bottom of the pool.	$L_2 - L_1 = (2.1$ m$)(\tan 42.6°)$
	$L_2 - L_1 = (2.1$ m$)(0.92)$
	$L_2 - L_1 = 1.9$ m

Part a. Step 5.	$L_2 = L_1 + 1.9 \text{ m} = 2.7 \text{ m} + 1.9 \text{ m}$
Determine the total horizontal distance (L_2).	$L_2 = 4.6 \text{ m}$

TEXTBOOK PROBLEM 39. A beam of light is emitted 8.0 cm beneath the surface of a liquid and strikes the surface 7.0 cm from the point directly above the source. If total internal reflection occurs, what can you say about the index of refraction of the liquid?

Part a. Step 1. Choose an appropriate scale factor and draw an accurate ray diagram.	Solution: (Section 23-6) 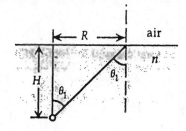
Part a. Step 2. Determine the angle (θ_1) the incident ray makes with the vertical.	Based on the diagram, $\tan \theta_1 = R/H = (7.0 \text{ cm})/(8.0 \text{ cm}) = 0.875$ $\theta_1 = \tan^{-1} 0.875 = 41.2°$
Part a. Step 3. Use Snell's law to determine the minimum index of refraction of the liquid.	$n_1 \sin \theta_1 = n_2 \sin \theta_2$ but at the critical angle $\theta_2 = 90°$ $n_1 \sin 41.2° = (1.00) \sin 90°$ $n_1 = (1.00)(1.00)/(\sin 41.2°) = (1.00)/(0.659) = 1.5$ The index of refraction of the liquid $n_1 \geq 1.5$

TEXTBOOK PROBLEM 47. A stamp collector uses a converging lens with focal length of 24 cm to view a stamp 18 cm in front of the lens. (a) Where is the image located? (b) What is the magnification?

Part a. Step 1. Choose an appropriate scale factor and draw an accurate ray diagram.	Solution: (Sections 23-7 and 23-8) 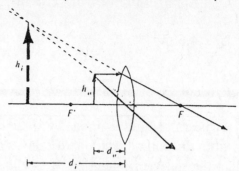
Part a. Step 2. State the characteristics of the image.	Based on the diagram, the image is virtual, erect, and magnified.
Part a. Step 3. Mathematically determine the image distance from the lens.	$1/d_o + 1/d_i = 1/f$ $1/(18 \text{ cm}) + 1/d_i = 1/(24 \text{ cm})$ $d_i = -72 \text{ cm}$
Part b. Step 1. Determine the magnification of the image.	$m = -d_i/d_o = -(-72 \text{ cm})/(18 \text{ cm}) = +4.0$

TEXTBOOK PROBLEM 57. A bright object and a viewing screen are separated by a distance of 66.0 cm. At what location(s) between the object and the screen should a lens of focal length 12.5 cm be placed in order to produce a crisp image on the screen? [Hint: first draw a diagram]

Part a. Step 1. Choose an appropriate scale factor and draw a ray diagram. Note: there are two possible locations.	Solution: (Section 23-5) 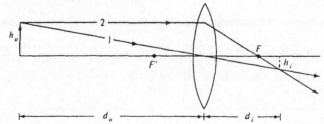

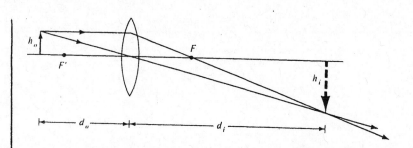

Part a. Step 2.	$1/d_o + 1/d_i = 1/f$ but $d_o + d_i = 66.0$ cm and $d_i = 66.00$ cm - d_o
Use the thin lens equation to solve for the object distance (d_o).	$1/d_o + 1/(66.0 \text{ cm} - d_o) = 1/(12.5 \text{ cm})$
	$[(66.0 \text{ cm} - d_o) + (d_o)]/[(d_o)(66.0 \text{ cm} - d_o)] = 1/(12.5 \text{ cm})$
	using algebra and rearranging gives
	$d_o^2 - (66.0 \text{ cm})d_o + 825 \text{ cm}^2 = 0$
	Using the quadratic formula results in two possible solutions,
	either $d_o = 49.2$ cm or $d_o = 16.8$ cm

CHAPTER 24

THE WAVE NATURE OF LIGHT

OBJECTIVES

After studying the material of this chapter, the student should be able to:

- use the wave model to explain reflection of light from mirrors and refraction of light as it passes from one medium into another.
- use the conditions for constructive and destructive interference of waves to explain the interference patterns observed in the Young's double slit experiment, single slit diffraction, diffraction grating, and thin film interference.
- solve problems involving a single slit, double slit, and diffraction grating for m, λ, d, D, and angular separation (θ), when the other quantities are given.
- solve problems involving thin film interference for m, λ, n, or t when the other quantities are given.
- explain how the Michelson interferometer can be used to determine the wavelength of a monochromatic light source and solve problems to determine the wavelength of the light from the source.
- use the wave model to explain plane polarization of light, polarization by reflection, and polarization by double refraction.
- calculate the angle of maximum polarization for reflected light.

KEY TERMS AND PHRASES

interference patterns are produced when two (or more) coherent sources producing waves of the same frequency and amplitude superimpose.

sources are coherent when there is a fixed phase between the waves emitted by the sources.

crest is the highest point of that portion of a transverse wave above the equilibrium position.

trough is the lowest point of that portion of a transverse wave below the equilibrium position.

destructive interference occurs if the amplitude of the resultant of two interfering waves is smaller than the displacement of either wave.

constructive interference occurs if the amplitude of the resultant of two interfering waves is larger than the displacement of either wave.

diffraction occurs when waves bend and spread out as they pass an obstacle or narrow opening. The amount of diffraction depends on the wavelength of the waves and the size of the obstacle. In the case of a narrow opening, the amount of bending increases as the size of the opening decreases.

diffraction grating consists of a large number of closely spaced parallel slits which diffract light incident on the grating. The diffracted light will exhibit a pattern of constructive interference and destructive interference.

thin film interference patterns occur when light waves are reflected at both the top and bottom surfaces of the film. The pattern of dark and bright lines observed results from light reflected from the top surface interfering with the light reflected from the bottom surface.

polarization is a property of light that indicates light is a transverse wave phenomenon. A transverse wave is a wave in which the particles of the medium move at right angles to the direction of motion of the wave. An electromagnetic wave in which the electric vector is vibrating in only one plane is said to be plane polarized.

Brewster's angle is the angle of incidence at which maximum polarization of the reflected light occurs.

double refraction occurs when light passes through certain transparent crystals and the incident ray is separated into two refracted rays. One ray, known as the ordinary ray (or o ray), follows Snell's law. The second ray, known as the extraordinary ray (or e ray), does not follow Snell's law.

SUMMARY OF MATHEMATICAL FORMULAS

speed of light and the wavelength in a medium other than a vacuum.	$v = c/n$	The speed of light (v) in a medium is the ratio of the speed of light in a vacuum (c) to the medium's index of refraction (n).
	$\lambda_1 = \lambda_{vacuum}/n_1$	The wavelength of the light in a medium (λ_1) is the ratio of the wavelength in a vacuum to the index of refraction of the medium (n_1).

double slit interference	$m\lambda = d \sin\theta$	The relationship between the wavelength (λ), the distance between the slits (d), and θ is the grating angle for constructive interference. m is the order of the interference fringe, m = 0, 1, 2, 3, etc.
	$(m + \frac{1}{2})\lambda = d \sin\theta$	The relationship between the wavelength (λ), the distance between the slits (d), and θ is the grating angle for destructive interference.
single slit diffraction bright fringes dark fringes	$(m + \frac{1}{2})\lambda = D \sin\theta$ $m\lambda = D \sin\theta$	The pattern for bright or dark fringes in a single slit diffraction pattern depends on the wavelength (λ), the slit width (D), and the the diffraction angle (θ). m = 1, 2, 3, etc.
thin film interference	$t = (2m + 1)\lambda_{film}/4$ or $4t = (2m + 1)\lambda_{film}$ where $\lambda_{film} = \lambda_{air}/n_{film}$ $t = m\,\lambda_{film}/2$ or $2t = m\,\lambda_{film}$	If the indices of refraction of the media above and below the thin film are lower (or higher) than the index of refraction of the medium of the thin film, then maximum reflection of light of a particular wavelength occurs if the thickness (t) of the film is an odd-number multiple of quarter wavelengths. Note: m = 0, 1, 2, etc. Minimum reflection of the light occurs if the thickness of the film is a whole-number multiple of half wavelengths. Note: m = 1, 2, etc. Note: if the media above and below the thin film are such that a phase change occurs at both the top and bottom surfaces, then the equation for maximum reflection is given by $2t = m\,\lambda_{film}$ and for minimum reflection by $4t = (2m + 1)\lambda_{film}$.
Michelson interferometer	$t = m\,\lambda/2$	The Michelson interferometer is a device which uses wave interference to determine the wavelength of light. By measuring the distance that one of the mirrors moves and the number of interference fringes (m) that pass the observer's field of view, the wavelength (λ) of the light can be determined.

Brewster's angle	$\tan \theta_p = n_2/n_1$	The angle of incidence at which maximum polarization (θ_p) of the reflected light occurs is known as Brewster's angle. The tangent of Brewster's angle (θ_p) is related to the ratio of the index of refraction (n_2) of the substance from which the light is reflected to the index of refraction (n_1) of the substance in which the light is initially traveling.

CONCEPT SUMMARY

Reflection

In Chapter 11 it was observed that a **wave front** striking a straight barrier follows the law of reflection (see page 11-13). This law was found to apply to light in Chapter 23.

Refraction

Water waves undergo **refraction** as they travel from deep to shallow water because the speed of the wave changes (see page 11-14). Refraction of light occurs as light travels from one medium to another and this was discussed in terms of Snell's law in Chapter 23. The wave theory proposed by Christian Huygens (1629-1695) predicts that the speed of light is less in water or glass than in air. Measurements of the speed of light in various materials agree with the wave theory. The speed of light in a medium (v) is inversely related to the medium's index of refraction (n), i.e., $v = c/n$ where c is the speed of light in a vacuum and $c = 3.0 \times 10^8$ m/s.

For light traveling from medium 1 into medium 2

$n_2/n_1 = v_1/v_2 = \lambda_1/\lambda_2$ where λ is the wavelength of the light in the medium.

EXAMPLE PROBLEM 1. Light of wavelength 600 nm passes from a vacuum into glass of index of refraction 1.50. Determine the a) speed of light in the glass and b) wavelength of the light in the glass.

Part a. Step 1.	Solution: (Section 24-2)
Determine the speed of light in glass.	The speed of light in a vacuum is 3.0×10^8 meters per second and the index of refraction of a vacuum is 1.00. Assume that medium 1 is the vacuum and medium 2 is the glass. $n_2/n_1 = v_1/v_2$ $1.50/1.00 = (3.0 \times 10^8 \text{ m/s})/v_2$

	$v_2 = 2.0 \times 10^8$ m/s
Part b. Step 1. Determine the wavelength of the light in glass.	1 nm = 1 x 10⁻⁹ m; however, it is appropriate to express the wavelength of light in nanometers (nm). There is no need to convert the wavelength to meters. $n_2/n_1 = \lambda_1/\lambda_2$ $1.50/1.00 = 600 \text{ nm}/\lambda_2$ $\lambda_2 = 400$ nm

Interference

Two **coherent** sources producing waves of the same frequency and amplitude produce an interference pattern. The following diagram was used in Chapter 12 (page 12-9) to demonstrate the pattern produced by sound waves. s_1 and s_2 represent the sources of the waves while point A is a point of **constructive interference** and point B is a point of **destructive interference**.

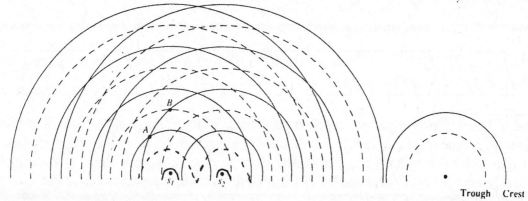

Trough   Crest

Young's Double Slit Experiment

Young's double slit experiment shows that mono-chromatic light passing through two openings also produces an interference pattern. The position of lines of constructive interference can be determined from the following equation:

$$m\,\lambda = d \sin \theta$$

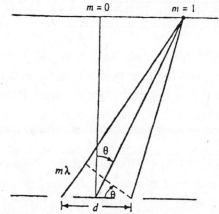

where d is the distance between the slits, and θ is the grating angle. This angle is equal to the angle formed by a line drawn from the center point between the slits to the region of constructive interference and the normal drawn to the center of the line which connects the slits. m is the order of the interference fringe. m is a dimensionless number which takes on integer values starting with zero, i.e., m = 0, 1, 2, 3, etc. λ is the wavelength of the monochromatic light incident on the double slit. $m\lambda$ is the path difference between the sources and the region of constructive interference. The path difference is a whole number of wavelengths for constructive interference.

Regions of destructive interference alternate with regions of constructive interference in an interference pattern. The position of lines of destructive interference can be found from

$(m + \frac{1}{2})\lambda = d \sin \theta$

where m, λ, d, and θ are the same as described previously and $(m + \frac{1}{2})\lambda$ is the path difference between the sources to the points of destructive interference.

TEXTBOOK QUESTION 7. Two rays of light from the same source destructively interfere if their path lengths differ by how much?

ANSWER: In order for destructive interference to occur, the path length must differ by an odd-numbered multiple of one-half wavelength, i.e., path difference = $(m + \frac{1}{2})\lambda$ where m = 0, 1, 2, etc.

TEXTBOOK QUESTION 5. If Young's double-slit experiment were submerged in water, how would the fringe pattern be changed?

Answer: The interference pattern depends on the wavelength. For constructive interference, $\sin \theta = m \lambda/d$ where $\sin \theta \approx \Delta X/L$ and ΔX is the separation between the bright fringes. The wavelength of light in water (λ_{water}) is shorter than in air, i.e., $\lambda_{water} = \lambda/n_{water}$. As a result, the bright fringes would be closer together if the experiment was conducted underwater.

TEXTBOOK QUESTION 10. Why doesn't the light from the two headlights of a distant car produce an interference pattern?

ANSWER: The light from the two headlamps is not coherent. The light produced by one light is produced at random with respect to the second light. In order to produce an interference pattern, the light waves must be coherent, i.e., the emitted waves maintain the same phase relationship to one another at all times.

EXAMPLE PROBLEM 2. A laser beam is incident on two slits separated by 0.63 mm. The interference pattern is formed on a screen 2.0 m from the slits and the first bright fringe is found to be 0.20 cm to the right of the central maximum. a) Calculate the wavelength of the light in nanometers (nm) and b) determine the maximum number of bright fringes that can be observed.

Part a. Step 1. Convert all of the data to SI units.	Solution: (Section 24-3) 0.63 mm = 6.3×10^{-4} m 0.20 cm = 2.0×10^{-3} m.
Part a. Step 2. Refer to the upper diagram on page 24-5 and determine the angular deflection for m = 1.	$\tan \theta = \Delta X / L = (2.0 \times 10^{-3}$ m$)/(2.0$ m$)$ $\tan \theta = 1.0 \times 10^{-3}$ The angle is small, and if the angle is expressed in radians, then $\sin \theta \approx \tan \theta \approx \theta$ and $\theta = 1.0 \times 10^{-3}$ rad
Part a. Step 3. Determine the wave-length of the light.	$m \lambda = d \sin \theta_1$ $(1)(\lambda) = (6.3 \times 10^{-4}$ m$)(1.00 \times 10^{-3})$ $\lambda = 6.3 \times 10^{-7}$ m or 630 nm
Part b. Step 1. Determine the maximum number of bright fringes that can be observed.	The maximum value that θ can have is 90°. $m \lambda = d \sin \theta$ $m (6.3 \times 10^{-7}$ m$) = (6.3 \times 10^{-4}$ m$)(\sin 90°)$ $m = (6.3 \times 10^{-4}$ m$)(1.00)/(6.30 \times 10^{-7}$ m$)$ $m = 1000$ 1000 fringes are formed on each side of the central maximum. Thus, counting the central fringe, the theoretical maximum number of lines that could be observed would be 1000 + 1000 + 1 = 2001. However, it is necessary to round the number to two significant figures, thus the answer would be 2000.

EXAMPLE PROBLEM 3. A laser beam is incident on two slits separated by 0.63 mm. The wavelength of the light produced by the laser is 630 nm. The interference pattern is formed on a screen 2.0 m from the slits. Determine the a) angular deflection of the third order dark fringe (dark line), b) distance from the central maximum to this fringe.

Part a. Step 1.	Solution: (Section 24-3)
Determine the angular deflection of the third order dark fringe.	$(m + \frac{1}{2})\lambda = d \sin \theta_3$
	$(3 + \frac{1}{2})(6.3 \times 10^{-7} \text{ m}) = (6.3 \times 10^{-4} \text{ m}) \sin \theta_3$
	$\sin \theta_3 = 3.5 \times 10^{-3}$
	The angle is small, $\sin \theta_3 \approx \tan \theta_3$
	$\theta_3 = 3.5 \times 10^{-3}$ radians
Part b. Step 1.	$\tan \theta_3 = \Delta X/L$
Determine the distance from the central maximum to the third order dark fringe.	$\Delta X = L \tan \theta_3 = (2.0 \text{ m})(3.5 \times 10^{-3} \text{ rad})$
	$\Delta X = 7.0 \times 10^{-3} \text{ m} = 7.0 \text{ mm}$

Diffraction

The wave theory predicts and experiments confirm that waves will bend and spread out as they pass an obstacle or narrow opening. In the case of a narrow opening, the amount of bending increases as the size of the opening decreases.

Because the wavelength of light is much shorter than that of water waves or sound waves, the **diffraction** of light is only noticeable when the size of the opening is comparable to the wavelength of the light.

Single Slit Diffraction

Due to the combined effects of diffraction and interference, monochromatic light passing through a single slit of width D produces an interference pattern of alternating bright and dark lines. The positions of lines of destructive interference can be determined by the following equation:

$m\lambda = D \sin \theta$

where D is the width of the opening and θ is the angle between the normal to the center of the slit and the center of the dark fringe. m is the order of the dark fringe, m = 1, 2, etc. For m = 0, the wave theory predicts that constructive rather than destructive interference will occur. λ is the wavelength of the light incident on the slit. mλ is the path difference between the two edges of the slit and the position of the dark fringe.

EXAMPLE PROBLEM 4. A laser beam is incident on a single slit of width 0.21 mm. The wavelength of the light produced by the laser is 630 nm. The diffraction pattern is formed on a screen 2.0 m from the slit.   Calculate the distance from the center of the central maximum to the first order dark fringe.

Part a. Step 1.	Solution: (Section 24-5)
Refer to the diagram on page 24-8 and determine the angular deflection (θ_1) of the first order dark fringe.	$m\lambda = D \sin \theta_1$ $(1)(6.3 \times 10^{-7} \text{ m}) = (0.21 \times 10^{-3} \text{ m})(\sin \theta_1)$ $\sin \theta_1 = (6.3 \times 10^{-7} \text{ m})/(0.21 \times 10^{-3} \text{ m})$ $\sin \theta_1 = 3.0 \times 10^{-3}$ θ is small, then $\sin \theta \approx \tan \theta$, and $\theta_1 = 3.0 \times 10^{-3}$ radians
Part a. Step 2.	$\tan \theta_1 = \Delta X/L$
Determine the distance (ΔX) from the center of the central maximum to the first order dark fringe.	$\Delta X = L \tan \theta_1 = (2.0 \text{ m})(3.0 \times 10^{-3} \text{ rad})$ $\Delta X = 6.0 \times 10^{-3}$ m or 6.0 mm

Diffraction Grating

   A **diffraction grating** consists of a large number of closely spaced parallel slits which diffract light incident on the grating. The diffracted light will exhibit constructive interference at points given by the equation

$$m\lambda = d \sin \theta$$

where m is the order of the bright fringe, m = 0, 1, 2, etc. λ is the wavelength of the incident light.   d is the distance between the centers of the adjacent slits.   The grating angle (θ) is measured from a line drawn normal to the center of the grating and a line drawn from the center of the grating to the position of the bright fringe.

The grating equation has the same form as the equation found in the Young double slit experiment. In the double slit experiment the distance d was measured between the centers of the two slits. In the diffraction grating equation d refers to the distance between the centers of adjacent slits. The amount of light passing through the diffraction grating is much greater than in the case of the double slit. As a result, the intensity of the bright lines is much greater.

A diffraction grating is particularly useful in separating the component wavelengths of the light incident on the grating. Because of this, it is frequently used in the analysis of spectrum produced by various gases, e.g., mercury, hydrogen, and helium.

EXAMPLE PROBLEM 5. A laser beam is incident on a 7500 groove per inch diffraction grating. The wavelength of the light produced by the laser is 600 nm and the interference pattern is observed on a screen 1.00 m from the grating. Determine the a) angular deflection of the first-order bright fringe, b) angular deflection of the second-order bright fringe, c) distance from the central maximum to the second-order bright fringe, and d) maximum number of bright fringes that can be observed.

Part a. Step 1.	Solution: (Section 24-6)
Determine the distance between adjacent grooves in the diffraction grating in meters.	There are 7500 grooves per inch; then the distance between adjacent grooves is 1/7500 inch. d = (1 inch/7500 grooves)(2.54 cm/1 inch)(1 m/100 cm) d = 3.39×10^{-6} m
Part a. Step 2.	$m \lambda = d \sin \theta_1$
Determine the first-order angular deflection.	$(1)(6.00 \times 10^{-7}$ m$) = (3.39 \times 10^{-6}$ m$) \sin \theta_1$ $\sin \theta_1 = 0.177$ and $\theta_1 = 10.2°$
Part b. Step 1.	$m \lambda = d \sin \theta_2$
Determine the second-order angular deflection.	$(2)(6.00 \times 10^{-7}$ m$) = (3.39 \times 10^{-6}$ m$) \sin \theta_2$ $\sin \theta_2 = 0.354$ and $\theta_2 = 20.7°$

24-10

Part c. Step 1.	$\tan \theta_2 = \Delta X / L$
Determine the distance from the central maximum to the second-order bright fringe.	$\Delta X = L \tan \theta_2 = (1.00 \text{ m})(\tan 20.7°)$ $\Delta X = 0.378 \text{ m}$
Part d. Step 1. Determine the maximum number of bright fringes that can be observed.	$m \lambda = d \sin \theta$ The maximum possible value of θ is 90°; thus $m (6.00 \times 10^{-7} \text{ m}) = (3.39 \times 10^{-6} \text{ m})(\sin 90°)$ $m = 5.65$ Since m must be an integer, only five bright fringes can be observed on either side of the central bright fringe. The maximum number of fringes that can be observed is $5 + 5 + 1 = 11$.

Interference by Thin Films

On page 11-14, it was noted that a transverse wave pulse traveling through a light section of rope will undergo partial transmission and partial reflection at a point where the pulse enters the heavier section. If the wave pulse is a crest, then the transmission will be a crest but the reflection will be a trough (see diagram). A 180° phase change occurs for the reflected part of the wave. If the pulse is traveling in a heavy section and enters a light section, no phase change occurs for either the transmitted or reflected wave pulse.

Light waves traveling from one medium to another undergo partial reflection and partial transmission at the interface of the two mediums. By analogy with the rope of Chapter 11, the medium with the lower index of refraction is analogous to the light section of the rope. The medium with the higher index of refraction is analogous to the heavy section of rope.

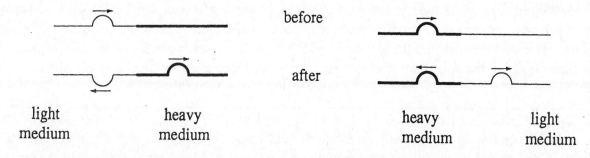

| light medium | heavy medium | | heavy medium | light medium |

As shown in the diagram at the top of the next page, reflection and transmission of light waves in **thin films** occurs at both the top and bottom surfaces of the film. The light reflected back to the observer is the result of light reflected from the top surface interfering with the reflection from the bottom surface.

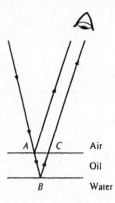

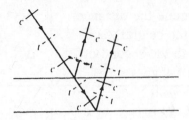

A   C   Air

Oil

B   Water

Light waves reflecting from
both the top and bottom
surfaces of a thin film.

Crests and troughs are represented by a solid line for a crest
and a dotted line for a trough.
Pulses are shown reflecting
from both top and bottom
surfaces of the thin film.

If the indices of refraction of the media above and below the thin film are lower (or higher)
than the index of refraction of the medium of the thin film, then **maximum** reflection of light
of a particular wavelength occurs if the thickness (t) of the film is an odd-number multiple of
quarter wavelengths. The equation for thickness of the film in terms of wavelength is

$$t = (2m + 1)\lambda_{film}/4 \quad \text{or} \quad 4t = (2m + 1)\lambda_{film} \quad \text{where} \quad \lambda_{film} = \lambda_{air}/n_{film} \quad \text{and} \quad m = 0, 1, 2, \text{etc.}$$

Minimum reflection of the light occurs if the thickness of the film is a whole number
multiple of half wavelengths. The equation for minimum reflection is given by

$$t = m\,\lambda_{film}/2 \quad \text{or} \quad 2t = m\,\lambda_{film} \quad \text{where} \quad m = 0, 1, 2, \text{etc.}$$

If the media above and below the thin film are such that a phase change occurs at both the
top and bottom surfaces, then the equation for **maximum reflection** is given by $2t = m\,\lambda_{film}$ and
for **minimum reflection** by $4t = (2m + 1)\,\lambda_{film}$.

Since ordinary white light is made up of colors of varying wavelengths (400 nm - 700 nm),
then maximum reflection of certain colors will occur while there will be no reflection of other
colors. This and the fact that the thickness of the thin film usually varies across the surface of
the film results in the spectrum of colors seen reflected from soap bubbles and oil films.

EXAMPLE PROBLEM 6. A ring of wire is dipped into a soap solution and then held so that
the soap film on the ring is vertical. As the soap film gradually drains toward the bottom, a dark
band appears at the top with alternating bright and dark bands of light appearing along the length
of the film. Determine the thickness of the film at the first three bright bands, as counted from
the top, if the incident light has a wavelength of 546 nm and is directed perpendicular to the
surface of the film. The index of refraction of the soap solution is 1.33.

Part a. Step 1.	Solution: (Section 24-8)
Assume that a wave crest (c_1) strikes the top interface. Determine whether each subsequent transmission and reflection will be a crest or a trough.	Since $n_{film} > n_{air}$, then a phase change occurs and the reflection of c_1 at the top interface will be a trough (t_1). No phase change occurs for transmitted light so the crest (c_1) will strike the lower interface as a crest. At the lower interface the crest (c_1) will reflect as a crest (c_1'). This is because $n_{film} > n_{air}$. When c_1' strikes the top interface, it will be transmitted as a crest.

Part a. Step 2.	
Draw a diagram using c for crest and t for trough for the situation described in part a.	air soap film air 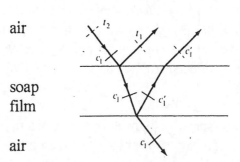

Part a. Step 3.	
Derive a formula for the film thickness that results in maximum reflection of the incident light.	If the thickness of the film is $\frac{1}{4}\lambda$, then c_1' will have traveled $\frac{1}{4}\lambda_{film} + \frac{1}{4}\lambda_{film} = \frac{1}{2}\lambda_{film}$ and c_1' will meet trough t_2 which is reflecting at the top interface as a crest c_2. The superposition of c_1' and c_2 will result in constructive interference and maximum reflection. Thus, maximum reflection occurs if the thickness of the film follows the equation $t = (2m + 1)\lambda_{film}/4 \quad or \quad 4t = (2m + 1)\lambda_{film}$ where $\lambda_{film} = \lambda_{air}/n_{film} = 546$ nm$/1.33 = 411$ nm

---

Part a. Step 4.	
Determine the thickness of the film at the position of the first three bright bands as counted from the top of the film.	The minimum thickness for maximum reflection occurs if $m = 0$. Since the film drains toward the bottom, the minimum thickness is located at the position of the first bright band observed near the top of the film. $4t = (2m + 1) \lambda_{film}$ If $m = 0$, then $4t = [2(0) + 1](411$ nm$)$ and $t = 103$ nm $= 1.03 \times 10^{-7}$ m Successive bright bands occur for $m = 1$, $m = 2$.

If m = 1, then 4t = [2 (1) + 1](411 nm) and

t = 308 nm = 3.08 x 10⁻⁷ m

If m = 2, then 4t = [2 (2) + 1](411 nm) and

t = 513 nm = 5.13 x 10⁻⁷ m

Michelson Interferometer

The **Michelson interferometer** is a device which uses wave interference to determine the wavelength of light. As shown in the diagram, the light from source S strikes a half-silvered mirror (M_s). Part of the light is reflected to a movable mirror (M_1) while part of the light is transmitted and reflects from a fixed mirror (M_2). After reflection, the beams arrive at the observer's eye.

Because the speed of light is less in glass than in air, a glass plate called the compensator (C) is placed in the path of beam 2. Thus, both beams travel through the same thickness of glass before arriving at the observer's eye.

Assume that light reaching the observer's eye from mirrors M_1 and M_2 are in phase. Constructive interference occurs and the observer sees light. If mirror M_1 is moved ¼λ toward the observer, the path difference changes by ¼λ + ¼λ = ½λ and destructive interference occurs. If the mirror is moved another ¼λ, then the total path difference has changed by ½λ + ½λ = 1λ from its original setting and constructive interference again occurs.

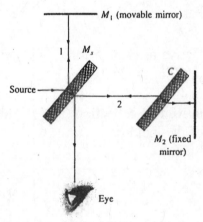

The distance (t) that the mirror M_1 moves can be determined by a device known as a micrometer. The wavelength of the monochromatic light from source S can be determined by counting the number of bright (or dark) fringes (m) which pass the field of view as M_1 is moved. The equation used is t = m λ/2.

EXAMPLE PROBLEM 7. The bright green line produced by the mercury spectrum is used as the source in an experiment involving the Michelson interferometer. When the movable mirror is moved 0.273 mm, 1000 bright fringes are counted. Determine the wavelength of the light.

Part a. Step 1.	Solution: (Section 24-9)
Determine the wavelength of the light. Note: refer to the diagram on the previous page.	Each time mirror M_1 travels $\frac{1}{2}\lambda$, another bright fringe passes the field of view. Therefore $t = m\,\lambda/2$ and rearranging gives $\lambda = 2t/m = 2(0.273\text{ mm})/1000$ $\lambda = 2(0.273 \times 10^{-3}\text{ m})/1000$ $\lambda = 0.546 \times 10^{-6}\text{ m} = 5.46 \times 10^{-7}\text{ m} = 546\text{ nm}$

Polarization of Light

The interference phenomena previously studied in this chapter can be produced by longitudinal as well as transverse waves. **Polarization** is a property of light that indicates light is a **transverse wave** phenomena.

An electromagnetic wave in which the electric vector is vibrating in only one plane is said to be **plane polarized**. Ordinary light is not polarized, which means that its electric vector is vibrating in many planes at the same time.

There are several ways to polarize light. One method is to use a material such as polaroid that removes all of the electric vectors except those in a particular plane. The polaroid filter that causes the electric vector to vibrate in only one plane is known as the polarizer. The axis of a second polaroid filter, known as the analyzer, can be crossed with the axis of the polarizer and block all light to the observer.

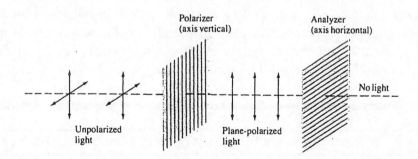

A second method to polarize unpolarized light is to reflect ordinary light from materials such as glass or water. The angle of incidence at which maximum polarization of the reflected light occurs is known as **Brewster's angle**. Brewster's angle θ_p can be found by using the equation

$\tan \theta_p = n_2/n_1$

where n_1 is the index of refraction of the medium in which the light is initially traveling and n_2 is the index of refraction of the medium from which the light is reflected.

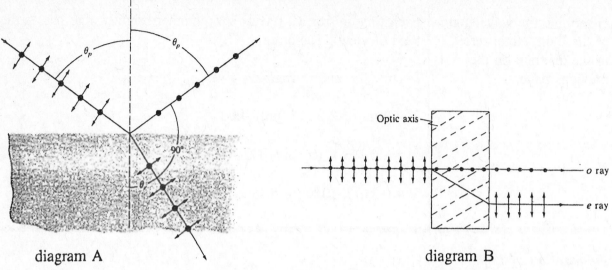

diagram A   diagram B

As shown in diagram A, at Brewster's angle the reflected light is plane polarized parallel to the surface. The angle between the refracted ray and the reflected ray is 90°.

A third method to polarize unpolarized light is by **double refraction**. When light passes through a crystal such as calcite, the incident ray is separated into two refracted rays (see diagram B). One ray, known as the ordinary ray (or o ray), follows Snell's law. The second ray, known as the extraordinary ray (or e ray), does not follow Snell's law. The two emerging beams are both plane polarized. The plane of polarization of the o ray is perpendicular to the plane of polarization of the e ray.

TEXTBOOK QUESTION 27. What does polarization tell us about the nature of light?

ANSWER: Polarization indicates that light must be a transverse wave rather than a longitudinal wave. As discussed in Chapter 12, sound waves are longitudinal waves which exhibit interference when emitted by two sources which are in phase. Sound waves are longitudinal waves; the molecules of the medium vibrate only along the direction of motion of the wave. There is no way that a longitudinal wave can be plane polarized.

TEXTBOOK QUESTION 29. How can you tell if a pair of sunglasses is polarizing or not?

ANSWER: Look for glare reflecting from a waxed floor, a lake, or a non-metallic reflecting surface. Make sure that the angle between the normal to the surface and your eye is 45° or more. Hold one lens in front of your eye and rotate the lens. If the lens is made of polarizing material, the intensity of the light reaching your eye will change as the lens rotates.

Part a. Step 1.	Solution: (Section 24-10)
Determine Brewster's angle for an air-glass interface.	$\tan \theta_p = n_2/n_1$ $\tan \theta_p = 1.65/1.00$ $\theta_p = 58.8°$
Part b. Step 1.	$\tan \theta_p = 1.65/1.33$
Determine Brewster's angle for a water-glass interface.	$\tan \theta_p = 1.24$ $\theta_p = 51.1°$

PROBLEM SOLVING SKILLS

For problems involving double slit interference, single slit diffraction, or a diffraction grating:

1. Determine whether the light is passing through a double slit, a single slit or a diffraction grating.
2. Draw an accurate diagram labeling d or D, ΔX, L, the order of the image (m) and the path difference $m\lambda$.
3. Complete a data table based on the information given in the problem.
4. Note whether the problem involves constructive interference (bright fringes) or destructive interference (dark fringes).
5. Choose the appropriate formula and solve the problem.

For problems involving thin film interference:

1. Assume a crest is incident on the top surface of the film.
2. Determine whether the subsequent reflections and transmissions from each interface will be a crest or trough.
3. Draw a diagram placing a t for trough and c for crest at each position where the light is reflected and transmitted.
4. Determine the minimum thickness of the film in terms of a fraction of a wavelength of the incident light required to produce a) maximum reflection of the light and b) minimum reflection of the light.
5. Based on step 4, write a formula for film thicknesses that will produce a) maximum reflection

of the light and b) minimum reflection of the light.

6. If the wavelength of the light in air is given, it is necessary to determine the wavelength of the light in the film.

7. Complete a data table, choose the appropriate formula, and solve the problem.

For problems involving the angle of maximum plane polarization of reflected light:

1. Determine the relative index of refraction of the two mediums, i.e., the ratio of the index of refraction of the medium from which the light is reflected to the index of refraction of the medium in which the light is incident.

2. Use $\tan \theta_p = n_2/n_1$ to solve for Brewster's angle (θ_p).

SOLUTIONS TO SELECTED TEXTBOOK PROBLEMS

TEXTBOOK PROBLEM 5. Light of wavelength 680 nm falls on two slits and produces an interference pattern in which the fourth-order fringe is 38 mm from the central fringe on a screen 2.0 m away. What is the separation of the two slits?

Part a. Step 1.	Solution: (Section 24-3)
Complete a data table. Convert all of the data to meters.	$\lambda = 680$ nm $= 6.80 \times 10^{-7}$ m $\qquad \Delta X = 38$ mm $= 3.8 \times 10^{-2}$ m $m = 4 \qquad L = 2.0$ m $\qquad d = ?$

Part a. Step 2.	$m \lambda = d \sin \theta_4$
Refer to the diagram on page 24-5. Determine the distance between the slits.	The angle is small, then $\sin \theta \approx \tan \theta$ where $\tan \theta = \Delta X/L$ $m \lambda = d \Delta X/L$ and rearranging gives $d = (m \lambda L)/\Delta X$ $d = [(4)(6.80 \times 10^{-7}$ m$)(2.0$ m$)]/(3.8 \times 10^{-2}$ m$)$ $d = 1.4 \times 10^{-4}$ m or 0.14 mm

TEXTBOOK PROBLEM 15. A light beam strikes a piece of glass at a 60.00° incident angle. The beam contains two wavelengths, 450.0 nm and 700.0 nm., for which the index of refraction of the glass is 1.4820 and 1.4742, respectively. What is the angle between the two refracted beams?

Part a. Step 1.	Solution: (Section 23-5)
Use Snell's law to determine the angle of refraction of the 450.0-nm beam. Note: let n_2 represent the index of refraction of glass for the 450.0-nm beam.	$n_1 \sin \theta_1 = n_2 \sin \theta_2$ $(1.00) \sin 60.00° = (1.4820) \sin \theta_2$ $\sin \theta_2 = (1.00)(0.866)/(1.4820)$ $\sin \theta_2 = 0.584$ $\theta_2 = 35.75°$
Part a. Step 2.	$n_1 \sin \theta_1 = n_3 \sin \theta_3$
Use Snell's law to determine the angle of refraction of the 700.0-nm beam. Note: let n_3 represent the index of refraction of glass for the 700.0-nm beam.	$(1.00) \sin 60.00° = (1.4742) \sin \theta_3$ $\sin \theta_3 = (1.00)(0.866)/(1.4742)$ $\sin \theta_3 = 0.587$ $\theta_3 = 35.98°$
Part a. Step 3.	The angle between the two beams = $\theta_3 - \theta_2$ = 35.98° - 35.75°
Determine the angle between the two beams.	$\theta_3 - \theta_2 = 0.22°$

TEXTBOOK PROBLEM 18. Monochromatic light falls on a slit that is 2.60 x 10⁻³ mm wide. If the angle between the first dark fringes on either side of the central maximum is 35.0° (dark fringe to dark fringe), what is the wavelength of the light used?

Part a. Step 1.	Solution: (Section 24-5)	
Draw an accurate diagram and complete a data table listing information both given and implied.	$m = 1 \qquad \lambda = ?$ $D = 2.60 \times 10^{-3}$ mm $\theta = \frac{1}{2}(35.0°) = 17.5°$	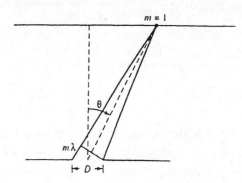

24-19

Part a. Step 2.	$m \lambda = D \sin \theta$
Determine the wavelength of the light used.	$(1)\lambda = (2.60 \times 10^{-3} \text{ mm})(\sin 17.5°)$
	$\lambda = (7.82 \times 10^{-4} \text{ mm})[(1\text{m})/(1000 \text{ mm})[(1 \text{ nm})/10^{-9} \text{ m})]$
	$\lambda = 782$ nm

TEXTBOOK PROBLEM 34. What is the highest spectral order that can be seen if a grating with 6000 lines per cm is illuminated with 633-nm laser light? Assume normal incidence.

Part a. Step 1.	Solution: (Section 24-6)
Determine the distance between adjacent grooves in the diffraction grating in meters.	There are 6000 lines per centimeter; then the distance between adjacent grooves is 1/6000 cm.
	$d = (1 \text{ cm}/6000 \text{ lines})(1 \text{ m}/100 \text{ cm})$
	$d = 1.67 \times 10^{-6}$ m

Part a. Step 2.	The maximum possible value of θ is 90°.
Determine the highest spectral order that can be observed.	$m \lambda = d \sin \theta$
	$m (633 \text{ nm})(1.00 \times 10^{-9} \text{ m})/(1 \text{ nm}) = (1.67 \times 10^{-6} \text{ m})(\sin 90°)$
	$m = 2.64$
	Since m must be an integer, only the first two orders can be observed. The maximum number of fringes that can be observed is $2 + 2 + 1 = 5$.

TEXTBOOK PROBLEM 42. A lens appears greenish yellow (λ = 570 nm is strongest) when white light reflects from it. What minimum thickness of coating (n = 1.25) do you think is used on such a glass (n = 1.52) lens, and why?

Part a. Step 1.	Solution: (Section 24-8)
Assume that a wave crest (c) strikes the top interface. Determine whether each subsequent transmission and reflection will be a crest or a trough.	Since $n_{film} > n_{air}$, then a phase change occurs and the reflection of c at the top interface will be a trough (t). No phase change occurs for transmitted light so the crest (c) will strike the lower interface as a crest. At the lower interface the crest (c) will reflect as a trough (t). This is because $n_{glass} > n_{film}$. When t reaches the top interface, it will be transmitted as a trough.

Part a. Step 2.

Draw a diagram using c for crest and t for trough for the situation described in part a.

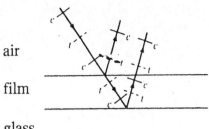

air

film

glass

Part a. Step 3.

Derive a formula for film thickness that results in maximum reflection of the incident light.

If the thickness of the film is $\frac{1}{2}\lambda$, then the initial transmission will have traveled $\frac{1}{2}\lambda_{film} + \frac{1}{2}\lambda_{film} = 1\lambda_{film}$ before emerging from the top of the lens as a trough. The trough will meet the reflection of the the next crest which is reflecting at the top interface as a trough. The superposition of the two troughs will result in constructive interference and maximum reflection. Thus, maximum reflection occurs if the thickness of the film follows the equation

$$t = (m\ \lambda_{film})/2 \quad \text{or} \quad 2t = m\ \lambda_{film}$$

where $\lambda_{film} = \lambda_{air}/n_{film} = (570\ nm)/(1.25) = 456.nm$

Part a. Step 4.

Determine the minimum thickness of the film that reflects light of wavelength 570 nm.

If m = 0, the thickness of the film is zero. The first non-zero thickness of the film occurs for m = 1.

$$t = (m\ \lambda_{film})/2$$

$$t = (1)(456\ nm)/2 = 228\ nm = 2.28 \times 10^{-7}\ m$$

TEXTBOOK PROBLEM 51. How far must the mirror M_1 in a Michelson interferometer be moved if 850 fringes of 589-nm light are to pass by a reference line?

Part a. Step 1.

Determine the distance that M_1 moves in nm and mm.

Solution: (Section 24-9)

A wavelength of 589 nm corresponds to yellow light. The distance (t) that M_1 moves from one yellow fringe to the next equals $\frac{1}{2}\lambda$. Therefore, the distance (t) that M_1 moves can be found by using

$$t = m\ \lambda/2 \quad \text{where}\quad m = 850\ \text{fringes}$$

$$t = (850)(589\ nm)/2 = 2.5 \times 10^5\ nm$$

$$t = (2.5 \times 10^5\ nm)[(1\ m)/(10^9\ nm)][(1000\ mm)/(1\ m)] = 0.250\ mm$$

CHAPTER 25

OPTICAL INSTRUMENTS

OBJECTIVES

After studying the material of this chapter, the student should be able to:

- identify the major components of a simple camera and explain how these components combine to produce a clear image.
- identify the major components of the human eye and explain how an image is formed on the retina of the eye.
- explain the causes of myopia and hyperopia, describe how these conditions may be corrected, and solve word problems related to corrective lenses for these conditions.
- explain how a magnifying glass can be used to produce an enlarged image, and solve word problems involving the magnifying glass.
- explain how two convex lenses can be arranged in order to form an astronomical telescope, and solve word problems related to this type of telescope.
- explain the operation of the Newtonian focus telescope, Cassegrainian focus telescope, and Galilean type telescope.
- explain how two convex lenses can be arranged in order to form a compound microscope, and solve word problems related to this type of microscope.
- distinguish between spherical aberration and chromatic aberration and explain how each type of aberration can be corrected.
- describe the factors which affect resolution of an image and limit the effective magnification of a telescope or microscope.
- explain how the phenomena of X-ray diffraction can be used either to determine the distance between the atoms of a crystal or the wavelength of the incident X-rays. Use the Bragg equation to solve word problems involving X-ray diffraction.

KEY TERMS AND PHRASES

simple camera consists of a convex lens, light-tight box, photographic plate or film, shutter, and an iris diaphragm or stop. The image of the object is formed on the film. The image is real, inverted, and diminished in size.

human eye functions in a manner similar to a simple camera. Light passes through a transparent outer membrane called the cornea, a clear liquid called the aqueous humor, and the lens into a cavity which is filled with a second clear liquid called the vitreous humor. The image is formed on the back of the eye on the retina and transmitted to the brain via the optic nerve.

rods on the retina are sensitive to the intensity level of light while the **cones** distinguish color.

fovea is the region of the retina where the cones are most densely packed. Vision is most acute in the fovea.

accommodation refers to the ability of the lens of the eye to change shape in order to bring objects into focus.

myopia or nearsightedness is a condition where a person can see nearby objects clearly but objects at a distance are blurred.

hyperopia or farsightedness is a condition where a person can see distant objects clearly but nearby objects are blurred.

diopter is a measure of the refractive power of a lens. The power in diopters is inversely related to the focal length expressed in meters.

magnifying glass is a converging lens which is used to produce a virtual, upright, and enlarged image of an object.

near point is the closest distance from the eye that an object can be placed and still be focused on the retina.

astronomical telescope, a refraction type, consists of a large-diameter convex lens with a long focal length (f_o) called the objective and a small-diameter convex lens with a short focal length (f_e) as the eyepiece. A magnified, inverted, virtual image of an object, e.g., a planet, is observed.

compound microscope consists of two convex lenses, an objective, and an eyepiece separated by a distance. The object is placed just beyond the focal point of the objective. The image observed is virtual, inverted, and greatly magnified.

spherical aberration occurs when rays of light from a point on the axis of the lens do not produce a point image after refraction. Instead, they produce a small circular patch of light.

chromatic aberration occurs because the index of refraction of transparent materials varies with wavelength. As a result, white light from a point source produces an image containing colors spread out over a small region.

achromatic doublet is a lens combination of two lenses used to correct chromatic aberration. One lens is a convex lens and the other a concave lens. The lenses have different indices of refraction and radii of curvature.

Rayleigh criterion states that two images are just resolvable when the center of the diffraction disk of one image is directly over the first minimum of the diffraction pattern of the other.

X-rays are electromagnetic waves of very short wavelength emitted when high energy electrons strike a metal target.

X-ray diffraction patterns are produced as a result of reflection of an X-ray beam from the surface of a crystal. The subsequent reflection of the X-rays from the planes of atoms results in a pattern of constructive and destructive interference.

SUMMARY OF MATHEMATICAL CONCEPTS

f-stop	f-stop = f/D	The f-stop refers to the adjustment to the size of the opening necessary to compensate for the outside brightness and also the shutter speed. f is the focal length of the lens and D is the diameter of the opening.
angular magnification	$M = \theta'/\theta$	The angular magnification or magnifying power (M) is the ratio of the angle the object subtends with the eye (θ') when the magnifier is used to the angle the object subtends with the eye (θ) without the magnifier at a distance of 25 cm.
magnification of a magnifying glass with the image at the near point	$M = 1 + N/f$	The magnification depends on the near point (N) and the focal length of the lens. N is usually taken to be 25 cm.
magnification of a magnifying glass with the image at infinity	$M = N/f$	If the eye is relaxed, the image is at infinity and the object is then precisely at the focal point. The magnification (M) equals the ratio of the near point to the focal length.
magnification of a refracting telescope	$M = - f_o/f_e$	The magnification (M) equals the ratio of the focal length of the objective lens (f_o) to the focal length of the eyepiece (f_e). The negative sign indicates an inverted image.

magnification of a compound microscope	$M = M_e M_o$ or $M \approx N\ell/(f_e\, f_o)$	The total magnification (M) of a compound microscope equals the product of the magnification of the eyepiece (M_e) and the magnification of the objective (M_o). The approximation is accurate when f_e and f_o are small compared to the distance between the lenses.
Rayleigh criterion	$\theta_{min} = 1.22\ \lambda/D$	The Rayleigh criterion is used to determine when two images can be resolved. The minimum angle (θ_{min}) for resolution depends on the wavelength (λ) of the light and the diameter of the objective lens (D).
resolving power of a microscope	$RP = s = f\,\theta = 1.22\lambda f/D$	For a microscope, the resolution is defined in terms of the resolving power (RP) where s is the minimum separation of two objects that can just be resolved and f is the focal length of the objective lens.
X-ray diffraction Bragg equation	$m\lambda = 2\,d \sin\theta$	If an X-ray beam is incident on the crystal, the subsequent reflection of the X-rays from the planes of atoms results in a diffraction pattern. Constructive interference occurs when $m\lambda = 2d \sin\theta$ where $m\lambda$ is a whole number of wavelengths and m = 1, 2, 3, etc. d is the distance between the layers of atoms and θ is the angle of incidence of the x-ray beam with the surface of the crystal.

CONCEPT SUMMARY

The Camera

As shown in the diagram at the top of the next page, a simple **camera** consists of a convex lens, light-tight box, photographic plate or film, shutter, and an iris diaphragm or stop.

Three main adjustments must be made in order to produce a clear image of the object on the film: shutter speed, f-stop, and distance from lens to film. The shutter speed determines the length of time that the shutter is open to allow light to pass through to the film. Shutter speed varies from a second or more for a long exposure to 1/1000 second or less to capture the image of a moving object and avoid a blurred picture.

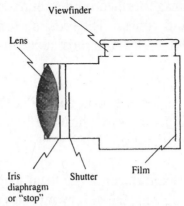

The amount of light reaching the film is controlled by the iris diaphragm or stop. The stop blocks light which has passed through the lens from reaching the film. The size of the opening is adjusted to compensate for the outside brightness and also the shutter speed and is specified in terms of the f-stop. The f-stop is given by the equation

f-stop = f/D   where f is the focal length of the lens and D is the diameter of the opening

Depending on the distance from the object to the lens, it may be necessary to adjust the distance between the lens and the film in order to produce a sharp image on the film. This is accomplished by turning a ring on the lens which moves the lens toward or away from the film. On inexpensive cameras, the optics are arranged so that all objects beyond a certain distance, usually 6 feet, are in focus and no adjustment is necessary.

TEXTBOOK QUESTION 3. Why must a camera lens be moved farther from the film to focus on a closer object?

ANSWER: The relationship between the object distance, image distance, and focal length of the lens is given by the lens equation: $1/d_o + 1/d_i = 1/f$. The focal length of the lens is fixed and the image distance is measured from the lens to the film where the image is formed. If the object is close to the lens, the image distance must be increased so that a clear image will be formed on the film (see page 23-15, diagram C). If the object is far away from the lens, the image distance between the lens and film must be decreased in order to form a clear image (see page 23-15, diagram A). Inexpensive cameras often are made so that the distance from the lens to the film is fixed. In this type of camera an object distance of 6 feet to infinity ensures that a clear image will be formed on the film.

The Human Eye

The basic structure of the **human eye** is shown in the diagram at the top of the next page. Light passes through a transparent outer membrane called the **cornea**, a clear liquid called the **aqueous humor**, and the lens into a cavity which is filled with a second clear liquid called the **vitreous humor**. The image is formed on the back of the eye on the **retina** and transmitted to the brain via the **optic nerve**. The image formed on the retina is received by millions of light-sensitive receptors known as **rods and cones**. The rods are sensitive to the intensity level of light while the cones distinguish color. Vision is most acute at the **fovea**, a region where the cones are most densely packed.

The amount of light passing through the eye is controlled by a diaphragm called the **iris** which opens and closes automatically as the eye adjusts to light intensity. The opening in the center of the iris is called the pupil of the eye.

The lens of the eye changes shape and therefore its focal length in order to focus the image of an object on the retina. For distant objects, the **ciliary muscles** relax, the lens becomes thinner and the focal length greater, and the image is focused on the retina. For nearby objects, the muscles contract, causing the curvature of the lens to increase and decreasing the focal length. Again the image comes to a focus on the retina. The ability of the eye to adjust in this manner is called **accommodation**.

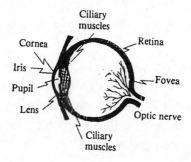

Common defects of the eye include nearsightedness or **myopia** and farsightedness or **hyperopia**. In nearsightedness, a person can see nearby objects clearly but objects at a distance are blurred. This is usually because the eyeball is too long and the image comes to a focus in front of the retina. This condition can be corrected by a concave or diverging lens which causes the rays to come to focus at the retina.

A farsighted person can see distant objects clearly but a nearby object, e.g., the print on this page, is blurred. This is usually caused by an eyeball that is too short. The rays from the object have not yet come to a focus when they strike the retina. A convex or converging lens is used to converge the rays so that the image comes to a focus at the retina. When prescribing eyeglasses, the power (P) of a lens, expressed in **diopters**, is used in place of the focal length. $P = 1/f$ where the focal length (f) is expressed in meters. P is positive for a converging lens and negative for a diverging lens.

TEXTBOOK QUESTION 4. Why are bifocals needed mainly by older persons and not generally by younger people?

ANSWER: Nearsightedness, or myopia, is usually caused by an eyeball that is too long. As a result, images of distant objects come to a focus in front of the retina. This problem is usually corrected by wearing glasses which contain lenses which are slightly concave, i.e., diverging lenses. If the focal length is correct, the image comes to a focus on the retina.

With age, the eye loses the ability to focus on nearby objects as well as distant objects. This condition is known as presbyopia, literally "old eyes." In older people this is usually caused by the inability of the lens of the eye to change shape sufficiently to shorten its focal length in order to read newsprint or a book. The lower part of the eye glasses are made slightly convex in order to accommodate to the changing eye. Since each lens has an upper portion for distance viewing and a lower portion for close-up viewing, the term bifocal is commonly used.

The Magnifying Glass

A simple **magnifier** or magnifying glass is a converging lens which is used to produce an enlarged image of an object on the retina of the eye. The diagram shown below shows the virtual image of an object produced when the lens is used as a magnifier compared to the object viewed by the unaided eye focused at its near point.

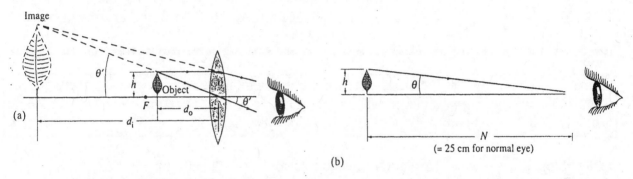

The **angular magnification** or **magnifying power M** is given by $M = \theta'/\theta$. θ' is the angle the object subtends with the eye when the magnifier is used and θ is the angle the object subtends with the eye without the magnifier at a distance of 25 cm. The angular magnification can be written in terms of the focal length of the lens as follows:

$$M = 1 + N/f = 1 + 25 \text{ cm}/f$$

N is the near point, which is the closest distance from the eye that an object can be placed and still be focused on the retina. N is usually taken to be 25 cm. If the eye is relaxed when using the magnifying glass, the image is then at infinity and the object is then precisely at the focal point. In this instance $M = N/f$.

EXAMPLE PROBLEM 1. A student requires reading glasses of +1.25 diopters to read the print on this page when the page is held 25.0 cm from her eyes. Determine the minimum distance that she would have to hold the newspaper in order to read the print without her eyeglasses. Assume that the distance from the lens to the eye is negligible.

Part a. Step 1.	Solution: (Section 25-2)
Determine the focal length of the lens.	The power of the lens is related to the focal length by the equation $P = 1/f$. The power of the lens is +1.25 diopters, and the focal length of the lens is
	$f = 1/{+}1.25 \text{ diopters} = +\, 0.800 \text{ meters} = 80.0 \text{ cm}$

Part a. Step 2. Use the lens equation determine the image distance.	With her glasses on, the print is held 25.0 cm from her eyes and the image is formed at the near point. Thus $d_o = 25.0$ cm while d_i is the distance from the lens to the near point of her vision. $1/d_o + 1/d_i = 1/f$ $1/(25.0 \text{ cm}) + 1/d_i = 1/(80.0 \text{ cm})$ $d_i = -36.4 \text{cm}$ Without her glasses, the student would have to hold the page at least 36.4 cm from her eyes in order to read the print.

EXAMPLE PROBLEM 2. The maximum magnification produced by a particular converging lens is 3.5 times when used by a person whose near point is 22 cm. Determine the a) focal length of the lens and b) magnification when the person's eye is relaxed.

Part a. Step 1. Determine the focal of the lens.	Solution: (Section 25-3) The maximum magnification is related to the focal length as length follows: $M = 1 + N/f$ $3.5 = 1 + 22/f$ and rearranging $22/f = 3.5 - 1, \quad 22/f = 2.5 \quad f = 22/2.5$ $f = 8.8 \text{ cm}$
Part b. Step 1. Determine the magnification when the eye is relaxed.	When the eye is relaxed, the image is seen at infinity and the object is at the focal point; thus $M = N/f = 22 \text{ cm}/8.8 \text{ cm} = 2.5$

Astronomical Telescope

A refraction-type astronomical **telescope** consists of a large-diameter convex lens with a long focal length (f_o) called the **objective** and a small-diameter convex lens with a short focal length (f_e) as the **eyepiece**. As shown in the diagram, distant objects such as stars and planets can be considered to be an infinite distance from the telescope. As a result, the image (I_1) is located at the focal point of the objective. The eyepiece is positioned so that the image produced by the objective lens is at or just inside the focal point of the eyepiece. A magnified, inverted virtual image (I_2) of the object is observed.

For an object at infinity, the distance between the lenses is $f_o + f_e$ and the magnifying power of the telescope is $M = \theta'/\theta = -f_o/f_e$ where the negative sign indicates an inverted image.

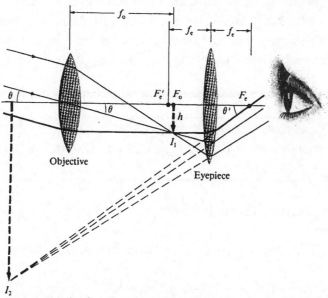

The largest astronomical telescopes employ a concave mirror as the objective. In a **Newtonian focus telescope**, the light from the objective is reflected to a plane mirror. The plane mirror reflects the light to a convex lens which acts as the eyepiece.

In a **Cassegrainian focus**, the light is reflected from the objective to a small convex mirror which then reflects the light through a hole in the objective.

A concave mirror can be used as the telescope of an astronomical telescope. Either a lens or a mirror can be used as the eyepiece. Arrangement (a) is called the Newtonian focus and (b) the Cassegrainian focus. Other arrangements are also possible.

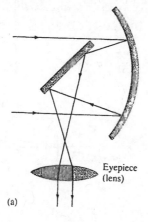

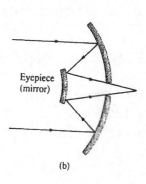

ANSWER: A large objective allows in more light and this is necessary to produce bright images of distant objects. However, the construction and grinding of large lenses is difficult and they tend to sag under their own weight. A mirror has only one surface to be ground and can be supported along the entire area of the non-reflecting side. Also, since mirrors reflect light they do not exhibit chromatic aberration.

Terrestrial Telescope

A **terrestrial telescope** is designed so that an erect image is observed. The **Galilean telescope** uses a convex lens of long focal length as the objective and a concave lens of short focal length as the eyepiece. The eyepiece is a concave lens, its focal length is negative, and the angular magnification is positive. The angular magnification is given by $M = - f_o / f_e$.

EXAMPLE PROBLEM 3. An astronomical telescope consists of two convex lenses, an eyepiece of focal length 100 cm and an objective of focal length 2.5 cm. Determine the magnification of the telescope if used to view a distant object.

Part a. Step 1.	Solution: (Section 25-4)
Determine the magnification of the telescope. Hint: assume that the object is at infinity.	The magnifying power can be found as follows: $M = - f_o/f_e = - 100 \text{ cm}/2.5 \text{ cm}$ $M = -40$

EXAMPLE PROBLEM 4. Two converging lenses are 40.0 cm apart. The focal lengths of the lenses are 10.0 cm and 20.0 cm, respectively. An object 3.0 cm high is placed 30.0 cm in front of the 10.0-cm lens. a) Draw an accurate ray diagram and locate the final image formed by the combination. b) Mathematically determine the position and size of the final image.

Part a. Step 1.	Solution: (Section 25-4)
Draw an accurate diagram and locate the final image formed by the combination.	

Part b. Step 1.	$1/d_o + 1/d_i = 1/f_1$
Determine the position and size of the image produced by the lens which is closest to the object.	$1/30.0 \text{ cm} + 1/d_i = 1/10.0 \text{ cm}$
	$d_i = 15.0 \text{ cm}$
	$h_i/h_o = -d_i/d_o$
	$h_i/3.0 \text{ cm} = -15.0 \text{ cm}/30.0 \text{ cm}$
	$h_i = -1.5 \text{ cm}$ The image is real, inverted, and diminished.

Part b. Step 2.	The image produced by the first lens now becomes the object for the second lens. The distance from the second lens to the image produced by the first lens is equal to the distance between the lenses minus the distance from the first lens to the image distance, i.e.,
Determine the distance from the second lens to the image and also the height of the image.	
	$40.0 \text{ cm} - 15.0 \text{ cm} = 25.0 \text{ cm}.$
State the characteristics of the final image.	$1/d_o + 1/d_i = 1/f_2$
	$1/25.0 \text{ cm} + 1/d_i = 1/20.0 \text{ cm}$
	$d_i = 100 \text{ cm}$
	The height of the final image is given by
	$h_i/h_o = -d_i/d_o$
	$h_i/(-1.5 \text{ cm}) = -(100 \text{ cm})/(25.0 \text{ cm})$
	$h_i = +6.0 \text{ cm}$
	The final image is real, upright, and magnified.

Compound Microscope

As shown at the top of the next page, a **compound microscope** consists of two convex lenses, an objective, and an eyepiece or ocular separated by a distance. The object is placed just beyond the focal point (f_o) of the objective. The image formed (I_1) is real, inverted, and enlarged. I_1 is formed inside the focal point (f_e) of the eyepiece and the viewer observes an image (I_2) which is virtual, inverted, and greatly magnified.

The magnification produced by the objective is given by $M_o = d_i/d_o = (\ell - f_e)/d_o$ where d_i is the distance from the lens to the image, d_o is the distance from the lens to the object, and ℓ is the distance between the two lenses.

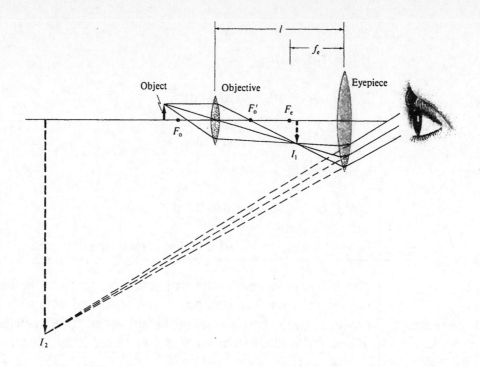

The magnification produced by the eyepiece is $M_e = N/f_e$ where N is the near point, which is usually taken to be 25 cm. The total magnification is given by

$$M = M_e / M_o = N/f_e \ \text{x} \ (\ell - f_e)/d_o \approx N\ell/(f_e \, f_o)$$

The approximation is accurate when f_e and f_o are small compared to ℓ.

EXAMPLE PROBLEM 5. During a laboratory experiment, a student positions two lenses to form a compound microscope. The objective and eyepiece have focal lengths of 0.40 cm and 3.0 cm, respectively. The lenses are 4.9 cm apart and an object is placed 0.50 cm in front of the objective lens. a) Draw a ray diagram and locate the position of the final image. b) Mathematically determine the position of the final image.

Part a. Step 1.

Draw an accurate diagram and locate the position of the final image.

Solution: (Section 25-5)

The following diagram is not drawn to scale. This is necessary in order to show the final image on the diagram.

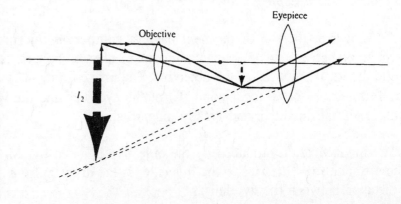

Part b. Step 1. Mathematically determine the position of the image produced by the eyepiece.	Let f_o = focal length of the objective lens. $1/d_o + 1/d_i = 1/f_o$ $1/0.50 \text{ cm} + 1/d_i = 1/0.40 \text{ cm}$ $d_i = 2.0 \text{ cm}$
Part b. Step 2. Mathematically determine the position of the final image.	The image produced by the objective lens becomes the object for the eyepiece. The two lenses are 4.9 cm apart; thus the image produced is 2.9 cm from the eyepiece. Therefore, $1/d_o + 1/d_i = 1/f_e$ where f_e = focal length of the eyepiece $1/2.9 \text{ cm} + 1/d_i = 1/3.0 \text{ cm}$ $d_i = -87 \text{ cm}$ The final image is a virtual image formed 87 cm in front of the eyepiece.

Lens Aberrations

The ray diagrams drawn to locate images produced by thin lenses are only approximately correct. Only an "ideal" lens gives an undistorted image, the formation of images by real lenses are limited by what are referred to as **lens aberrations**.

In spherical aberrations (diagram a) rays of light from a point on the axis of the lens, which pass through different sections of the lens, do not produce a point image after refraction. Instead, they produce a small circular patch of light. The point at which the circle has its smallest diameter is referred to as the circle of least confusion. Spherical aberrations can be approximately corrected by the expensive method of grinding a nonspherical lens or, more frequently. by using a combination of two or more lenses.

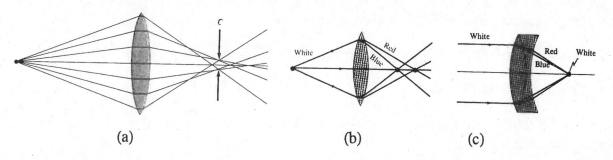

(a)   (b)   (c)

Chromatic aberration (diagram b) occurs because the index of refraction of transparent materials varies with wavelength. For example, blue light is bent more by glass than red light. As a result, white light from a point source produces an image containing colors spread out over a small region.

Chromatic aberration (diagram c) is approximately corrected through use of a combination of two lenses called an **achromatic doublet**. One lens is a convex lens and the other a concave lens. The lenses have different indices of refraction and radii of curvature.

TEXTBOOK QUESTION 13. Which aberrations present in a simple lens are not present (or are greatly reduced) in the human eye?

ANSWER: The cornea of the human eye is less curved at the edges than at the center and the lens is less dense at the edges than at the center. Both effects cause rays at the outer edges to be bent less strongly and thus help reduce spherical aberration. Also, the retina of the eye is curved and this compensates for curvature of field which occurs due to the flat film used in simple cameras. In addition, wide-angle lenses commonly show distortion but the small lens of the eye plus the curved retina compensate for this aberration.

There is significant absorption of the shorter wavelengths by the lens of the eye and the retina is less sensitive to the blue and violet wavelengths. This is the region of the spectrum where chromatic aberration is the greatest and as a result chromatic aberration is greatly reduced in the eye.

Limits of Resolution

The ability of a lens to produce distinct images of two point objects which are very close together is called the **resolution** of the lens. Two principal factors limit the resolution of a lens. One factor is lens aberration and the second is the wave nature of light.

The magnification produced by a microscope or telescope is limited by diffraction. Magnification beyond a certain point does not lead to an increase in sharpness or resolution of images. The **Rayleigh criterion** states that two images are just resolvable when the center of the diffraction disk of one image is directly over the first minimum of the diffraction pattern of the other. The first minimum is at an angle $\theta = 1.22 \, \lambda/D$ from the central maximum and therefore the objects are considered to be "just" resolvable at this angle. λ is the wavelength of the incident light and D is the diameter of the objective lens.

For a microscope, the resolution is defined in terms of the **resolving power (RP)** where

$$RP = s = f \, \theta = 1.22 \, \lambda f/D$$

where s is the minimum separation of two objects that can just be resolved and f is the focal length of the objective lens.

EXAMPLE PROBLEM 6. The objective lens of a compound microscope is 1.50 cm in diameter and has a 10.0-cm focal length. Light of wavelength 600 nm is used to illuminate objects to be viewed. a) What is the angular separation of nearby objects when they are "just" resolvable? b) What is the resolving power of this lens?

Part a. Step 1. What is the angular separation of nearby objects when they are "just" resolvable?	Solution: (Section 25-7) The minimum angular separation is given by $\theta = 1.22\ \lambda/D = (1.22)(600\ nm)/(1.50\ cm)$ $\quad = (1.22)(600 \times 10^{-9}\ m)/(1.50 \times 10^{-2}\ m)$ $\theta = 4.88 \times 10^{-5}$ radians or 2.8×10^{-3} degrees This is the limit on resolution for this lens due to diffraction.
Part b. Step 1. What is the resolving power of this lens?	resolving power = RP = s = f θ = 1.22 $\lambda f/D$ RP = (1.22)(600 nm)(10.0 cm)/(1.50 cm) RP = 4.88×10^{-6} m or 4.88×10^{-3} mm This distance is the minimum separation of two point objects that can "just" be resolved.

X-ray Diffraction

X-rays were discovered in 1895 by Wilhelm Roentgen. The nature of X-rays was not determined until 1913 when it was shown by Max Von Laue that X-rays exhibit properties of electromagnetic waves of very short wavelength.

The wavelength of X-rays is comparable to the spacing of atoms in crystals such as sodium chloride (NaCl). If an X-ray beam is incident on the crystal, the subsequent reflection of the X-rays from the planes of atoms results in a diffraction pattern. Based on the diagram, constructive interference will occur if the extra distance, known as the path difference, that ray I travels before rejoining ray II equals a whole number of wavelengths. The path difference can be shown to be equal to 2 d sin θ. Constructive interference occurs when mλ = 2 d sin θ (Bragg equation).

mλ is a whole number of wavelengths and m = 1, 2, 3, etc., d is the distance between the layers of atoms and θ is the angle of incidence of the X-ray beam with the surface of the crystal.

EXAMPLE PROBLEM 7. X-rays having a wavelength of 0.150 nm are scattered by a crystal. Determine the angles for all diffraction maxima produced by one set of layers of ions spaced 0.412 nm apart.

Part a. Step 1.	Solution: (Section 25-11)
Use the Bragg equation to determine the possible angles.	θ cannot exceed 90°; therefore, the possible angles are 90° or less.

$m\lambda = 2d \sin \theta_1$ If $m = 1$, then

$(1)(0.150 \text{ nm}) = 2(0.412 \text{ nm}) \sin \theta_1$

$\sin \theta_1 = 0.182$

$\theta_1 = 10.4°$

If $m = 2$, then

$(2)(0.150 \text{ nm}) = 2(0.412 \text{ nm}) \sin \theta_2$

$\sin \theta_2 = 0.364$

$\theta_2 = 21.4°$

If $m = 3$, then

$(3)(0.150 \text{ nm}) = 2(0.412 \text{ nm}) \sin \theta_3$

$\sin \theta_3 = 0.546$

$\theta_3 = 33.1°$

If $m = 4$, then

$(4)(0.150 \text{ nm}) = 2(0.412 \text{ nm}) \sin \theta_4$

$\sin \theta_4 = 0.728$

$\theta_4 = 46.7°$

If $m = 5$, then

$(5)(0.150 \text{ nm}) = 2(0.412 \text{ nm}) \sin \theta_4$

$\sin \theta_5 = 0.910$

$\theta_5 = 65.5°$

If $m = 6$, then $\sin \theta_6 > 1$, therefore, only five diffraction maxima are produced.

PROBLEM SOLVING SKILLS

For problems related to power of a lens:

1. Use the equation $P = 1/f$ to determine either the power or the focal length of the lens.
2. Use the Gaussian form of the lens equation to determine the position of the image.
3. Use the equation $M = 1 + N/f$ to determine the magnification of a converging lens.

For problems related to the final image produced by two converging lenses:

1. Use the Gaussian form of the lens equations to determine the position and size of the image produced by the lens closest to the object.
2. Determine the distance from the second lens to the image produced by the first lens. This distance represents the object distance from the second lens.
3. Use the Gaussian form of the lens equation to determine the position and size of the image produced by the second lens.

To determine the limit of resolution of a lens for two point objects which are close together:

1. Apply the Rayleigh criterion to determine the minimum angle at which the objects are resolvable.
2. If the objects are being viewed through a microscope, use the formula for resolving power to determine the minimum separation of the objects which allows resolution.

For problems related to X-ray diffraction:

1. Apply the Bragg equation to determine either the distance between atoms in the crystal or the wavelength of the incident X-rays.

SOLUTION TO SELECTED TEXTBOOK PROBLEMS

TEXTBOOK PROBLEM 5. If an f = 135-mm telephoto lens is designed to cover object distances from 1.2 m to ∞, over what distance must the lens move relative to the plane of the film?

Part a. Step 1.	Solution: (Section 25-1)
Determine the distance from the lens to the film if the object is 1.2 m from the lens.	$1/d_o + 1/d_i = 1/f$
	$1/(1200 \text{ mm}) + 1/d_i = 1/(135 \text{ mm})$
	$1/d_i = 0.00657$
	$d_i = 152 \text{ mm}$

Part a. Step 2.	$1/d_o + 1/d_i = 1/f$
Determine the distance from the lens to the film if the object is an infinite distance from the lens.	$1/(\infty \text{ mm}) + 1/d_i = 1/(135 \text{ mm})$ $1/d_i = 1/(135 \text{ mm})$ $d_i = 135 \text{ mm}$
Part a. Step 3. Determine the relative distance the lens must be moved in order to cover the required object distances.	The lens must be 152 mm from the film in order to clearly focus objects 1.2 m from the lens. The lens must be 135 mm from the film in order to clearly focus objects at infinity. Therefore, the range of distance from the lens to the film is from 135 mm to 152 mm or a distance of 17 mm.

TEXTBOOK PROBLEM 14. A person struggles to read by holding a book at arm's length, a distance of 45 cm away (= near point). What power of reading glasses should be prescribed for him, assuming they will be placed 2.0 cm from the eye and he wants to read at the normal near point of 25 cm?

Part a. Step 1. Complete a data table based on the information provided.	Solution: (Section 25-2) Wearing glasses, an object placed 25 cm from the eye, 23 cm from the lens, produces a virtual image at the actual near point of the person. However, the actual near point of the person is 45 cm which will be 43 cm from the lens of the glasses. $d_o = 25 \text{ cm} - 2.0 \text{ cm} = 23 \text{ cm}$ $d_i = 45 \text{ cm} - 2.0 \text{ cm} = 43 \text{ cm}$ Note: $d_i = -43$ cm because the image is virtual.
Part a. Step 2. Use the lens equations from Chapter 23 to solve for the focal length.	$1/d_o + 1/d_i = 1/f$ $1/(23 \text{ cm}) + 1/(-43 \text{ cm}) = 1/f$ $0.0435 \text{ cm}^{-1} + -0.0233 \text{ cm}^{-1} = 1/f$ $1/f = 0.0202 \text{ cm}^{-1}$ $f = 49.4 \text{ cm} = 0.494 \text{ m}$

Part a. Step 3.	$P = 1/f$
Determine the power of the lens in diopters.	$= 1/(+0.494\ m)$
	$P = +2.0\ D$

TEXTBOOK PROBLEM 24 Sherlock Holmes is using a 9.00-cm focal length lens as his magnifying glass. To obtain maximum magnification, where must the object be placed (assume a normal eye), and what will be the magnification?

Part a. Step 1.	Solution: (Section 25-3)
Determine the object distance from the lens.	The lens is used as a magnifying glass. The image will be located on the other side of the lens from Sherlock's eye. The normal eye has a near point of 25.0 cm, therefore, $d_i = -25.0$ cm.
	$1/d_o + 1/d_i = 1/f$
	$1/d_o + 1/(-25.0\ cm) = 1/(9.00\ cm)$
	$1/d_0 = 0.151$
	$d_0 = 6.62$ cm The lens must be placed 6.62 cm from the object.
Part a. Step 2.	$M = 1 + N/f$
Determine the magnification of the image.	$= 1 + (25.0\ cm)/(9.00\ cm)$
	$M = 3.78$

TEXTBOOK PROBLEM 35. What is the magnifying power of an astronomical telescope using a reflecting mirror whose radius of curvature is 6.0 m and an eyepiece whose focal length is 3.2 cm?

Part a. Step 1.	Solution: (Section 25-4)
Determine the focal length of the objective.	The reflecting mirror acts as the objective.
	$f_o = r/2 = (6.0\ m)/2 = 3.0\ m$
Part a. Step 2.	$M = -f_o/f_e$
Determine the magnification of the telescope.	$M = -(300\ cm)/(3.2\ cm)$
	$M = -94$

TEXTBOOK PROBLEM 40. A 620x microscope uses a 0.40-cm focal length objective lens. If the tube length is 17.5 cm, what is the focal length of the eyepiece? Assume a normal eye and that the final image is at infinity.

Part a. Step 1.	Solution: (Section 25-5)
Complete a data table. Note: for a normal eye, the near point N = 25 cm.	$M = 620$ $\ell = 17.5$ cm $N = 25$ cm $f_o = 0.40$ cm $f_e = ?$

Part a. Step 2. Determine the focal length of the eyepiece.	Since $f_e << \ell$ then, to a good approximation $M \approx N\ell/(f_e\, f_o)$ $620 \approx [(25\ \text{cm})(17.5\ \text{cm})]/[(f_e)(0.40\ \text{cm})]$ $f_e \approx [(25\ \text{cm})(17.5\ \text{cm})]/[(620)(0.40\ \text{cm})]$ $f_e \approx 1.8$ cm

TEXTBOOK PROBLEM 55. X-rays of wavelength 0.0973 nm are directed at an unknown crystal. The second diffraction maximum is recorded at 23.4° relative to the crystal surface. What is the spacing between crystal planes?

Part a. Step 1. Use the Bragg equation to determine the spacing.	Solution: (Section 25-11) $m\lambda = 2d \sin \theta_1$ If $m = 2$, then $(2)(0.0973\ \text{nm}) = 2\ d \sin 23.4°$ $d = 0.244$ nm

CHAPTER 26

THE SPECIAL THEORY OF RELATIVITY

OBJECTIVES

After studying the material of this chapter, the student should be able to:

- state in the student's own words the postulates of the special theory of relativity.
- explain what is meant by a frame of reference and distinguish between an inertial and non-inertial frame of reference.
- explain what is meant by the principle of simultaneity and explain in your own words the "thought" experiment described in the text.
- explain what is meant by proper time and relativistic time and solve word problems involving time dilation.
- explain what is meant by proper length and relativistic length and solve word problems involving length contraction.
- explain what is meant by proper mass and relativistic mass and solve word problems related to the mass of a moving object as measured by an observer at rest relative to the object.
- use the principle of special relativity to determine the relative velocity of an object as measured by an observer moving with respect to the object.
- explain what is meant by rest energy and total energy and solve word problems involving Einstein's mass-energy equation.

KEY TERMS AND PHRASES

postulates of the special theory are 1) all inertial frames of reference are equivalent, and 2) observers, regardless of their relative velocity or the velocity of the source, must measure the same value for the speed of light in a vacuum.

frame of reference is a point in space with respect to which the motion of objects can be measured.

inertial frame of reference is one which is either at rest or is moving in a straight line at a constant speed. In all inertial frames of reference, the laws of physics are the same and hold in the same way.

noninertial frame of reference is one which is undergoing accelerated motion. This means a frame in which the speed is changing or the direction of motion is changing or both the speed and the direction of motion are changing.

principle of simultaneity Two events which are observed to occur simultaneously at different points in one frame of reference will not occur simultaneously according to an observer in a second frame of reference which is in motion relative to the first.

proper time is measured by an observer at rest with respect to the timing device.

relativistic time dilation refers to the difference in the measurement of time by an observer in motion relative to the timing device.

proper length is measured by an observer at rest with respect to the object measured.

relativistic length contraction refers to the decrease in measured length of an object when the object is in motion relative to an observer.

proper mass is measured by an observer at rest with respect to the object measured.

relativistic mass refers to the mass of an object, as measured by an observer, when the object is in motion relative to an observer.

rest energy is the energy an object has due to its mass alone.

total energy of an object equals the sum of the object's mechanical energy and rest energy. The total energy is expressed by the formula $E = mc^2$ where m is the relativistic mass and c is the speed of light. According to the equation, it is possible to convert mass into energy and vice versa.

SUMMARY OF MATHEMATICAL FORMULAS

time dilation equation	$\Delta t = \Delta t_0 / [1 - (v/c)^2]^{1/2}$	Δt_o is the proper time measured by the observer at rest with the timing device. Δt is the relativistic time measured by the observer in motion relative to the timing device. v is the speed of the timing device relative to observer measuring Δt while c is the speed of light in a vacuum.

length contraction equation	$L = L_o[1 - (v/c)^2]^{\frac{1}{2}}$	L is the relativistic length measured by the observer in motion relative to the object. L_o is the proper length measured by the observer at rest with respect to the object.
relativistic mass	$m = m_o/[1 - (v/c)^2]^{\frac{1}{2}}$	m is the relativistic mass and is measured by an observer in motion relative to the object. The proper mass (m_o) is measured by an observer at rest with respect to the object.
relativistic addition of velocities	$u = (v + u')/(1 + vu'/c^2)$	u is the speed of the object as measured by an observer in motion relative to the frame of reference of the moving object. v is the speed of the moving frame of reference. u' is the speed of the object relative to an observer at rest in the moving frame of reference. If u' is in the same direction as v, then u' is positive. If u' is opposite from v, then u' is negative.
rest energy	$E_o = m_o c^2$	An object's rest energy (E_o) equals the product of the rest mass (m_o) and the speed of light (c) squared.
relativistic total energy	$E = mc^2$	The total energy (E) equals the product of the relativistic mass (m) and the speed of light (c) squared. The total energy of an object equals the sum of the object's mechanical energy and rest energy.
relativistic kinetic energy	$KE = E - E_o$ $KE = (m - m_o)c^2$	For a moving object, the kinetic energy equals the difference between an object's total energy (E) and its rest energy (E_o).

CONCEPT SUMMARY

Postulates of the Special Theory

The **special theory of relativity** is based on two postulates formulated by Albert Einstein:

1) All inertial frames of reference are equivalent.

2) Observers, regardless of their relative velocity or the velocity of the source, must measure the same value for the speed of light in a vacuum.

TEXT QUESTION 5. If you were on a spaceship traveling at 0.5 c away from a star, at what speed would the starlight pass you?

ANSWER: Second postulate of special relativity: Light propagates through empty space with a definite speed c independent of the speed of the source or observer. Based on this postulate, you would measure the speed of the passing light to be c.

Inertial Frames of Reference

A **frame of reference** is a point in space with respect to which the motion of objects can be measured. An **inertial frame of reference** is one which is either at rest or is moving in a straight line at a constant speed. In all inertial frames of reference, the laws of physics are the same and hold in the same way.

A **noninertial frame of reference** is one which is undergoing accelerated motion. This means a frame in which the speed is changing or the direction of motion is changing or both the speed and the direction of motion are changing.

As a consequence of the first postulate, it is possible to determine relative motion between objects traveling at constant velocity with respect to one another, but it is not possible to determine absolute motion.

Principle of Simultaneity

Two events which are observed to occur simultaneously at different points in one frame of reference will not occur simultaneously according to an observer in a second frame of reference which is in motion relative to the first.

The following "thought" experiment will be used to clarify this principle. Two railroad cars are directly opposite one another when lightning bolts strike points shown in the diagram located below. The cars are moving relative to one another with observer 0_1 in car 1 assuming that he is at rest while car 2 is moving to the left at speed v. Likewise, observer 0_2 in car 2 assumes that he is at rest and sees car 1 moving to the right at speed v.

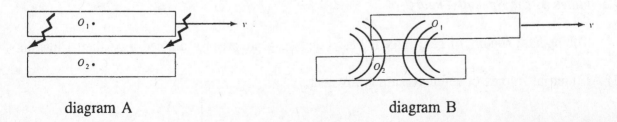

diagram A   diagram B

As shown in diagram A, the lightning bolts strike the front and rear ends of each car simultaneously. As shown in diagram B, a moment later the light from the events simultaneously reach 0_2's position. However, in 0_1's reference frame, the light from the front end has already reached 0_1, while the light from the rear has not yet reached 0_1. So from 0_1's reference frame the lightning strike at the front of the car must have occurred before the strike at the rear of the car.

Predictions from the Special Theory

Time Dilation

Einstein's two postulates plus the principle of simultaneity lead to the prediction that time is measured differently in frames of reference moving relative to one another.

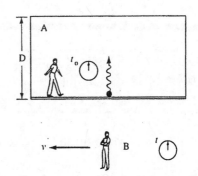

Observer A on a spaceship is moving toward the right at a speed v relative to an Earthbound observer B. A pulse of laser light is fired vertically upward from the floor of the spacecraft just as the spacecraft is passing B's position.

A is moving at a constant speed in a straight line, he is an inertial frame of reference. From A's point of view he is **not** moving but instead B is moving toward the left at speed v. A sees the light travel vertically upward toward the ceiling, a distance D, at the speed of light (c). The time required for the light to travel from the floor to the ceiling, from A's viewpoint, can be determined from the equation

time = distance/velocity or $\Delta t_o = D/c$ and $D = c\, \Delta t_o$

The time measured by observer A is the proper time Δt_o. D is the distance from the floor to the ceiling. c is the speed of light in a vacuum.

As shown in the diagram 1 at the top of the next page, from B's frame of reference he is at rest and it is A who is moving toward the right at speed v. B sees the beam of light travel from the floor to the ceiling, but A is traveling toward the right, and B observes that the beam does not travel vertically upward. B sees the beam travel at an angle to the vertical. According to the second postulate, both A and B must measure the speed of light to be c. Thus, B sees the beam travel at speed c at an angle to the vertical.

As shown in diagram 2, B uses the Pythagorean theorem and determines that the component of the light's speed in the vertical direction is $(c^2 - v^2)^{1/2}$. The time required for the light to travel from the floor to the ceiling, the vertical component of motion, is given by
$\Delta t = D/(c^2 - v^2)^{1/2}$ and $D = (c^2 - v^2)^{1/2}\, \Delta t$ where Δt is the relativistic time.

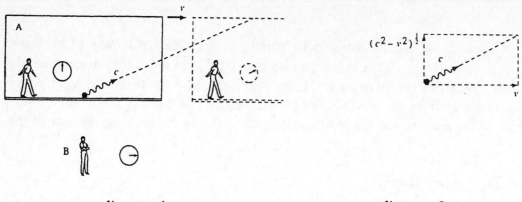

diagram 1   diagram 2

Both observers measure D to be the same because there is no relative motion between their frames of reference in the vertical direction. Therefore, $c \, \Delta t_o = (c^2 - v^2)^{1/2} \, \Delta t$ and rearranging gives $\Delta t = \Delta t_o / [1 - (v/c)^2]^{1/2}$. This equation is known as the time dilation equation.

EXAMPLE PROBLEM 1. Observer A in the diagram shown above measures the passage of 5.0 hours in the spacecraft. Determine the amount of time that has passed according to observer B's clock if the spacecraft is moving at a) 0.80c and b) 0.98c.

Part a. Step 1. Use the time dilation equation to determine the amount of time that passes on B's clock if $v = 0.8c$.	Solution: (Sections 26-4 and 26-5) Observer A measures the proper time (Δt_o) while B measures the relativistic time (Δt) and $v = 0.80c$. Then $[1 - (v/c)^2]^{1/2} = [1 - (0.8c/c)^2]^{1/2} = (1 - 0.64)^{1/2} = 0.60$ and $\Delta t = \Delta t_o / (1 - (v/c)^2)^{1/2}$ $\Delta t = (5.0\ \text{h})/(0.60) = 8.3\ \text{h}$
Part b. Step 1. Use the time dilation equation to determine the amount of time that passes on B's clock if $v = 0.98\ c$.	If $v = 0.98c$, then $[1 - (v/c)^2]^{1/2} = [1 - (0.98c/c)^2]^{1/2} =$ $(1 - 0.96)^{1/2} = 0.20$ and $\Delta t = \Delta t_o \, (1 - (v/c)^2)^{1/2}$ $\Delta t = (5.0\ \text{h})/(0.20) = 25.0\ \text{h}$

Length Contraction

Not only would observer B measure time differently than observer A, but he would measure lengths differently as well. Objects in the spacecraft are at rest relative to observer A. Observer A measures the **proper length** (L_o) of objects in the spacecraft.

The spacecraft shown in the diagram on the previous page is moving in the x direction relative to observer B. Observer B will see the length of objects in the spacecraft shortened in the x direction but not in the y and z directions. The length of the object (L) as measured by observer B will differ from the length measured by observer A according to the following equation:

$$L = L_o[1 - (v/c)^2]^{\frac{1}{2}}$$

L is the **relativistic length** and is measured by the observer in motion relative to the object. The proper length L_o is the length measured by the observer at rest relative to the object.

Mass Increase

Special relativity leads to the prediction that the mass of an object will increase as its speed increases. If the object is moving at speed v relative to an observer, the observer will measure its mass to be m where

$$m = m_o/[1 - (v/c)^2]^{\frac{1}{2}}$$

m is the **relativistic mass** and is measured by an observer in motion relative to the object. The **proper mass** (m_o) is measured by an observer at rest relative to the object.

It should be noted that each of the equations discussed thus far, time dilation, length contraction, and mass increase, predict that there is an ultimate speed. As v approaches c, $[1 - (v/c)^2]^{\frac{1}{2}}$ approaches zero. Thus as v approaches c, the relativistic time and relativistic mass approach infinity while the relativistic length approaches zero. Since an infinite mass would require infinite energy to accelerate it, the principle of special relativity predicts that it is impossible for an object to travel at speeds equal to or greater than the speed of light.

TEXTBOOK QUESTION 9. If you were traveling away from Earth at a speed of 0.5c, would you notice a change in your heartbeat? Would your mass, height, or waistline change? What would observers on Earth using telescopes say about you?

ANSWER: You are at rest with respect to objects in the spacecraft and of course you are at rest with respect to your own body. Because of this, you would not notice any change in your heartbeat, mass, height, or waistline. However, you are in motion relative to an Earthbound observer and

$[1 - (v/c)^2]^{1/2} = [1 - (0.5c/c)^2]^{1/2} = [1 - 0.25]^{1/2} = 0.87$

Using the relativity equations, it can be shown that according to the Earthbound observer, your heartbeat and waistline are reduced by the factor 0.87 while your mass has increased by $1/0.87 = 1.15$. Your height does not change because there is no relative motion between the Earthbound observer and your height.

EXAMPLE PROBLEM 2. Observer A (diagram 1 on page 26-6) measures the length of a piece of wood to be 3.0 meter and its mass to be 2.0 kg. The spacecraft is moving at 0.98c relative to observer B. Determine the a) length of the wood and b) mass of the wood as observed by B.

Part a. Step 1.	Solution: (Sections 26-4 and 26-5)
Determine which observer measures the proper length and mass and which observer measures the relativistic length and mass.	Observer A is at rest relative to the piece of wood; therefore, observer A measures the proper length and proper mass. $L_o = 3.0$ m and $m_o = 2.0$ kg The piece of wood is in motion relative to observer B; then B measures the relativistic length (L) and relativistic mass (m).
Part a. Step 2. Determine the relativistic length of the wood.	In problem 1, it was determined that if $v = 0.98c$, then $[1 - (v/c)^2]^{1/2} = 0.20$. $L = (1 - v/c)^2)^{1/2} L_o$ $L = (0.20)(3.0 \text{ m}) = 0.60 \text{ m}$
Part b. Step 1. Determine the relativistic mass of the wood.	$m = m_o /[1 - (v/c)^2]^{1/2}$ $m = (2.0 \text{ kg})/(0.20) = 10 \text{ kg}$

$E = mc^2$; *Mass and Energy*

When mechanical work is done on an object, the object's energy must change. However, according to the special theory, as the object's speed increases some of the energy is found in the increasing mass of the object instead of its speed and $KE = \frac{1}{2} mv^2$ no longer applies. The total energy that the object possesses is given by

$E = mc^2$

E is the object's total energy in joules or electron volts (eV) where $1.0 \text{ eV} = 1.6 \times 10^{-19}$ J. m is the object's relativistic mass in kg and $c^2 = (3.0 \times 10^8 \text{ m/s})^2 = 9.0 \times 10^{16}$ m^2/s^2.

The **total energy** of an object can be shown to be equal to the sum of the object's mechanical energy and **rest energy**. The object's rest energy is the energy which is the product of the rest mass and the speed of light squared, i.e., $E_0 = m_0 c^2$. For a moving object the total energy can be expressed as follows:

total energy = KE + rest energy

$E = KE + E_0$

The object's kinetic energy can be determined by rearranging the above formula:

$KE = E - E_0 = mc^2 - m_0 c^2 = (m - m_0) c^2$

According to the theory, even an object at rest has energy and it is possible to convert mass into energy and vice versa.

Although the relation $E = mc^2$ applies to all processes involving energy transfer, it is usually detected only in certain nuclear processes, e.g., nuclear fission and fusion. However, any situation where an object's energy changes should result in a change in its mass. For example, lifting a book and placing it on a table should result in an increase in the book's mass and heating a rod in a fire should result in an increase in its mass.

TEXTBOOK QUESTION 19. It is not correct to say that "matter can neither be created nor destroyed." What must we say instead?

ANSWER: The formula $E = mc^2$ relates mass and energy and can be used to determine how much energy can be obtained from mass and vice versa. As stated in the textbook: "Mass and energy are interconvertible. The law of conservation of energy must include mass as a form of energy."

EXAMPLE PROBLEM 3. Determine the a) relativistic mass and b) total energy of an object of rest mass 10.0 kg which is traveling at 0.80 c.

Part a. Step 1.	Solution: (Section 26-9)
Determine the relativistic mass.	$m = m_0 / [1 - (v/c)^2]^{\frac{1}{2}}$
	$= (10.0 \text{ kg}) / [1 - (0.80c/c)^2]^{\frac{1}{2}}$
	$m = (10.0 \text{ kg})/(0.60) = 16.7 \text{ kg}$

26-9

Part b. Step 1.	$E = mc^2 = (16.7 \text{ kg})(3.0 \times 10^8 \text{ m/s})^2$
Determine the total energy.	$E = 1.50 \times 10^{18} \text{ J}$

EXAMPLE PROBLEM 4. An electron is accelerated from rest through a potential difference of 5.00×10^5 volts. Determine the electron's a) kinetic energy, b) rest energy, c) total energy, and d) speed after passing through the potential difference.

Part a. Step 1.	Solution: (Section 26-9)
Determine the electron's kinetic energy in electron volts and joules after passing through the potential difference.	As the electron accelerates from rest through the potential difference, work is done on it and potential energy is converted to kinetic energy. $W = -\Delta PE = +\Delta KE$ where $\Delta PE = PE_f - PE_i$. $\Delta PE = 0 \text{ eV} - (1 \text{ electron})(5.00 \times 10^5 \text{ V})$ $\Delta PE = -5.00 \times 10^5 \text{ eV}$ $-\Delta PE = \Delta KE, \quad \Delta KE = KE_f - KE_i \quad \text{and} \quad KE_i = 0 \text{ J}$ $-(-5.00 \times 10^5 \text{ eV}) = KE_f - 0$ $KE_f = 5.00 \times 10^5 \text{ eV} = 0.500 \text{ MeV}$ Express the electron's kinetic energy in joules. $KE_f = (5.00 \times 10^5 \text{ eV})(1.6 \times 10^{-19} \text{ J}/1.0 \text{ eV}) = 8.00 \times 10^{-14} \text{ J}$

Part b. Step 1.	The rest energy (E_o) of an electron is given by
Determine the electron's rest energy in joules and MeV.	$E_o = m_o c^2$ $\quad = (9.1 \times 10^{-31} \text{ kg})(9.0 \times 10^{16} \text{ m}^2/\text{s}^2)$ $E_o = 8.19 \times 10^{-14} \text{ J}$ $1.0 \text{ MeV} = 1.0 \times 10^6 \text{ eV} = 1.6 \times 10^{-13} \text{ J}$

Therefore, the rest energy of the electron may be expressed in MeV as follows:

$$E_o = (8.19 \times 10^{-14} \text{ J})(1.0 \text{ MeV}/1.6 \times 10^{-13} \text{ J})$$

$$E_o = 0.512 \text{ MeV}$$

Part c. Step 1. Determine the electron's total energy in joules and MeV.	$E = KE + E_o$ $= 8.00 \times 10^{-14} \text{ J} + 8.19 \times 10^{-14} \text{ J}$ $E = 1.62 \times 10^{-13} \text{ J}$ or expressing the answer in MeV, $E = 0.500 \text{ MeV} + 0.512 \text{ MeV} = 1.01 \text{ MeV}$
Part d. Step 1. Determine the electron's relativistic mass.	$E = mc^2$ and $m = E/c^2$ $m = (1.62 \times 10^{-13} \text{ J})/(9.0 \times 10^{16} \text{ m}^2/\text{s}^2)$ $m = 1.80 \times 10^{-30} \text{ kg}$
Part d. Step 2. Use the relativistic mass equation to determine the electron's velocity.	$m = m_o/[1 - (v/c)^2]^{\frac{1}{2}}$ and $[1 - (v/c)^2]^{\frac{1}{2}} = m_o/m$ $[1 - (v/c)^2]^{\frac{1}{2}} = (9.1 \times 10^{-31} \text{ kg})/(1.80 \times 10^{-30} \text{ kg})$ $[1 - (v/c)^2]^{\frac{1}{2}} = 0.506$ Squaring both sides of the equation gives $1 - (v/c)^2 = 0.256$ and $- (v/c)^2 = -1 + 0.256 = -0.744$ $(v/c)^2 = 0.744$ $v/c = 0.863$ and $v = 0.863\ c = 2.59 \times 10^8 \text{ m/s}$

Relativistic Addition of Velocities

Suppose observer A in the spacecraft shown in the diagram located at the top of page 26-6 throws an object in the positive x direction with a speed of u'. Observer B will not measure the relative speed of the object to be u where u = v + u'. The reason for this is because the addition of the two velocities might give a value larger than c. Instead, the speed of the object as measured by B is given by

$$u = (v + u')/(1 + vu'/c^2)$$

where u is the speed of the object as measured by observer B and v is the speed of observer A relative to observer B. u' is the speed of the object relative to observer A. If u' is in the same direction as v, then u' is positive. If u' is opposite from v, then u' is negative.

EXAMPLE PROBLEM 5. The spacecraft shown in diagram 1 on page 26-6 is traveling at 0.90 c relative to observer B. Observer A throws an object in the same direction that the spacecraft is moving. The object's velocity relative to observer A is 0.80c. Determine the velocity of the object as measured by observer B.

Part a. Step 1.	Solution: (Section 26-10)
Use the equation for the relativistic addition of velocities.	$u = (v + u')/(1 + v\,u'/c^2)$ $= (0.90c + 0.80c)/(1 + (0.90c)(0.80c)/c^2)$ $u = 0.99c$

PROBLEM SOLVING SKILLS

For problems related to time dilation, length contraction, and mass increase:

1. Determine which observer is at rest relative to the objects to be measured. This observer measures the proper time, proper length, and proper mass. The observer who is moving relative to the objects measures the relativistic time, length, and mass.
2. Use the equations for time dilation, length contraction, and mass increase to solve the problem.

For problems related to the relativistic addition of velocities:

1. Determine the speed of the object as measured by one of the observers.
2. Determine the relative speed of the two observers.
3. Use the equation for relativistic addition of velocities to determine the speed of the object relative to the second observer.

For problems related to the total energy of an object:

1. Determine the object's mechanical energy in electron volts and joules.
2. Determine the object's rest energy ($E = m_oc^2$) in electron volts and joules.
3. Determine the object's total energy ($E = mc^2$) in electron volts and joules.
4. If requested, use the total energy equation to determine the object's relativistic mass and the mass increase equation to determine the object's velocity.

SOLUTIONS TO SELECTED TEXTBOOK PROBLEMS

TEXTBOOK PROBLEM 1. A spaceship passes you at a speed of 0.750c. You measure its length to be 28.2 m. How long would it be when at rest?

Part a. Step 1.	Solution: (Sections 26-4 and 26-5)
Use the length contraction equation to measure the length when at rest.	When the spacecraft passes the observer measures the contracted length (L). The length measured when at rest is L_o.

$$L = L_o(1 - (v/c)^2)^{1/2}$$

$$28.2 \text{ m} = L_o[1 - (0.750c/c)^2]^{1/2}$$

$$28.2 \text{ m} = L_o (0.661)$$

$$L_o = 42.6 \text{ m}$$

TEXTBOOK PROBLEM 11. Suppose a news report stated that the Starship *Enterprise* had just returned from a 5-year voyage while traveling at 0.84c. (a) If the report meant 5.0 years of *Earth time*, how much time elapsed on the ship? (b) If the report meant 5.0 years of *ship time*, how much time passed on the Earth?

Part a. Step 1.	Solution: (Sections 26-4 and 26-5)
Use the time dilation equation to measure the amount of time that passes on the ship's clock.	Earth observers measure 5.0 years while the crew of the *Enterprise* measures a shorter time.

$$\Delta t = \Delta t_o/(1 - (v/c)^2)^{1/2}$$

$$5.0 \text{ years} = \Delta t_o/[1 - (0.84c/c)^2]^{1/2}$$

$$5.0 \text{ years} = \Delta t_o/(0.294)^{1/2}$$

$$\Delta t_o = (5.0 \text{ years})(0.543) = 2.7 \text{ years}$$

Part b. Step 1.	The crew of the *Enterprise* measures 5.0 years while an Earth observer measures a longer time.
Use the time dilation equation to measure the amount of time that passes on an Earth observer's clock.	$\Delta t = \Delta t_o/(1 - (v/c)^2)^{1/2}$ $\Delta t = (5.0 \text{ years})/[1 - (0.84c/c)^2]^{1/2}$ $\Delta t = (5.0 \text{ years})/(0.294)^{1/2}$ $\Delta t_o = (5.0 \text{ years})/(0.542) = 9.2 \text{ years}$

TEXTBOOK PROBLEM 15. What is the momentum of a proton traveling at $v = 0.85c$?

Part a. Step 1.	Solution: (Section 26-8)
Determine the relativistic mass of the proton.	$m = m_o /[1 - (v/c)^2]^{1/2}$ $= (1.67 \times 10^{-27} \text{ kg})/[1 - (0.85c/c)^2]^{1/2}$ $= (1.67 \times 10^{-27} \text{ kg})/[1 - 0.723]^{1/2}$ $m = 3.2 \times 10^{-27} \text{ kg}$
Part a. Step 2.	$p = m v = (3.2 \times 10^{-27} \text{ kg})[(0.85)(3.0 \times 10^8 \text{ m/s})]$
Determine the proton's momentum.	$p = 8.1 \times 10^{-19} \text{ kg m/s}$

TEXTBOOK PROBLEM 35. Calculate the speed of a proton ($m_o = 1.67 \times 10^{-27}$ kg) whose kinetic energy is exactly half (a) its total energy, (b) rest energy.

Part a. Step 1.	Solution: (Section 26-9)
Determine the relativistic mass of the proton.	total energy = kinetic energy + rest energy $mc^2 = KE + m_oc^2$ but $KE = \frac{1}{2} mc^2$ $mc^2 = \frac{1}{2} mc^2 + m_oc^2$ Note: c^2 cancels $m = \frac{1}{2} m + m_o$ $m = 2 m_o = 2(1.67 \times 10^{-27} \text{ kg}) = 3.34 \times 10^{-27} \text{ kg}$

Part a. Step 2.	$m = m_0/(1 - (v/c)^2)^{1/2}$
Use the mass increase formula to determine the speed.	$2m_0 = m_0/(1 - (v/c)^2)^{1/2}$ Note: m_0 cancels giving
	$2 = 1/(1 - (v/c)^2)^{1/2}$ and rearranging
	$(1 - (v/c)^2)^{1/2} = 1/2$
	$1 - (v/c)^2 = 0.250$
	$-(v/c)^2 = 0.250 - 1 = -0.750$
	$(v/c)^2 = 0.750$
	$v = 0.866c$

Part b. Step 1	total energy = kinetic energy + rest energy but $KE = \frac{1}{2} m_0 c^2$
Determine the relativistic mass of the proton.	$mc^2 = \frac{1}{2} m_0 c^2 + m_0 c^2$ Note: c^2 cancels
	$m = \frac{1}{2} m_0 + m_0$
	$m = 1.50 m_0 = (1.50)(1.67 \times 10^{-27} \text{ kg}) = 2.51 \times 10^{-27} \text{ kg}$

Part b. Step 2.	$m = m_0/(1 - (v/c)^2)^{1/2}$
Use the mass increase formula to determine the speed.	$1.50\, m_0 = m_0/(1 - (v/c)^2)^{1/2}$ Note: m_0 cancels giving
	$1.5 = 1/(1 - (v/c)^2)^{1/2}$ and rearranging
	$(1 - (v/c)^2)^{1/2} = 2/3$
	$1 - (v/c)^2 = 0.444$
	$-(v/c)^2 = 0.444 - 1 = -0.556$
	$(v/c)^2 = 0.556$
	$v = 0.745c$

TEXTBOOK PROBLEM 40. An electron ($m_0 = 9.11 \times 10^{-31}$ kg) is accelerated from rest to a speed v by a conservative force. In this process, its potential energy decreases by 6.60×10^{-14} J. Determine the electron's speed, v.

Part a. Step 1.	Solution: (Section 26-9)
Determine the electron's kinetic energy in joules.	As the electron accelerates from rest, it loses potential energy and gains kinetic energy. $+ \Delta KE = - \Delta PE$ where $\Delta KE = KE_f - KE_i$ and $\Delta PE = PE_f - PE_i$. $KE_f - 0 \text{ J} = - (0 - 6.60 \times 10^{-14} \text{ J})$ $KE_f = 6.6 \times 10^{-14} \text{ J}$
Part a. Step 2.	The rest energy (E_o) of an electron is given by
Determine the electron's rest energy in joules.	$E_o = m_o c^2$ $= (9.1 \times 10^{-31} \text{ kg})(9.0 \times 10^{16} \text{ m}^2/\text{s}^2)$ $E_o = 8.19 \times 10^{-14} \text{ J}$
Part a. Step 3.	$E = KE + E_o$
Determine the electron's total energy in joules.	$= 6.6 \times 10^{-14} \text{ J} + 8.19 \times 10^{-14} \text{ J}$ $E = 1.48 \times 10^{-13} \text{ J}$
Part a. Step 4.	$E = mc^2$ and $m = E/c^2$
Determine the electron's relativistic mass.	$m = (1.48 \times 10^{-13} \text{ J})/(9.0 \times 10^{16} \text{ m}^2/\text{s}^2)$ $m = 1.64 \times 10^{-30} \text{ kg}$
Part a. Step 5.	$m = m_o/[1 - (v/c)^2]^{1/2}$ and $[1 - (v/c)^2]^{1/2} = m_o/m$
Use the relativistic mass equation to determine the electron's velocity.	$[1 - (v/c)^2]^{1/2} = (9.1 \times 10^{-31} \text{ kg})/(1.64 \times 10^{-30} \text{ kg})$ $[1 - (v/c)^2]^{1/2} = 0.554$ Squaring both sides of the equation gives $1 - (v/c)^2 = 0.307$ and $- (v/c)^2 = - 1 + 0.307 = - 0.693$ $(v/c)^2 = 0.693$

$v/c = 0.833$ and

$v = 0.833c = 2.50 \times 10^8$ m/s

TEXTBOOK PROBLEM 43. A person on a rocket traveling at 0.50c (with respect to the Earth) observes a meteor come from behind and pass her at a speed she measures as 0.50c. How fast is the meteor moving with respect to the Earth?

Part a. Step 1.	Solution: (Section 26-10)
Use the equation for the relativistic addition of velocities.	Let u represent the speed of the meteor as measured by an observer on Earth.

$u = (v + u')/(1 + v\,u'/c^2)$

$\quad = (0.50c + 0.50c)/(1 + (0.50c)(0.50c)/c^2)$

$u = (1.0c)/(1 + 0.25)$

$u = 0.80c$

CHAPTER 27

EARLY QUANTUM THEORY AND MODELS OF THE ATOM

OBJECTIVES

After studying the material of this chapter, the student should be able to:

- describe the method used by J. J. Thomson to determine the ratio of the charge on an electron to its mass.
- describe the apparatus used and solve problems involving a charged particle passing undeflected through a velocity selector.
- use Wien's law to determine the peak wavelength emitted by a black body at a given temperature.
- describe Planck's quantum hypothesis and calculate the energy of a photon at a given frequency or wavelength.
- state the experimental results of the photoelectric effect and use the photon theory to explain these results.
- use the photon theory to determine the maximum kinetic energy of photons emitted from the surface of a metal or the threshold wavelength for the metal.
- use the photon theory and Compton's hypothesis to calculate the wavelength of a photon after it has been scattered as a result of a collision with an electron.
- use $E = mc^2$ to determine the minimum energy required for pair production.
- explain the significance of the Principle of Complementarity.
- use de Broglie's hypothesis to determine the wavelength of moving particle.
- describe the apparatus used in the Rutherford scattering experiment and describe the experimental results.
- describe Rutherford's model of the atom and list two problems with the model.
- write the Balmer equation and use the equation to determine the wavelength of a photon emitted as an electron drops from a higher energy level to a lower energy level. Determine the frequency and energy of this photon.
- list Bohr's postulates and use these postulates to explain the emission spectra produced by the hydrogen atom.
- determine the Bohr radius and angular momentum of an electron in a given energy level.

KEY TERMS AND PHRASES

e/m experiment showed that cathode rays consist of charged particles now known as electrons. As a result of this experiment, the ratio of the charge on an electron (e) to its mass (m) was determined.

Planck's quantum hypothesis predicts that the molecules in a heated object can vibrate only with discrete amounts of energy. Thus the energy of the vibrating atom is quantized.

Millikan oil drop experiment determined that the charge on a microscopically small oil drop is always a small whole-number multiple of 1.6×10^{-19} C. This value equals the charge on the electron. Once the value of the charge on an electron was determined, the accepted value for the mass of the electron was determined to be 9.1×10^{-31} kg.

blackbody radiation refers to the intensity of spectral radiation emitted by a "perfectly" radiating object.

photoelectric effect indicates that light has characteristics of particles. Light particles are called photons.

threshold frequency is the minimum frequency at which electrons are ejected from a surface.

work function (W_o) is the minimum energy required to break the electron free from the attractive forces which hold the electron to the surface of a metal.

pair production occurs when a high-energy photon known as a gamma ray traveling near the nucleus of an atom disappears and an electron and a positron may appear in its place.

positron has the same mass as an electron and carries the same magnitude of electric charge; however, the electron is negatively charged while the positron carries a positive charge.

wave-particle duality refers to the phenomena where both particles, such as electrons and protons, and light exhibit both the properties of waves and the properties of particles.

Compton effect shows that the interaction of a photon with an electron can be viewed as a two-particle collision.

de Broglie wavelength is the wavelength of a particle of mass m traveling at speed v. The wavelength is given by $\lambda = h/(mv)$ where λ is the wavelength of the particle.

emission spectra are produced by a high voltage placed across the electrodes of a tube containing a gas under low pressure. The light produced can be separated into its component colors by a diffraction grating. Such analysis reveals a spectra of discrete lines and not a continuous spectrum.

ionization energy refers to the energy required to remove an electron from an atom.

SUMMARY OF MATHEMATICAL FORMULAS

Wien's law	$\lambda_p T = 2.90 \times 10^{-3}$ m K	the relationship between absolute temperature (T) and the peak wavelength (λ_p) in blackbody radiation
photon energy	$E = h f$	The energy (E) of a photon is related to the frequency (f) of the light.
photoelectric effect	$KE_{max} = hf - W_o$	The maximum kinetic energy (KE_{max}) of the emitted photoelectrons equals the difference between the energy of the incident photon (hf) and the work function (W_o) of the metal surface.
Compton effect	$\lambda' - \lambda = (h/mc)(1 - \cos \phi)$	A collision between a photon and an electron results in a change of wavelength for the photon.
de Broglie wavelength	$\lambda = h/mv$	λ represents the wavelength of a particle of mass m traveling at speed v.
Balmer equation	$1/\lambda = R\, Z^2(1/n'^2 - 1/n^2)$	An electron dropping from a higher energy level to a lower energy level emits a photon of wavelength λ.
ionization energy for an electron from a hydrogen-like atom	$E = (-13.6\text{ eV})Z^2/n^2$	ionization energy (E) of an electron located in the nth level of the hydrogen-like atom
Bohr radius	$r = (0.53 \times 10^{-10}\text{ m})(n^2)$	radius (r) of the orbit of an electron in the hydrogen atom, where n = 1, 2, etc.
angular momentum of an electron	$L = m v\, r_n = n\, h/2\pi$	the angular momentum (L) of an electron orbiting the hydrogen nucleus

CONCEPT SUMMARY

The Electron

In 1897, J. J. Thomson performed the e/m experiment which showed that **cathode rays** consist of charged particles now known as electrons. As a result of his experiment, he was able to determine that the ratio of the charge on an electron (e) to its mass (m) is

$$e/m = v/Br$$

where v is the electron's velocity, B is the magnetic field strength which is directed perpendicular to the electron's path, and r is the radius of the circular path in which the particle travels.

B and r are readily measured while the velocity of the particle can be determined by using a device known as a **velocity selector**. In a velocity selector (see example problem 1), an electric field is arranged so that the electric force $F = q\,E$ balances the force exerted on the charged particle by the magnetic field $F = q\,v\,B \sin 90°$. As a result, $q\,E = q\,v\,B$ and $v = E/B$. The particle passes through the selector undeflected. Since $e/m = v/Br$ and $v = E/B$, then $e/m = E/B^2 r$. Substituting experimental values for E, B, and r, Thomson obtained a value for the electron's charge to mass ratio close to the modern value of 1.76×10^{11} C/kg.

In 1909, R. A. Millikan was able to determine the charge on an electron in an experiment referred to as the **Millikan oil drop experiment**. Millikan found that the charge on a microscopically small oil drop is always a small whole-number multiple of 1.6×10^{-19} C. Once the charge on the electron became known it was possible to determine its mass. The accepted value for the mass of the electron is 9.1×10^{-31} kg.

EXAMPLE PROBLEM 1. In a device known as a "velocity selector" a beam of protons is directed between the plates of a parallel plate capacitor as shown in the diagram. The electric field between the plates has a magnitude of 1000 N/C. Determine the magnitude and direction of the magnetic field required to allow protons traveling at 5.0×10^4 m/s to pass between the plates undeflected.

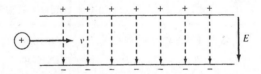

Part a. Step 1.	Solution: (Section 27-1)
Locate the forces acting on the particle. Determine the net force acting on the particle.	In order for the beam to pass undeflected at constant speed, the net force on the particle must be zero. Based on the diagram, the electric force is directed downward. The magnetic force must be equal in magnitude to the electric force but directed upward in order that the net force equal zero.

$$\text{net } F = 0 \quad + \quad \begin{array}{c} \uparrow F_{magnetic} \\ \\ \downarrow F_{electric} \end{array}$$

$$F_{elec} - F_{mag} = 0$$

Part a. Step 2.	$F_{elec} - F_{mag} = 0$
Determine the magnitude of the magnetic field required to allow the particle to pass undeflected.	$F_{elec} = F_{mag}$
	$q\,E = q\,v\,B\,\sin\theta$
	q cancels from both sides of the equation and solving for B gives
	$B = E/v\,\sin\theta$
	$= (1000\ N/C)/(5.0 \times 10^4\,m/s)(\sin 90^\circ)$
	$B = 0.020\ N/C\ m/s = 0.020\ T$

Part a. Step 3.	The magnetic force is directed upward. Using the right-hand rule, the fingers point toward the right of the page while the thumb is directed towards the top of the page. When you bend your fingers they point into the paper. Therefore, the magnetic field is directed into the paper.
Use the right-hand rule to determine the direction of the magnetic field.	

Planck's Quantum Hypothesis

The intensity of spectral radiation emitted by a "perfectly" radiating object, known as **blackbody radiation,** when graphed as a function of wavelength, produces curves as shown at the top of the next page.

The wavelength at which peak intensity occurs decreases as the Kelvin temperature of the object increases. The relationship between the peak wavelength and the absolute temperature is given by **Wien's law,**

$\lambda_p\,T = 2.90 \times 10^{-3}\ m\ K$ where λ_p is the **peak wavelength**

The peak wavelength is the wavelength of the emitted light at the point where the intensity of the emitted light is a maximum. T is the temperature of the blackbody in degrees Kelvin.

Max Planck produced a theory which agreed with the experimental data by assuming that the molecules in the heated object can vibrate only with discrete amounts of energy. Thus the energy of the vibrating atom is quantized.

The energy is related to the natural frequency of vibration of the molecules of the radiating object by the formula $E = n\,h\,f$.

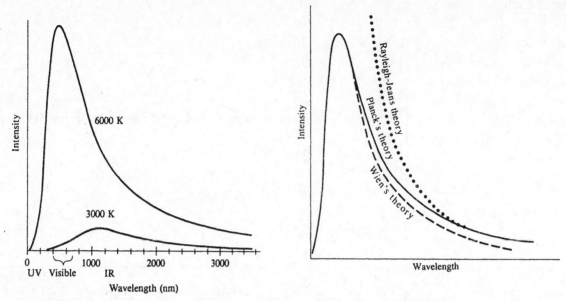

h is **Planck's constant**, h = 6.626 x 10^{-34} J s.   E is the energy in joules, n is a whole number, n = 1, 2, 3, etc., while f is the object's natural frequency of vibration in hertz.

TEXTBOOK QUESTION 2. If energy is radiated by all objects, why can't we see them in the dark? (See also Section 14-8).

ANSWER: The human eye is sensitive only to visible light, i.e., violet (400 nm) through red (700 nm). The energy emitted by objects such as this book or your body is far below that required to produce visible light. For example, let us assume that the skin temperature of your body is approximately 34°C (307 K). Using Wien's law to determine the peak wavelength of light λ_p (307 K) = 2.90 x 10^{-3} m K gives a peak wavelength of 9.45 x 10^{-6} m or 9450 nm. This radiation is in the infrared portion of the electromagnetic spectrum.

EXAMPLE PROBLEM 2. a) Use Wien's law to determine the peak wavelength of light emitted by a blackbody whose temperature is 6000 K. b) Is this wavelength in the visible portion of the electromagnetic spectrum?

Part a. Step 1.	Solution: (Section 27-2)
Determine the peak wavelength.	λ_p T = 2.90 x 10^{-3} m K
	λ_p (6000 K) = 2.90 x 10^{-3} m K
	λ_p = 4.83 x 10^{-7} m = 483 nm

| Part b. Step 1.

Is this wavelength the visible spectrum? | The visible spectrum extends from 400 to 700 nm. The peak wavelength is 483 nm; therefore, it is in the visible range. 483 nm is located in the blue portion of the electromagnetic spectrum. |

EXAMPLE PROBLEM 3. Show that the energy of the photon, E = h f, can be written as E = 1240 eV nm/λ.

| Part a. Step 1.

Express the formula for the energy of a photon in terms of joule meter divided by the wavelength. | Solution: (Section 27-3)

$E = h f$ but $f = c/\lambda$

$E = hc/\lambda$

$\quad = (6.63 \times 10^{-34} \text{ J s})(3.00 \times 10^{8} \text{ m/s})/\lambda$

$E = 1.99 \times 10^{-25} \text{ J m}/\lambda$ |
| Part a. Step 2.

Use the conversion from joules to eV and from meters to nm to show that E = hf can be written as E = 1240 eV nm/λ. | $E = (1.99 \times 10^{-25} \text{ J m}/\lambda)(1.0 \text{ eV}/1.6 \times 10^{-19} \text{ J})$

$\quad = (1.243 \times 10^{-6} \text{ ev m})(1.0 \text{ nm}/1.0 \times 10^{-9} \text{ m})/\lambda$

$E = 1240 \text{ ev nm}/\lambda$ |

Photoelectric Effect

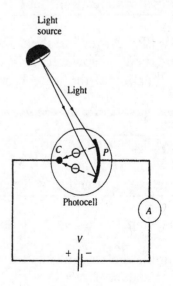

Light source

Light

Photocell

When light is incident on a metal surface, electrons are emitted. In the photocell shown in the diagram, the incident light causes the electrons to be emitted from the **plate** (P) and the difference in potential causes the electrons to travel to the **collector** (C).

The experimental results indicate that 1) the number of electrons emitted per second (electron current) increases as the light intensity increases, 2) the maximum kinetic energy of the electrons is not affected by the intensity of the light, and 3) below a certain frequency, called the **threshold frequency,** no electrons are ejected no matter how intense the light beam.

The wave theory agrees with the first result but cannot explain the second and third experimental results. In 1905, Albert Einstein proposed an extension of the quantum theory to explain the results of the **photoelectric effect**. His theory says that light is emitted in particles which are now called **photons**. Each photon has energy which is related to the frequency of the light according to the formula

$$E = h f$$

The experimental results of the photoelectric effect can be explained by the photon theory:

1) An electron in the metal making up the plate of the photocell absorbs the energy of the incident photon. If the photon energy is large enough, the electron escapes from the metal. As the intensity of the light beam increases, the number of photons increases and the number of electrons emitted each second increases. However, the intensity of the light does not affect the energy of the incident photons.

2) The maximum kinetic energy of the emitted electrons is related to the energy of the incident photon by the equation $hf = KE_{max} + W_o$ where hf is the energy of the incident photon in joules, KE_{max} is the kinetic energy of the most energetic electrons emitted from the metal in joules. The **work function** (W_o) is the minimum energy required to break the electron free from the attractive forces which hold the electron to the metal.

3) If the energy of the incident photon is below that of the work function, no electrons are emitted. The minimum frequency required to eject an electron is called the threshold frequency or "cut off" frequency.

TEXTBOOK QUESTION 5. If the threshold wavelength in the photoelectric effect increases when the emitting metal is changed to a different metal, what can you say about the work functions of the two metals?

ANSWER: At the threshold wavelength, no electrons are ejected from the surface of the metal and $hf_o = W_o$. However, $\lambda_o = c/f_o$ and therefore, $W_o = hc/\lambda_o$. If the wavelength increases, the work function of the second metal must be less than the work function of the original metal.

EXAMPLE PROBLEM 4. The maximum kinetic energy of electrons emitted from a surface coated with sodium is 4.00 eV. If the work function is 2.28 eV, determine the a) energy of the incident photons and b) wavelength of the incident photons.

Part a. Step 1.

Determine the energy of the incident photons.

Solution: (Section 27-3)

$$E = KE_{max} + W_o$$

$$E = 4.00 \text{ eV} + 2.28 \text{ eV} = 6.28 \text{ eV}$$

Part b. Step 1.	$E = h f$ and $f = E/h$
Determine the frequency of the incident photons.	$f = (6.28 \text{ eV})/(6.63 \times 10^{-34} \text{ J s}) \times (1.6 \times 10^{-19} \text{ J})/(1 \text{ eV})$
	$f = 1.52 \times 10^{15} \text{ Hz}$

Part b. Step 2.	$\lambda = c/f = (3.0 \times 10^8 \text{ m/s})/(1.52 \times 10^{15} \text{ Hz})$
Determine the wavelength of the incident photons.	$\lambda = 1.98 \times 10^{-7} \text{ m} = 198 \text{ nm}$
	alternate solution:
	$E = 1240 \text{ eV nm}/\lambda$ and $\lambda = 1240 \text{ eV nm}/E$
	$\lambda = (1240 \text{ eV nm})/(6.28 \text{ eV}) = 198 \text{ nm}$

EXAMPLE PROBLEM 5. The threshold wavelength for incident photons to eject electrons from the surface of gold is 257 nm. If light of wavelength 100 nm shines on the metal, determine the a) work function of the metal and b) energy of the most energetic electrons.

Part a. Step 1.	Solution: (Section 27-3)
Determine the work function of the metal.	At the threshold wavelength, the energy absorbed by the electron is used to break free from the attractive forces which hold it to the metal. As a result, it has no kinetic energy when it finally breaks free. Therefore,
	$E = KE_{max} + W_o$ but $KE_{max} = 0 \text{ J}$
	$hf = 0 \text{ J} + W_o$ where $hf = hc/\lambda$
	$W_o = (6.63 \times 10^{-34} \text{ J s})(3.0 \times 10^8 \text{ m/s})/(2.57 \times 10^{-7} \text{ m})$
	$W_o = 7.74 \times 10^{-19} \text{ J}$ or 4.83 eV
	alternate solution:
	$W_o = hf = hc/\lambda = 1240 \text{ eV nm}/\lambda$
	$W_o = (1240 \text{ eV nm})/(257 \text{ nm}) = 4.83 \text{ eV}$

Part b. Step 1.	$E = KE_{max} + W_o$
Determine the energy of the most energetic electrons.	where $E = h f = (1240 \text{ eV})/(100 \text{ nm}) = 12.4 \text{ eV}$
	$12.4 \text{ eV} = KE_{max} + 4.83 \text{ eV}$
	$KE_{max} = 7.57 \text{ eV}$

Compton Effect

In 1922, Arthur Compton used the photon theory to explain why x-rays scattered from certain materials have different wavelengths than the incident x-rays. According to Compton, an incident photon transferred some of its energy to an electron in the material and was scattered with lower energy and therefore longer wavelength.

According to the theory, the incident photon carries both energy ($E = h f$) and momentum ($p = E/c = h f/c = h/\lambda$). The photon interaction with the electron can be considered to be a two-particle collision as shown in the diagram.

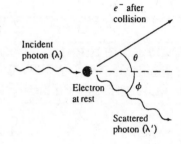

Applying the laws of conservation of energy and momentum to the collision, the theory correctly predicts a wavelength change given by the following equation:

$$\lambda' - \lambda = (h/mc)(1 - \cos \phi)$$

λ is the wavelength of the incident light, λ' is the wavelength of the scattered photon, m is the mass of the recoil electron, and ϕ is the angle between the direction of the incident photon and the scattered photon.

TEXTBOOK QUESTION 8. If an X-ray photon is scattered by an electron, does its wavelength change? If so, does it increase or decrease?

ANSWER: The incident photon transfers some of its energy to the electron. The result is a scattered photon which has less energy and a different wavelength than the incident photon. Since $E = hf = hc/\lambda$, if the energy is less, the wavelength must increase.

EXAMPLE PROBLEM 6. An X-ray of wavelength 0.2000 nm is scattered by an electron. In the resulting collision, the scattered photon is reflected directly backward while the electron travels in the direction of the incident photon. Determine the a) wavelength of the scattered photon and b) energy of the recoil electron.

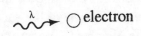 electron

before collision   after collision

Part a. Step 1.	Solution: (Section 27-5)
Determine the wavelength of the scattered photon.	The collision is head on and the angle between the incident photon and the scattered photon is 180°. Therefore,

$$\lambda' - \lambda = (h/mc)(1 - \cos \phi)$$

where $\cos \phi = \cos 180° = -1$ and

$$m\,c = (9.1 \times 10^{-31} \text{ kg})(3.0 \times 10^{8} \text{ m/s})$$

$$m\,c = 2.7 \times 10^{-22} \text{ kg m/s}$$

$$\lambda' - 0.2000 \text{ nm} = [(6.63 \times 10^{-34} \text{ J s})/(2.7 \times 10^{-22} \text{ kg})](1 - -1)$$

$$\lambda' - 0.2000 \text{ nm} = 4.86 \times 10^{-12} \text{ m}$$

But 4.86×10^{-12} m $= 4.86 \times 10^{-3}$ nm.

$$\lambda' = 0.2000 \text{ nm} + 4.86 \times 10^{-3} \text{ nm} = 0.2049 \text{ nm}$$

Part b. Step 1.	The kinetic energy of the recoil electron can be determined by using the law of conservation of energy. The energy of the recoil photon equals the sum of the energy of the scattered photon and the kinetic energy of the recoil electron.
Determine the kinetic energy of the recoil electron.	

$$(1240 \text{ eV nm})/\lambda = (1240 \text{ eV nm})/\lambda' + KE_{electron}$$

$$(1240 \text{ eV nm})/(0.2000 \text{ nm}) = (1240 \text{ eV nm})/(0.2049 \text{ nm}) + KE_{electron}$$

$$6200 \text{ eV} = 6050 \text{ eV} + KE_{electron}$$

$$KE_{electron} = 150 \text{ eV}$$

Pair Production

The equation $E = m\,c^2$ implies that it is possible to convert mass into energy and vice versa. One example of the conversion of energy to mass is **pair production**. In this process, a high-energy photon known as a gamma ray traveling near the nucleus of an atom may disappear and an electron and a positron may appear in its place. The electron and the positron have the same mass and carry the same magnitude of electric charge; however, while the electron is negatively charged the positron carries a positive charge. Thus, in addition to the laws of conservation of energy and momentum, the law of conservation of electric charge also holds true.

The minimum energy of a gamma ray required for the pair production of electron and positron can be shown to be about 1.02 Mev. If the energy of the gamma ray is above this amount, then excess energy is shared equally between the particles in the form of kinetic energy.

Wave Particle Duality; the Principle of Complementarity

Experiments such as the Young's interference experiment, polarization, and single slit diffraction indicate that light is a wave. The photoelectric effect and the Compton effect indicate that light is a particle. Light is a phenomena which exhibits both the properties of waves and the properties of particles. This is known as **wave-particle duality.**

Niels Bohr proposed the **principle of complementarity** which says that for any particular experiment involving light, we must either use the wave theory or the particle theory, but not both. The two aspects of light complement one another.

EXAMPLE PROBLEM 7. What is the minimum energy of a photon required to produce a proton-antiproton pair? Note: the antiproton has the same rest mass as a proton but is negatively charged.

Part a. Step 1.	Solution: (Section 27-6)
Use the law of conservation of energy to determine the minimum energy of the photon.	Using the law of conservation of energy, the minimum energy is equal to the sum of the rest energies of the two particles.

$$E = m c^2 = m_{proton} c^2 + m_{antiproton} c^2$$

$$E = (1.67 \times 10^{-27} \text{ kg})(9.0 \times 10^{16} \text{ m}^2/\text{s}^2)$$
$$+ (1.67 \times 10^{-27} \text{ kg})(9.0 \times 10^{16} \text{ m}^2/\text{s}^2)$$

$$E = 3.01 \times 10^{-10} \text{ J} = 1.88 \times 10^9 \text{ eV} = 1.88 \text{ BeV}$$

Wave Nature of Matter

Just as light exhibits properties of both particles and waves, particles such as electrons, protons, and neutrons also exhibit wave properties. Thus the wave-particle theory extends to matter as well as light. In 1923, Louis de Broglie suggested that the wavelength of a particle of mass m traveling at speed v is given by

$\lambda = h/(m\ v)$ where λ is the **de Broglie wavelength** of the particle.

In 1927, two Americans, Davisson and Germer, produced diffraction patterns by scattering electron beams from the surface of a metal crystal. The calculated wavelength of the electron waves agreed with de Broglie's prediction. It was later shown that protons, and neutrons as well as other particles exhibit wave properties as well as particle properties.

Rutherford's Model of the Atom

In 1909, Ernest Rutherford suggested to two of his students, Hans Geiger and Ernest Marsden, that they bombard a piece of thin gold foil with high-energy alpha particles.

The positively charged alpha particles are emitted spontaneously from a radioactive source. Because of their high energy, Rutherford expected that the alpha particles would pass through the gold foil without being significantly deflected from their original direction. However, as shown in the diagram, Geiger and Marsden found that while most of the alpha particles did pass through undeflected, some were deflected by 30° or more and a few were deflected by 90° or more.

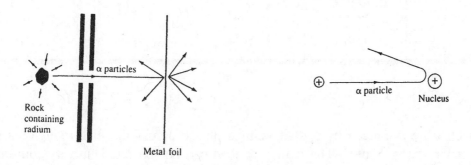

Based on observations, Rutherford concluded that most of the mass of the atom resided in a tiny region called the **nucleus**. The nucleus contains the positively charged protons while the electrons travel in orbits around the nucleus.

Rutherford's model had two problems: 1) since protons repel one another, the positively charged nucleus should not exist, and 2) the orbiting electrons are undergoing centripetal acceleration and, according to electromagnetic theory, accelerated electric charges give rise to electromagnetic waves. Thus the electrons should give up their energy in the form of a continuous spectrum as they spiral into the nucleus.

EXAMPLE PROBLEM 8. Suppose in the Rutherford scattering experiment that an alpha particle traveling at 5.00×10^7 m/s struck a gold nucleus in a head-on collision. Determine the closest approach of the alpha particle to the gold nucleus if the nucleus did not recoil.

Part a. Step 1.	Solution: (Section 27-10)
Use the law of conservation of energy to determine the closest approach of the alpha particle to the gold nucleus.	The collision is head on; therefore, all of the kinetic energy of the alpha particle will be converted into the form of electrical potential energy at the point of closest approach.

Note: $q_\alpha = +2e$ $q_{gold} = +79e$

$e = 1.6 \times 10^{-19}$ C $m_\alpha = 6.64 \times 10^{-27}$ kg

$q_\alpha q_{gold} = (+2 \times 1.6 \times 10^{-19}$ C$)(+79 \times 1.6 \times 10^{-19}$ C$)$

$q_\alpha \, q_{gold} = 4.04 \times 10^{-36}$ C^2

initial kinetic energy = electrical potential energy at closest approach

$\frac{1}{2} m v^2 = k \, q_\alpha \, q_{gold}/r$

$\frac{1}{2} (6.64 \times 10^{-27}$ kg$)(5.00 \times 10^7$ m/s$)^2 = (9 \times 10^9$ N m^2/C$^2)(4.04 \times 10^{-36}$ C$^2)/r$

Solving for r gives

$r = 4.38 \times 10^{-15}$ m

Atomic Spectra

Emission spectra are produced by a high voltage placed across the electrodes of a tube containing a gas under low pressure. The light produced can be separated into its component colors by a diffraction grating. Such analysis reveals a spectra of discrete lines and not a continuous spectrum.

In 1885, J. Balmer developed a mathematical equation which could be used to predict the wavelengths of the four visible lines in the hydrogen spectrum. Balmer's equation states

$1/\lambda = R \, Z^2 \, (1/n'^2 - 1/n^2)$

where Z is the atomic number of the atom, λ is the wavelength of the spectral line in meters, and $R = 1.097 \times 10^7$ m^{-1} is known as Rydberg's constant. n' is a whole number, 2 for lines in the visible spectrum, 3 for infrared, 1 for ultraviolet. n is a whole number; this number can be any number greater than n'. For example, for the visible spectrum n' = 2 while n = 3 (red light), n = 4 (blue light), n = 5 (violet light), and n = 6 (violet light).

Part a. Step 1.	Solution: (Sections 27-12)
Use the Balmer formula to determine the wavelength of the emitted photon.	$1/\lambda = R\ Z^2[(1/n'^2) - (1/n^2)]$
	$1/\lambda = (1.097 \times 10^7\ m^{-1})(1^2)[(1/2^2) - (1/4^2)]$
	$1/\lambda = (1.097 \times 10^7\ m^{-1})(0.25 - 0.063)$
	$1/\lambda = 2.06 \times 10^6\ m^{-1}$
	$\lambda = 4.86 \times 10^{-7}\ m = 486\ nm$

Part b. Step 1.	$c = f\ \lambda$
Determine the photon's frequency.	$3.00 \times 10^8\ m/s = f\ (4.86 \times 10^{-7}\ m)$
	$f = 6.17 \times 10^{14}\ Hz$

Part c. Step 1.	$E = h\ f$
Determine the photon's energy in joules and eV.	$E = (6.63 \times 10^{-34}\ J\ s)(6.17 \times 10^{14}\ Hz)$
	$E = 4.09 \times 10^{-19}\ J = 2.55\ eV$
	alternate solution:
	The energy of any particular level in the hydrogen atom is given by the formula
	$E = -\ 13.6\ eV/n^2$
	The difference in the energy between the third and second levels equals the energy of the emitted photon.
	$\Delta E = E_2 - E_3 = (-\ 13.6\ eV)/(2^2) - (-\ 13.6\ eV)/(4^2)$
	$\quad = (-\ 3.40\ eV) - (-\ 0.850\ eV)$
	$\Delta E = -\ 2.55\ eV$

> The electron loses 2.55 eV of energy as it drops from the third to the second level. The energy is in the form of a photon which has 2.55 eV of energy.

Bohr's Model of the Hydrogen Atom

Based on the Rutherford model of the atom, line spectra, and the Balmer formula, Niels Bohr in 1913 proposed the following postulates for the hydrogen atom:

1. The electron travels in circular orbits about the positively charged nucleus. However, only certain orbits are allowed. The electron does not radiate energy when it is in one of these orbits, thus violating classical theory.

2. The allowed orbits have radii (r_n) where

   $r_n = (0.53 \times 10^{-10} \text{ m}) n^2 = (0.53 \text{ nm}) n^2$ and   n = 1, 2, 3, etc.
   If n = 1, then the electron is in its smallest orbit, which is known as its **ground state**.

3. The orbits have angular momentum (L) given by

   $L = m v r_n = n h/2\pi$ where   n = 1, 2, 3, etc.   The angular momentum has only discrete values; thus it is **quantized**.

4. If an electron falls from one orbit, also known as **energy level**, to another, it loses energy in the form of a photon of light. The energy of the photon equals the difference between the energy of the orbits.

5. The energy level of a particular orbit is given by $E = -13.6 \text{eV}/n^2$ where n = 1, 2, 3, etc. If n = 1, the electron is in its lowest energy level and it would take 13.6 eV to remove it from the atom (**ionization energy**).

6. A hydrogen atom can absorb only those photons of light which will cause the electron to jump from a lower level to a higher level. Thus the energy of the photon must equal the difference in the energy between the two levels. Therefore, a continuous spectrum passing through a cool gas will exhibit dark absorption lines at the same wavelengths as the emission lines.

The Bohr model proved to be successful for the hydrogen atom and for one-electron ions; however, it did not work for multi-electron atoms.

EXAMPLE PROBLEM 10. Determine the energy required to ionize an electron which is in the ground state of an Li^{+2} ion.

Part a. Step 1.	Solution: (Section 27-12)
Determine the energy required. Hint: n = 1 for the ground state and n = ∞ for ionization.	The ionization energy is the energy required to completely remove the electron from the atom. The point at which it is completely free from the atom is referred to as the free state and at that point $E = 0$. The ionization energy equals the difference in the energy between the ground state (n = 1) and the free state (n = ∞).

The atomic number of lithium is 3; thus $Z = 3$. The energy of any level n in the lithium ion is given by

$$E = [(-13.6 \text{ eV}) Z^2]/n^2$$

$$\Delta E = E_\infty - E_1$$

$$= [(-13.6 \text{ eV})(3^2)]/\infty^2] - [(-13.6 \text{ eV})(3^2)]/1^2]$$

$$\Delta E = (0 \text{ eV}) - (-122 \text{ eV}) = 122 \text{ eV}$$ |

EXAMPLE PROBLEM 11. Determine the a) wavelength, b) frequency, and c) energy of a photon emitted from a Li^{+2} ion when an electron makes a transition from the second energy level to the first energy level.

Part a. Step 1.	Solution: (Section 27-12)
Use the Balmer formula to solve for the wavelength.	The atomic number of Li is 3; therefore, $Z = 3$.

$$1/\lambda = (1.097 \times 10^7 \text{ m}^{-1})(3^2)[1/1^2 - 1/2^2]$$

$$1/\lambda = 7.74 \times 10^7 \text{ m}^{-1}$$

$$\lambda = 1.35 \times 10^{-8} \text{ m}$$ |
| Part b. Step 1. | $c = f \lambda$ |
| Solve for the frequency. | $$3.00 \times 10^8 \text{ m/s} = f (1.35 \times 10^{-8} \text{ m})$$

$$f = 2.22 \times 10^{16} \text{ Hz}$$ |

Part c. Step 1.	The energy of the emitted photon equals the energy lost by the electron as it drops from the second level to the first level.
Solve for the energy of the emitted photon.	$\Delta E = E_1 - E_3$
	$= [(-13.6 \text{ eV})(3^2)]/1^2] - [(-13.6 \text{ eV})(3^2)]/2^2]$
	$\Delta E = (-122 \text{ eV}) - (-30.6 \text{ eV}) = -91.4 \text{ eV}$
	The electron loses 91.4 eV of energy as it drops from n = 2 to n = 1. Therefore, the emitted photon has 91.4 eV of energy.

De Broglie's Contribution to the Bohr Model

In 1923, de Broglie extended his idea of matter waves to the Bohr atom by stating that the electron wave must be a circular standing wave. The circumference ($2\pi \, r_n$) of the standing wave contains a whole number of wavelengths ($n\lambda$); thus $2\pi \, r_n = n\lambda$ where n = 1, 2, 3, etc.

$\lambda = 2\pi \, r_n/n$ but $\lambda = h/mv$. Then $2\pi \, r_n/n = h/mv$ and rearranging gives $m \, v \, r_n = n \, h/2\pi$.

This formula is Bohr's postulate concerning the quantization of the electron's angular momentum. de Broglie provided an explanation for the key postulate in Bohr's model of the hydrogen atom; that is, the idea that the angular momentum (L) of the electron is quantized, i.e., $L = m \, v \, r_n = n \, h/2\pi$. It is from this postulate that the equations for the discrete orbits and energy states are derived.

TEXTBOOK QUESTION 15. If an electron and proton travel at the same speed, which has the shorter wavelength? Explain.

ANSWER: Based on de Broglie's hypothesis, the wavelength of a particle is given by $\lambda = h/mv$. The wavelength is inversely proportional to the particle's mass. The mass of the proton is approximately 1840 times greater than the mass of the electron. Therefore, the wavelength of the proton is approximately 1840 times less than the wavelength of the electron. The proton has the shorter wavelength.

EXAMPLE PROBLEM 12. Three particles, an electron, a proton, and a 0.300-kg ball are traveling at 2.50×10^6 m/s. Determine the de Broglie wavelength of each particle.

Part a. Step 1.	Solution: (Section 27-13)
Use the de Broglie equation to determine the wavelength of each particle.	The rest mass of an electron = 9.1×10^{-31} kg and that of a proton is 1.67×10^{-27} kg. Use the de Broglie equation to solve for the wavelength.

electron: $\lambda = h/mv$

$\lambda = (6.63 \times 10^{-34} \text{ J})/[(9.1 \times 10^{-31} \text{ kg})(2.50 \times 10^6 \text{ m/s})]$

$\lambda = 2.91 \times 10^{-10}$ m = 0.291 nm

proton: $\lambda = h/mv$

$\lambda = (6.63 \times 10^{-34} \text{ J s})/[(1.67 \times 10^{-27} \text{ kg})(2.50 \times 10^6 \text{ m/s})]$

$\lambda = 1.59 \times 10^{-13}$ m

ball: $\lambda = h/mv$

$\lambda = (6.63 \times 10^{-34} \text{ J s})/[(0.300 \text{ kg})(2.50 \times 10^6 \text{ m/s})]$

$\lambda = 8.84 \times 10^{-40}$ m

The de Broglie wavelength of the electron described in this problem corresponds to the distance between the layers of atoms in a crystal. Therefore, it would be possible to detect the wave nature of the electrons by using a diffraction experiment. The wavelength of the proton and ball is such that the wave nature of these particles would not be detected.

EXAMPLE PROBLEM 13. Determine the Bohr radius of the n = 2 orbit of the hydrogen atom. Determine the angular momentum of an electron in this orbit.

Part a. Step 1.	Solution: (Section 27-12)
Determine the radius of the orbit.	The radius of orbit is given by

$r = (0.53 \times 10^{-10} \text{ m})(n^2)$ where n = 2

$r = (0.53 \times 10^{-10} \text{ m})(2^2)$

$r = 2.12 \times 10^{-10}$ m

Part a. Step 2.	$L = (n\ h)/(2\pi) = [(2)(6.63 \times 10^{-34}\ J\ s)]/(2\pi)$
Determine the angular momentum of the electron.	$L = 2.11 \times 10^{-34}\ J\ s$

PROBLEM SOLVING SKILLS

For problems involving the velocity selector:

1. The force exerted by the electric field on the particle is equal in magnitude to the force exerted by the magnetic field. The direction of the electric force may be determined by the conventions adopted in Chapter 16 and the right-hand rule described in Chapter 20.

2. Use the equation $v = E/B$ to solve for the unknown quantity.

For problems related to the photoelectric effect:

1. Construct a data table listing information both given and requested in the problem. For example, include the work function of the metal, threshold frequency and threshold wavelength for the metal, energy of the incident photons, wavelength of the incident photons, Planck's constant, and the speed of light.

2. Use the equation $E = KE_{max} + W_o$ where $W_o = h\ f_o$ to solve the problem.

For problems related to the Compton effect:

1. Construct a data table listing information both given and requested in the problem. For example, include the wavelength of the incident and scattered photons, the angle between the incident and scattered photons, and the energy of the recoil electron.
2. Use the equation $\lambda' - \lambda = (h/mc)\ (1 - \cos \phi)$ to solve for the unknown quantity.
3. Use the law of conservation of energy to solve for the energy of the recoil electron.

For problems related to pair production:

1. List the mass of the particle and its antiparticle.
2. Determine the minimum energy required for pair production.
3. Use the law of conservation of energy to determine the kinetic energy of the particles after they are produced.

For problems related to the de Broglie wavelength of a particle:

1. List the mass and velocity of the particle.
2. Use $\lambda = h/mv$ to solve for the de Broglie wavelength.

For problems related to emission spectra:

1. List information both given and requested in the problem.
2. Use the Balmer equation to solve for the wavelength of the light emitted when a electron transition occurs between two energy levels.
3. Use $f = c/\lambda$ to determine the frequency of the photon.
4. Use $E = hf$ to determine the energy of the photon.

SOLUTIONS TO SELECTED TEXTBOOK PROBLEMS

TEXTBOOK PROBLEM 3. An oil drop whose mass is determined to be 2.8×10^{-15} kg is held at rest between two large plates separated by 1.0 cm when the potential difference between them is 340 V. How many excess electrons does this drop have?

Part a. Step 1.	Solution: (Section 27-1)
Determine the electric field between the plates. Note: 1.0 cm = 0.010 m.	$V = E\,d$ $E = V/d = (340\ V)/(0.010\ m) = 3.4 \times 10^4$ V/m $= 3.4 \times 10^4$ N/C
Part a. Step 2. Determine the excess charge on the oil drop.	Since the oil drop is at rest, the net force on the drop equals zero. The electric force must be equal to the gravitational force but in the opposite direction. $F_{electric} = F_{gravitational}$ *diagram* *see chapter 17, problem 2* $q\,E = m\,g$ $q = (m\,g)/E$ $q = (2.8 \times 10^{-15}\ kg)(9.8\ m/s^2)/(3.4 \times 10^4\ N/C)$ $q = 8.1 \times 10^{-19}$ C
Part a. Step 3. Determine the number of excess electrons on the oil drop.	The magnitude of the charge on an electron (e) $= 1.60 \times 10^{-19}$ C $n\,e = 8.1 \times 10^{-19}$ C where n = the number of electrons $n\,(1.60 \times 10^{-19}$ C/electron) $= 8.1 \times 10^{-19}$ C $n \approx 5$ electrons

Part a. Step 1.	Solution: (Section 27-2)
Use Wien's law to determine the temperature.	$\lambda_p T = 2.90 \times 10^{-3}$ m·K and re-arranging gives $$T = (2.90 \times 10^{-3} \text{ m·K})/(\lambda_p)$$ $$= (2.90 \times 10^{-3} \text{ m·K})/[(440 \text{ nm})(1.0 \text{ m})/(1 \times 10^9 \text{ nm})]$$ $$T = 6.59 \times 10^3 \text{ K}$$

Part a. Step 1.	Solution: (Section 27-2)
Use Wien's law to determine the peak wavelength for ice.	$\lambda_p T = 2.90 \times 10^{-3}$ m·K 0°C = 273 K $$\lambda_p (273 \text{ K}) = 2.90 \times 10^{-3} \text{ m·K}$$ $$= (2.90 \times 10^{-3} \text{ m·K})/(273 \text{ K})$$ $$\lambda_p = 1.06 \times 10^{-5} \text{ m} = 10600 \text{ nm (infrared)}$$
Part b. Step 1. Determine the peak wavelength for the floodlamp.	$\lambda_p (3500 \text{ K}) = 2.90 \times 10^{-3}$ m·K $$\lambda_p = (2.90 \times 10^{-3} \text{ m·K})/(3500 \text{ K})$$ $$\lambda_p = 8.29 \times 10^{-7} \text{ m} = 829 \text{ nm (infrared)}$$
Part c. Step 1. Determine the peak wavelength for helium.	$\lambda_p (4 \text{ K}) = 2.90 \times 10^{-3}$ m·K $$\lambda_p = (2.90 \times 10^{-3} \text{ m·K})/(4 \text{ K})$$ $$\lambda_p = 7.3 \times 10^{-4} \text{ m} = 7.3 \times 10^5 \text{ nm} = 0.73 \text{ mm (microwave)}$$

Part d. Step 1.	λ_p (2.725 K) = 2.90 x 10^{-3} m·K
Determine the peak wavelength for the universe at 2.725 K.	λ_p = (2.90 x 10^{-3} m·K)/(2.725 K)
	λ_p = 1.06 x 10^{-3} m = 1.06 x 10^6 nm = 1.06 mm (microwave)

Note: visible light extends from 400 nm to 700 nm. Therefore, each wavelength is outside of the visible part of the EM spectrum.

TEXTBOOK PROBLEM 24. The threshold wavelength for emission of electrons from a given surface is 350 nm. What will be the maximum kinetic energy of ejected electrons when the wavelength is changed to (a) 280 nm, (b) 360 nm?

Part a. Step 1.	Solution: (Section 27-3)
Determine the work function of the metal. Note: 350 nm = 3.50 x 10^{-7} m	At the threshold wavelength, the energy absorbed by the electron is used to break free from the attractive forces which hold it to the metal. As a result, it has no kinetic energy when it finally breaks free. Thus
	$E = KE_{max} + W_o$ but $KE_{max} = 0$ J
	$hf = 0$ J $+ W_o$ where $hf = hc/\lambda$
	W_o = [(6.63 x 10^{-34} J s)(3.0 x 10^8 m/s)]/(3.50 x 10^{-7} m)
	W_o = 5.68 x 10^{-19} J or 3.54 eV
	alternate solution:
	$W_o = hf = hc/\lambda$ = 1240 eV nm/λ
	W_o = (1240 eV nm)/(350 nm) = 3.54 eV

Part a. Step 2.	$E = KE_{max} + W_o$
Determine the maximum kinetic energy of electrons emitted by light of wavelength 250 nm.	where E = h f = (1240 eV)/(280 nm) = 4.43 eV
	4.43 eV = KE_{max} + 3.54 eV
	KE_{max} = 0.889 eV

Part b. Step 1.	$E = KE_{max} + W_o$
Determine the maximum kinetic energy of electrons emitted by light of wavelength 350 nm.	where $E = h f = (1240 \text{ eV})/(360 \text{ nm}) = 3.44 \text{ eV}$
	3.44 eV is less than the work function of the metal. Therefore, no electrons will be emitted from the metal.

TEXTBOOK PROBLEM 32. How much total kinetic energy will an electron-positron pair have if produced by a 3.84 MeV photon?

Part a. Step 1.	Solution: (Section 27-6)
Use the law of conservation of energy to determine the minimum energy of the photon required to produce the electron-proton pair.	Using the law of conservation of energy, the minimum energy is equal to the sum of the rest energies of the two particles.
	$E = mc^2 = m_{electron}c^2 + m_{positron}c^2$
	$E = (9.1 \times 10^{-31} \text{ kg})(9.0 \times 10^{16} \text{ m}^2/\text{s}^2) +$
	$\qquad\qquad\qquad (9.1 \times 10^{-31} \text{ kg})(9.0 \times 10^{16} \text{ m}^2/\text{s}^2)$
	$E = (1.64 \times 10^{-13} \text{ J})/(1 \text{ MeV}/1.6 \times 10^{-13} \text{ J}) = 1.02 \text{ MeV}$
Part a. Step 2.	Total energy = kinetic energy + rest energy
Determine the total kinetic energy of the pair.	3.84 MeV = KE + 1.02 MeV
	$KE_{total} = 2.82 \text{ MeV}$

TEXTBOOK PROBLEM 46. What voltage is needed to produce electron wavelengths of 0.20 nm? (Assume that the electrons are nonrelativistic.)

Part a. Step 1.	Solution: (Section 27-9)
Use the de Broglie formula to determine the speed of the electrons in meters per second.	$\lambda = h/(mv)$
	$v = h/(m\lambda)$
	Note: nonrelativistic mass of an electron = 9.11×10^{-31} kg
	$v = (6.63 \times 10^{-34} \text{ J s})/[(9.11 \times 10^{-31} \text{ kg})(0.20 \times 10^{-9} \text{ m})]$
	$v = 3.63 \times 10^6 \text{ m/s}$

Part a. Step 2.	$W = q\,V = \Delta KE$
Determine the potential difference required to accelerate an electron to this energy.	$(1.6 \times 10^{-19} \text{ C})\,V = \tfrac{1}{2}\,(9.11 \times 10^{-31} \text{ kg})\,(3.63 \times 10^{6} \text{ m/s})^{2} - 0 \text{ J}$
	$V = 37.5$ volts

CHAPTER 28

QUANTUM MECHANICS OF ATOMS

OBJECTIVES

After studying the material of this chapter, the student should be able to:

- distinguish between the Bohr model and the quantum mechanical model of the atom.
- state two forms of the Heisenberg uncertainty principle and explain how the principle predicts an inherent unpredictability in nature. Use the uncertainty principle to compute the minimum uncertainty of a molecule's momentum or position.
- name the four quantum numbers required to describe the state of an electron in an atom. State the symbol used to represent each quantum number.
- given a value for the principle quantum number, list the range of values for the other quantum numbers.
- state the Pauli exclusion principle. Use this principle to determine the maximum number of electrons that fill the energy levels of atoms where n = 1 or n = 2.
- given the atomic number of a particular element, write the electronic configuration for the ground state of the atom.
- use a periodic table to identify an element whose outer electronic configuration is given.
- describe two ways X-ray photons can be produced.
- state the Bragg equation. Use the equation to determine either the wavelength of the X-rays incident on a crystal or the distance between the layers of atoms which make up the crystal.
- determine the cut-off frequency and wavelength of an X-ray photon produced by accelerating electrons through a known potential difference.
- determine the wavelength of a K_α X-ray of known energy. In addition, use Moseley's equation to determine the atomic number of the element which produced this X-ray.

KEY WORDS AND PHRASES

quantum mechanics or **wave mechanics** unified the wave-particle duality into a single consistent theory.

Heisenberg uncertainty principle is an important result of quantum mechanics. This principle results from the wave-particle duality and an intrinsic limit in our ability to make accurate measurements. One form of the uncertainty principle states that it is impossible to know simultaneously both the precise position and momentum of a particle.

quantum numbers   The state of an electron in the hydrogen atom is governed by four quantum numbers. The quantum numbers are n, ℓ, m_ℓ, and m_s.   n is called the principle quantum number, where n = 1, 2, 3, 4, etc. ℓ is the orbital quantum number. m_ℓ is the magnetic quantum number. m_s is the spin quantum number.

Pauli exclusion principle is used to explain the arrangement of electrons in multi-electron atoms. This principle states that "no two electrons in an atom can occupy the same quantum state." Thus each electron has a unique set of quantum numbers: n, ℓ, m_ℓ, and m_s.

electronic configuration of the elements listed in the periodic table of the elements can be specified using the n and ℓ quantum numbers. Electrons with the same value of ℓ are in the same subshell within the main shell designated by the letter n. The subshells are designated by the letters s, p, d, f, etc.

X-rays exhibit properties of electromagnetic waves of very short wavelength. X-rays can be produced in two ways. One method is for high-energy electrons to knock an electron out of an inner energy level of certain atoms. When an electron drops from a higher level to a lower level an X-ray photon is emitted. The second method is bremsstrahlung or braking radiation. In this method, the electron is deflected as it passes near the nucleus of the atom.

SUMMARY OF MATHEMATICAL FORMULAS

Heisenberg uncertainty principle	$(\Delta x)(\Delta p) \geq h/2\pi$	It is impossible to know simultaneously both the precise position (Δx) and momentum of a particle (Δp). The product of uncertainty of the position (Δx) and the uncertainty of the momentum (Δp) must be greater than or equal to Plank's constant divided by 2π.
	$(\Delta E)(\Delta t) \geq h/2\pi$	Another form of the uncertainty principle states that the product of the uncertainty of energy (ΔE) and the uncertainty in time (Δt) must be greater than or equal to Planck's constant divided by 2π.

principle quantum numbers	n = 1, 2, 3, etc.	n is called the principle quantum number, where n = 1, 2, 3, 4, etc.
	$\ell \leq n - 1$	ℓ is the orbital quantum number. ℓ can take on integer values up to n - 1. For example, if n = 3, then ℓ can have the following values: 0, 1, 2.
	$-\ell \leq m_\ell \leq +\ell$	m_ℓ is the magnetic quantum number. It is related to the direction of the electron's angular momentum. m_ℓ is an integer and can have values from $-\ell$ to $+\ell$.
	$m_s = +\frac{1}{2}$ or $-\frac{1}{2}$	m_s is the spin quantum number. It is related to the spin angular momentum
electronic configuration of the elements	The order of filling is as follows: $1s^2, 2s^2, 2p^6, 3s^2, 3p^6, 4s^2, 3d^{10}, 4p^6, 5s^2, 4d^{10}, 5p^6, 6s^2, 4f^{14}, 5d^{10}$, etc.	The s orbital holds up to 2 electrons, the p orbital up to 6 electrons, the d orbital up to 10 electrons, and the f orbital up to 14 electrons.
cut-off wavelength of X-ray photons	$\lambda_o = (hc)/(eV)$	The cut-off wavelength (λ_o) is the shortest wavelength X-ray produced, e is the charge on the electron, and V is the accelerating voltage.
Moseley's formula	$\lambda = (1.22 \times 10^{-7} \text{ m})/(Z - 1)^2$	The wavelength (λ) of the K_α line is related to the atomic number (Z) of the atoms of the metal target.

CONCEPTS AND EQUATIONS

Quantum Mechanics

About 1925, Erwin Schrodinger and Werner Heisenberg produced a new theory, called wave mechanics or **quantum mechanics**, which unified the wave-particle duality into a single consistent theory.

Applied to the atom, quantum mechanics pictures the electron as spread out in space in the form of a cloud of negative charge. The shape and size of the electron cloud can be mathematically determined for a particular state of an atom. For the hydrogen atom, the shape of the electron cloud is spherically symmetric about the nucleus. The cloud model can be interpreted as an electron wave spread out in space or as a probability distribution for electrons as particles.

Quantum mechanics has been used to explain phenomena such as spectra emitted by complex atoms, the relative brightness of spectral lines, and even how atoms form molecules. Quantum mechanics reduces to classical physics in instances where classical physics applies. Newtonian mechanics is a special case of quantum mechanics. Even Bohr's postulate on the quantization of angular momentum of the electron in the hydrogen atom can be shown to be a special case of the more general quantum mechanics.

Heisenberg Uncertainty Principle

Newtonian mechanics implies that if an object's position and momentum are known at a particular moment in time, and if the forces that are acting on it or will be acting on it are known, then its future position can be predicted. This idea is referred to as determinism.

However, an important result of quantum mechanics is the **uncertainty principle.** This principle results from the wave-particle duality and an intrinsic limit in our ability to make accurate measurements. One form of the uncertainty principle states that it is impossible to know simultaneously both the precise position and momentum of a particle. Expressed mathematically, the product of uncertainty of the position (Δx) and the uncertainty of the momentum (Δp) must be greater than or equal to Planck's constant divided by 2π, i.e., $(\Delta x)(\Delta p) \geq h/2\pi$.

Another form of the uncertainty principle states that the product of the uncertainty of energy (ΔE) and the uncertainty in time (Δt) must be greater than or equal to Plank's constant divided by 2π, i.e., $(\Delta E)(\Delta t) \geq h/2\pi$.

Thus, unlike Newtonian mechanics, quantum mechanics states that only approximate predictions are possible and that there is an inherent unpredictability in nature.

EXAMPLE PROBLEM 1. A nitrogen molecule travels at 1200 miles per hour (536 m/s) at 0°C. Suppose the uncertainty in an experimental measurement of its speed is ±25.0 m/s. Compute the minimum uncertainty in its position.

Part a. Step 1.	Solution: (Section 28-3)
Determine the mass nitrogen molecule.	Using the periodic table, the molecular weight of diatomic of a nitrogen is 28 grams/mole.
	$m = (28 \text{ g/mole})(1 \text{ mole}/6.02 \times 10^{23} \text{ molecules})$
	$m = 4.7 \times 10^{-23} \text{ g} = 4.7 \times 10^{-26} \text{ kg}$

Part a. Step 2.	The uncertainty in the measurement of its speed is ± 25.0 m/s. The uncertainty in the molecule's speed (Δv) can be found as follows $\Delta v = 25.0$ m/s - -25.0 m/s = 50.0 m/s
Determine the uncertainty in the molecule's momentum (Δp).	$\Delta p = m \, \Delta v$
	$\qquad = (4.7 \times 10^{-26} \text{ kg})(50.0 \text{ m/s})$
	$\Delta p = 2.35 \times 10^{-24}$ kg m/s

Part a. Step 3.	Using the Heisenberg uncertainty principle,
Determine the minimum uncertainty in the molecule's position (Δx).	$(\Delta x)(\Delta P) \geq h/2\pi$
	$(\Delta x)(2.35 \times 10^{-24} \text{ kg m/s}) \geq (6.63 \times 10^{-34} \text{ J s})/(2\pi)$
	$\Delta x \geq 4.49 \times 10^{-11}$ m.

Quantum Mechanics of the Hydrogen Atom; Quantum Numbers

The state of an electron in the hydrogen atom is governed by four quantum numbers: n, ℓ, m_ℓ, m_s. The energy of a particular level in the hydrogen atom is related to n by the equation $E = (-13.6 \text{ eV})/n^2$. n is retained in quantum mechanics and is called the **principle quantum number**, where $n = 1, 2, 3, 4$, etc.

ℓ is the **orbital quantum number**. The angular momentum (L) is related to the orbital quantum number by the formula $L = [\ell(\ell + 1)]^{1/2} (h/2\pi)$. ℓ can take on integer values up to $n - 1$. For example, if $n = 3$, then ℓ can have the following values: 0, 1, 2.

m_ℓ is the **magnetic quantum number**. It is related to the direction of the electron's angular momentum. m_ℓ is an integer and can have values from $-\ell$ to $+\ell$. The component of the angular momentum in an assigned direction, usually along the z axis, is given by $L_z = m_\ell (h/2\pi)$.

m_s is the **spin quantum number**. This quantum number only has values of $+\frac{1}{2}$ or $-\frac{1}{2}$. The spin angular momentum in an assigned direction equals $m_s(h/2\pi)$. The word "spin" was originally given to this quantum number because it was thought to be associated with the electron spinning on its axis as it revolves around the nucleus. However, this is an oversimplification because the electron exhibits properties of waves as well as particles.

Pauli Exclusion Principle

In order to explain the arrangement of electrons in multi-electron atoms, the **Pauli exclusion principle** is used. This principle states that "no two electrons in an atom can occupy the same quantum state." Thus, each electron has a unique set of quantum numbers: n, ℓ, m_ℓ, and m_s. For example, helium has two electrons. Both electrons have $n = 1$, $\ell = 0$, and $m_\ell = 0$. However, one of the electrons has $m_s = +\frac{1}{2}$ and the other has $m_s = -\frac{1}{2}$. Thus each electron has a different set of quantum numbers.

Electronic Configuration of the Elements

The **electronic configuration** of the elements listed in the periodic table of the elements can be specified using the n and ℓ quantum numbers. Electrons with the same value of n are in the same shell. The shells are given letter symbols as shown in the table shown at the top of the next page. If $n = 1$, then the electrons are in the K shell.

As shown in the table, electrons with the same value of ℓ are in the same subshell within the main shell designated by the letter n. The subshells are designated by the letters s, p, d, f, etc.

The number of electrons in the subshell can be found by using the formula $2(2\ell + 1)$. Thus, if $\ell = 3$, then $2[2(3) + 1] = 14$ and the f subshell can hold 14 electrons.

The designation of an electron involves both n and ℓ plus a superscript which designates the number of electrons in the subshell. For example, if $n = 2$ and $\ell = 1$ and there are three electrons in the orbital, then the designation would be $2p^3$.

value of n	symbol of subshell	value of ℓ	symbol of subshell	maximum number of electrons
1	K	0	s	2
2	L	1	p	6
3	M	2	d	10
4	N	3	f	14
.	.	4	g	18
.	.	5	h	32
.	.			

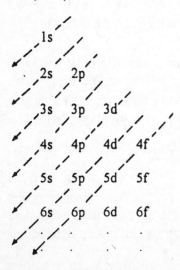

The order of filling is as follows: $1s^2$, $2s^2$, $2p^6$, $3s^2$, $3p^6$, $4s^2$, $3d^{10}$, $4p^6$, $5s^2$, $4d^{10}$, $5p^6$, $6s^2$, $4f^{14}$, $5d^{10}$, etc. A simplified way of determining the order of filling is shown in the diagram.

Write down the principle energy levels and their subshells and follow the diagonal lines. The diagonal lines follow the order of filling.

In filling the subshells, the lower energy subshells are filled first, as are the lower principle energy levels.

Each box in the periodic table contains the symbol of the element, its atomic number, its atomic mass, and the ground state electronic configuration of the outermost electrons. For example, the symbol for calcium is Ca, 20 is the atomic number, 40.08 is the atomic mass and $4s^2$ is the electronic configuration of the outermost electrons.

Ca	20
40.08	
$4s^2$	

TEXTBOOK QUESTION 15. Which of the following electron configurations are not allowed: a) $1s^2 2s^2 2p^4 3s^2 4p^2$; b) $1s^2 2s^2 2p^8 3s^1$; c) $1s^2 2s^2 2p^6 3s^2 3p^5 4s^2 4d^5 4f^1$? If not allowed, explain why.

ANSWER: a) Configuration (a) is forbidden because the 2p orbital fills to six electrons before the 3s orbital begins to fill. b) The 2p orbital contains a maximum of six electrons. Therefore, configuration (b) is forbidden. c) The 3p orbital must fill to six electrons before the 4s begins to fill; therefore, configuration (c) is forbidden.

TEXTBOOK QUESTION 17. In what column of the periodic table would you expect to find the atom with each of the following configurations: a) $1s^2 2s^2 2p^6 3s^2$; b) $1s^2 2s^2 2p^6 3s^2 3p^6$; c) $1s^2 2s^2 2p^6 3s^2 3p^6 4s^1$; d) $1s^2 2s^2 2p^5$?

ANSWER: The periodic table of the elements is given in the inside cover located in the back of the textbook. The table lists the outermost electrons (also known as valence electrons) for each atom. The outermost electrons for element a) is $3s^2$; therefore, the element is magnesium (Mg) which is in Group II. Element b) is $3s^2 3p^6$, which is the element argon (Ar). Argon is in Group VIII. Element c) is $4s^1$, which is the element potassium (K). Potassium is in Group I. Element d) is $2s^2 2p^5$, which is the element fluorine (F). Fluorine is in Group VII.

TEXTBOOK QUESTION 18. Why do chlorine and iodine exhibit similar properties?

ANSWER: Using the periodic table located in the text, it can be seen that the outer electron configuration of both elements consists of $s^2 p^5$ electrons. Chlorine is $3s^2 3p^5$ while iodine is $5s^2 5p^5$. The physical and chemical properties of atoms are primarily determined by the outer electronic configuration, also known as the valence electrons. Therefore, both elements exhibit similar properties and are said to belong to the same chemical family, in this case the halogen family.

EXAMPLE PROBLEM 2. List the possible quantum states for $_{10}Ne$.

Part a. Step 1.

Use the Pauli exclusion
principle to list the
possible quantum states.

Solution: (Sections 28-6 to 28-8)

The Pauli exclusion principle states, "no two electrons in an atom
can occupy the same quantum state." Each atom must have a
unique set of quantum numbers: n, ℓ, m_ℓ, and m_s.

Based on the exclusion principle it is possible to construct a table
of possible quantum states for $_{10}Ne$.

sub-shell	n	ℓ	m	m_s	sub-shell	n	ℓ	m	m_s
1s	1	0	0	+½	2p	2	1	-1	-½
1s	1	0	0	-½	2p	2	1	0	+½
2s	2	0	0	+½	2p	2	1	0	-½
2s	2	0	0	-½	2p	2	1	1	+½
2p	2	1	-1	+½	2p	2	1	1	-½

EXAMPLE PROBLEM 3. Determine the values of m_ℓ that are allowed for the a) 1s subshell and
b) 3d subshell.

Part a. Step 1.

Determine the values
for the 1s subshell.

Solution: (Section 28-6 to 28-8)

For the 1s subshell, $n = 1$, and since $\ell \leq n - 1$, then $\ell = 0$. m_ℓ can
have values from $-\ell$ to $+\ell$, but because $\ell = 0$, $m_\ell = 0$.

Part b. Step 1.

Determine the values
for the 3d subshell.

For the 3d subshell, $n = 3$; therefore, $\ell = 1$ or 2, and since m can
have values from $-\ell$ to $+\ell$, m_ℓ can have the values -2, -1, 0, +1, +2.

EXAMPLE PROBLEM 4. What is the range of values of the angular momentum of an electron
in the n = 4 state of the hydrogen atom?

Part a. Step 1.	Solution: (Section 28-6)
Determine the possible values of the orbital quantum number. Then determine the range of values of the angular momentum.	n = 4 and ℓ can take on integer values from 0 to n - 1; then ℓ can be 0, 1, 2, 3. The range of values for the angular momentum (L) can be determined by substituting the possible values of ℓ into the following equation:

$$L = [\ell(\ell + 1)]^{\frac{1}{2}} \, h/2\pi$$

$$\ell = 0, \quad L = [0(0 + 1)]^{\frac{1}{2}} \, h/2\pi = 0$$

$$\ell = 1, \quad L = [1(1 + 1)]^{\frac{1}{2}} \, (6.63 \times 10^{-34} \text{ J s})/2\pi = 1.49 \times 10^{-34} \text{ J s}$$

$$\ell = 2, \quad L = [2(2 + 1)]^{\frac{1}{2}} \, (6.63 \times 10^{-34} \text{ J s})/2\pi = 2.58 \times 10^{-34} \text{ J s}$$

$$\ell = 3, \quad L = [3(3 + 1)]^{\frac{1}{2}} \, (6.63 \times 10^{-34} \text{ J s})/2\pi = 3.66 \times 10^{-34} \text{ J s}$$

EXAMPLE PROBLEM 5. a) Write the electronic configuration for each of the following elements: $_{11}$Na, $_{17}$Cl, $_{20}$Ca, and $_{21}$Sc.

Part a. Step 1.	Solution: (Section 28-8)
Write the electronic configuration for each element.	The order of filling of the subshells is

$$1s^2 \; 2s^2 \; 2p^6 \; 3s^2 \; 3p^6 \; 4s^2 \; 3d^{10}, \text{ etc.}$$

Sodium (Na) has 11 electrons, and its electronic configuration is

$$1s^2 \; 2s^2 \; 2p^6 \; 3s^1$$

Using the same method, the other elements can be shown to have the following configurations:

$$_{17}\text{Cl} \quad 1s^2 \; 2s^2 \; 2p^6 \; 3s^2 \; 3p^5$$

$$_{20}\text{Ca} \quad 1s^2 \; 2s^2 \; 2p^6 \; 3s^2 \; 3p^6 \; 4s^2$$

$$_{21}\text{Sc} \quad 1s^2 \; 2s^2 \; 2p^6 \; 3s^2 \; 3p^6 \; 4s^2 \; 3d^1$$

EXAMPLE PROBLEM 6. Write the symbols for the elements whose outer electron configurations are as follows: $4s^2 \; 4p^3$, $3d^7 \; 4s^2$, and $5s^1$.

Part a. Step 1.	Solution: (Section 28-8)
Use the periodic table at the back of the text to determine the symbol for each element.	The periodic table in the textbook specifies the configuration of the outermost electrons and any other nonfilled subshells. Using the periodic table as a guide, it can be seen that arsenic (As) has an outer configuration of $4s^2 4p^3$.
	In the same manner, the symbol of each of the other elements can now be determined.
	$3d^7 4s^2$; the element is cobalt (Co).
	$5s^1$; the element is rubidium (Rb).

X-rays and X-ray Production

X-rays were discovered in 1895 by Wilhelm Roentgen. The nature of X-rays was not determined until 1913 when it was shown that X-rays exhibit properties of electromagnetic waves of very short wavelength.

X-rays can be produced in two ways. One method is for high-energy electrons to knock an electron out of an inner energy level of certain atoms. When an electron drops from a higher level to a lower level an X-ray photon is emitted. The second method is a continuous spectrum called **bremsstrahlung** or **braking radiation**. In this method, the electron is deflected as it passes near the nucleus of the atom. During the resulting deceleration, energy in the form of an X-ray is produced.

An X-ray tube produces a spectrum of wavelengths. The shortest wavelength X-ray is the result of the electron losing all of its kinetic energy during the collision. In this case,

energy lost by electron = energy gained by X-ray photon

$e V = h f_o$ and because $f_o = c/\lambda_o$, then $\lambda_o = (hc/e)V$. f_o is the cut-off frequency. f_o is the highest frequency X-ray produced. λ_o is the cut-off wavelength and is the shortest wavelength X-ray produced, e is the charge on the electron and V is the accelerating voltage.

EXAMPLE PROBLEM 7. The wavelength of the strongest X-ray spectral line from an X-ray tube with a copper target is 0.100 nm. Determine the minimum accelerating voltage required to give an electron enough energy to produce this X-ray.

Part a. Step 1.	Solution: (Section 28-9)
Determine the minimum accelerating voltage necessary for an electron to produce an X-ray photon of wavelength 0.100 nm.	Assume that all of the electron's energy is converted into the energy of the X-ray photon upon collision with the copper target.

kinetic energy of the electron $\quad=\quad$ energy of the X-ray photon

$$q\ V = h\ c/\lambda = 1240\ eV\ nm/\lambda$$

$$(1\ electron)\ V = (1240\ eV\ nm)/(0.100\ nm)$$

$$V = 12400\ volts$$

X-ray Diffraction

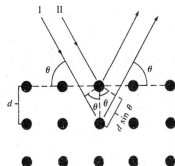

The wavelength of X-rays is comparable to the spacing of atoms in crystals such as sodium chloride (NaCl). If an x-ray beam is incident on the crystal, the subsequent reflection of the X-rays from the planes of atoms results in a diffraction pattern. Based on the diagram, constructive interference will occur if the extra distance, known as the path difference, that ray I travels before rejoining ray II equals a whole number of wavelengths. The path difference can be shown to be equal to $2\ d\ \sin \theta$ and therefore constructive interference occurs when $m\lambda = 2\ d\ \sin \theta$ (Bragg equation).

$m\lambda$ is a whole number of wavelengths and m = 1, 2, 3, etc., d is the distance between the layers of atoms, and θ is the angle of incidence of the X-ray beam with the surface of the crystal.

X-rays and Atomic Number

One method of producing X-rays is for high-energy electrons to knock an electron out of an inner shell of certain atoms. If the missing electron was in the K shell (n = 1) and is replaced by an electron which falls from the L shell (n = 2), then the X-ray is referred to as an K_α X-ray.

In 1914, Henry Moseley determined that the wavelength of the K_α x-ray line followed an empirical formula similar to Balmer's equation. Moseley's formula for the K_α line is as follows:

$$1/\lambda = R(Z - 1)^2 (1/1^2 - 1/2^2) \qquad \text{and upon rearranging this equation becomes}$$

$$\lambda = (1.22 \times 10^{-7}\ m)/(Z - 1)^2$$

λ is the wavelength of the K_α X-ray and Z is the atomic number of the atoms of the metal target.

Moseley determined that a graph of $(1/\lambda)^{\frac{1}{2}}$ vs. Z produces a straight line. Because of this, the atomic number of a number of elements could be determined and his research led to the arrangement of the periodic table on the basis of atomic number. Moseley enlisted in the British army and was sent to Turkey. He was killed at Gallipoli on August 10, 1915, at the age of 27.

EXAMPLE PROBLEM 8. The K_α X-ray of copper has an energy of 8000 eV. Determine the a) wavelength of this X-ray, and b) atomic number of Cu.

Part a. Step 1. Determine the wavelength of this X-ray.	Solution: (Section 28-9) $E = h\,f = h\,c/\lambda = 1240 \text{ eV nm}/\lambda$ $8000 \text{ eV} = 1240 \text{ eV nm}/\lambda;\ \lambda = 1240 \text{ eV nm}/8000 \text{ eV}$ $\lambda = 0.16 \text{ nm} = 1.6 \times 10^{-10} \text{ m}$
Part b. Step 1. Use Moseley's formula to determine the atomic number of the copper target.	The wavelength of a K_α X-ray is related to the atomic number of the target metal by the following equation: $\lambda = (1.22 \times 10^{-7} \text{ m})/(Z - 1)^2$ and rearranging gives $(Z - 1)^2 = (1.22 \times 10^{-7} \text{ m})/\lambda$ $(Z - 1)^2 = (1.22 \times 10^{-7} \text{ m})/(1.6 \times 10^{-10} \text{ m})$ $(Z - 1)^2 = 760$ and $Z - 1 = 28$ $Z = 29$. The atomic number of copper is 29.

PROBLEM SOLVING SKILLS

For problems involving the uncertainty principle:

1. Determine the object's mass.
2. If necessary, determine the uncertainty in the object's speed.
3. Determine the uncertainty in the molecule's momentum.
4. Use the uncertainty principle to determine the uncertainty in the object's position.
5. If the uncertainty in position is given, use the uncertainty principle to determine the uncertainty in the object's momentum and velocity.

6. If the problem involves uncertainty in energy, then note either the uncertainty in the energy or time and solve the problem using the equation $(\Delta E)(\Delta t) \geq h/2\pi$.

For problems involving the possible quantum states of an atom:

1. Note the value of the principle quantum number (n) for the atom. For example, if an atom is in the third period of the periodic table, then the value of n = 3.
2. Determine the possible values of ℓ. Remember that ℓ can have positive values up to n - 1.
3. Determine the possible values of m_ℓ. Remember that m_ℓ can have values of $-\ell$ to $+\ell$.
4. m_s can have values of $+\frac{1}{2}$ or $-\frac{1}{2}$.
5. Use the Pauli exclusion principle to construct a table for the possible quantum states.

For problems involving the range of values of the angular momentum of an electron:

1. Note the principle quantum number (n) of the particular orbital in which the electron is located.
2. Determine the range of values of the orbital quantum number (ℓ).
3. Use the equation $L = [\ell(\ell + 1)]^{1/2} \, h/2\pi$ to determine the possible values of the angular momentum (L).

For problems involving the electronic configuration of an element:

1. Write down the order of filling of the s, p, d, and f subshells and the maximum number of electrons in each subshell. Hint: rather than attempting to memorize the order of filling, commit the table on page 28-5 to memory.
2. Take note of the atomic number of the element.
3. Write down the electronic configuration until the number of electrons in the atom is reached.
4. If the outer electron configuration is given, the element can be easily identified by using a periodic table. For example, if the outer configuration is $4s^2 \, 4p^1$, then the element is located in period 4 and is a member of family IIIA. This element is gallium (Ga).

For problems involving x-ray spectra and atomic number:

1. If the accelerating voltage is given, then determine the electron's kinetic energy in electron volts ($W = \Delta KE = q \, \Delta V$).
2. The highest energy and shortest wavelength X-ray photon is produced if all of the electron's kinetic energy is used in producing the X-ray photon. Use the equation $q \, \Delta V = h \, c/\lambda$ to determine the wavelength of the X-ray.
3. If the wavelength of the shortest X-rays is given, then the equation $q \, \Delta V = h \, c/\lambda$ can be used to determine the voltage needed to give an electron the energy to produce the X-ray.
4. If the wavelength of the K_α X-ray is given, then the formula $\lambda = (1.22 \times 10^{-7} \, m)/(Z - 1)^2$ is used to determine the atomic number (Z) of the target element.

SOLUTIONS TO SELECTED TEXTBOOK PROBLEMS

TEXTBOOK PROBLEM 3. A proton is traveling with a speed of $(6.560 \pm 0.012) \times 10^5$ m/s. With what maximum accuracy can its positioned by ascertained? {*Hint*: $\Delta p = m \, \Delta v$.]

Part a. Step 1. Determine the uncertainty in the proton's speed.	Solution: (Section 28-3) The uncertainty in the proton's speed (Δv) is $\Delta v = 0.012 \times 10^5$ m/s
Part a. Step 2. Determine the uncertainty in the proton's momentum.	$\Delta p = m \, \Delta v$ $\quad = (1.67 \times 10^{-27}$ kg$)(0.012 \times 10^5$ m/s$)$ $\Delta p = 2.00 \times 10^{-24}$ kg m/s
Part a. Step 3. Determine the maximum accuracy the proton's position can be ascertained.	Using the Heisenberg uncertainty principle, $(\Delta x)(\Delta p) \geq h/2\pi$ $(\Delta x)(2.00 \times 10^{-24}$ kg m/s$) \geq (6.63 \times 10^{-34}$ J s$)/(2\pi)$ $\Delta x \geq 5.27 \times 10^{-11}$ m Therefore, the maximum accuracy of the proton's position is $\pm 5.27 \times 10^{-11}$ m

TEXTBOOK PROBLEM 8. A free neutron ($m = 1.67 \times 10^{-27}$ kg) has a mean life of 900 s. What is the uncertainty in its mass (in kg)?

Part a. Step 1. Use the Heisenberg uncertainty principle to determine the uncertainty in the neutron's energy.	Solution: (Section 28-3) $(\Delta E)(\Delta t) \geq h/2\pi$ $(\Delta E)(900$ s$) \geq (6.626 \times 10^{-34}$ J s$)/(2\pi)$ $\Delta E \geq (6.626 \times 10^{-34}$ J s$)/[(2\pi)(900$ s$)]$ $\Delta E \geq 1.17 \times 10^{-37}$ J

Part a. Step 2.	$\Delta E = (\Delta m)c^2$
Use the mass-energy equation to determine the uncertainty in the mass.	$\Delta m = (\Delta E)/c^2$
	$\Delta m = (1.17 \times 10^{-37} \text{ J})/(3.0 \times 10^8 \text{ m/s})^2$
	$\Delta m = 1.30 \times 10^{-54}$ kg

TEXTBOOK PROBLEM 15. How many electrons can be in the n = 6, ℓ = 3 subshell?

Part a. Step 1.	Solution: (Sections 26-6 to 28-8)
Use the Pauli exclusion principle to determine the maximum number of electrons in the n = 6, ℓ = 3 subshell.	According to the Pauli exclusion principle, the maximum number of electrons depends on the value of ℓ. For any value of ℓ, there are $2\ell + 1$ different m_ℓ values and for each m_ℓ value there are two values of m_s. Therefore, the maximum number of electrons in a ℓ subshell the maximum number of electrons = $2(2\ell + 1)$.
	Since ℓ = 3, then the maximum number of electrons equals $2[(2)(3) + 1] = 14$.

TEXTBOOK PROBLEM 30. What are the shortest wavelengths X-rays emitted by electrons striking the face of a 33.5-kV TV picture tube? What are the longest wavelengths?

Part a. Step 1.	Solution: (Section 28-9)
Determine the wavelength of the shortest wavelength X-rays produced.	The energy of the shortest wavelength X-rays equals the maximum kinetic energy of the electrons striking the picture tube. Therefore,
	$E = h f = h c/\lambda = 1240$ eV nm/λ
	$(33.5 \text{ KeV})[(1000 \text{ eV})/(1 \text{ KeV})] = 1240$ eV nm/λ
	$\lambda = (1240 \text{ eV nm})/(33500 \text{ eV})$
	$\lambda = 0.0370$ nm
Part a. Step 2.	An X-ray tube produces a spectrum of wavelengths. The shortest wavelength X-rays result when the electron loses all of its kinetic energy during the collision. This occurs when the high energy electron knocks an electron out of an inner energy level of the atom. When an electron drops from a higher level to a lower level an X-ray photon is emitted. A continuous spectrum called bremsstrahlung or braking radiation is produced when the electron is deflected as it passes near the nucleus of the atom. During the
Determine the wavelength of the longest wavelength X-ray produced.	

resulting deceleration, energy in the form of an X-ray is produced. The wavelength of the longest wavelength X-ray would be approximately 1 nm, which is the upper limit of the X-ray portion of the EM spectrum.

TEXTBOOK PROBLEM 34. Estimate the wavelength for an $n = 2$ to $n = 1$ transition in iron $(Z = 26)$.

Part a. Step 1.	Solution: (Section 28-9)
Use Moseley's formula to determine the wavelength of the X-ray.	$1/\lambda = R(Z - 1)^2 (1/1^2 - 1/2^2)$
	$1/\lambda = (1.097 \times 10^7 \text{ m}^{-1})(26 - 1)^2 (1/1^2 - 1/2^2)$
	$1/\lambda = (1.097 \times 10^7 \text{ m}^{-1})(625)(1 - 0.25)$
	$1/\lambda = 5.14 \times 10^9 \text{ m}^{-1}$
	$\lambda = 1.94 \times 10^{-10} \text{ m} = 0.194 \text{ nm}$

TEXTBOOK PROBLEM 43. What are the largest and smallest possible values for the angular momentum L of an electron in the $n = 5$ shell?

Part a. Step 1.	Solution: (Section 28-6)
Determine the possible values for the orbital quantum number ℓ.	ℓ can have values from 0 to $n - 1$. Therefore, since $n = 5$, then the possible values of ℓ are 0, 1, 2, 3, 4.

Part a. Step 2.	
Determine the smallest possible value for the angular moment L.	$L = [\ell(\ell + 1)]^{\frac{1}{2}} (h/2\pi)$. The smallest value of $\ell = 0$
	$\quad = [0(0 + 1)]^{\frac{1}{2}} (h/2\pi)$
	$L = 0 \text{ kg m}^2/\text{s}$

Part a. Step 3.	
Determine the largest possible value for the angular moment L.	$L = [\ell(\ell + 1)]^{\frac{1}{2}} (h/2\pi)$. The largest value of $\ell = 4$.
	$\quad = [4(4 + 1)]^{\frac{1}{2}} [(6.63 \times 10^{-34} \text{ J s})/(2\pi)]$
	$\quad = (4.47)(1.05 \times 10^{-34} \text{ kg m}^2/\text{s})$
	$L = 4.72 \times 10^{-34} \text{ kg m}^2/\text{s}$

CHAPTER 29

MOLECULES AND SOLIDS

OBJECTIVES

After studying the material of this chapter, the student should be able to:

- distinguish between a covalent bond, polar covalent bond, and an ionic bond.
- distinguish between a strong bond and a weak bond.
- explain what is meant by the bond length and the binding energy.
- calculate the potential energy between ions separated by a given distance and determine the magnitude of the electrostatic force acting between them.
- draw a potential energy vs. distance of separation diagram for point charges when the charges are either of the same sign or the opposite sign.
- draw a potential energy vs. distance of separation diagram for two atoms which come together to form a covalent bond or an ionic bond. The diagram should include a section to represent the activation energy.
- explain what is meant by the activation energy.
- identify the point on the potential energy vs. distance of separation curve where the potential energy is a minimum and based on the identification explain what is meant by bond length and binding energy.
- explain why molecular spectra appear in the form of band spectra rather than line spectra.
- determine the difference in energy between rotational and/or vibrational energy states of a diatomic molecule. Also, determine the moment of inertia and the bond length between the atoms of the molecule.
- use the band theory of solids to explain the classification of solids into conductors, insulators, and semiconductors.
- calculate the energy gap between the valence band and the conduction band for a semiconductor or insulator if the wavelength of the longest wavelength photon which causes a transition is given.
- explain why semiconductors become better electrical conductors as the temperature increases.
- use the idea of electronic configuration to explain how a "doped" semiconductor can become highly conducting.
- explain the principle of the junction diode and how the junction diode can be used as a half-wave rectifier.

KEY TERMS AND PHRASES

chemical bonds refer to the forces that hold the atoms of a molecule together.

pure covalent bond is the type of chemical bond in which the electrons are shared equally.

polar covalent bond refers to the type of bond in which the electrons which form the bond are not shared equally. One end of the molecule is charged positively while the other end is charged negatively; the molecule is called a **polar molecule**.

ionic bond is formed when one or more electrons are transferred from one atom to another. The bond formed is based on the electrostatic attraction of the negatively charged ion for the positively charged ion.

bond energy or **binding energy** refers to the energy required to break a chemical bond which holds the atoms of a molecule together. The bonds which hold atoms of a molecule together are called "**strong**" bonds.

"**weak**" **bond** or **Van der Waals bond** usually refers to electrostatic attraction between molecules. An example of a weak bond is between two dipoles and such a bond is often called a **dipole-dipole bond**.

activation energy refers to the energy which must be added in order to force atoms together to form a molecule.

molecular spectra or **band spectra** is exhibited by molecules. This is because molecules have additional energy levels due to the vibration of the atoms of the molecule with respect to each other and the rotational energy of the molecule.

n-type semiconductors are semiconductors where electrons (negative charge) carry the current.

p-type semiconductors are semiconductors where positive holes "appear" to carry the current.

semiconductor diode is produced when a p-type and an n-type semiconductor are joined. This combination is called a **pn junction diode**.

diode is forward biased if voltage from a battery connected to the diode is large enough to overcome the internal potential difference. The result is a current flow through the diode.

diode is reversed biased if voltage from a battery connected to the diode causes the holes and electrons to be separated, which means that the negative charge tends to be separated from the positive charge. The result is a diode that is essentially nonconducting.

half-wave rectifier allows current from an ac source to flow only in one direction. A pn junction diode acts as a half-wave rectifier. It can be used to change an ac current to a dc current.

SUMMARY OF MATHEMATICAL FORMULAS

potential energy between two point charges	$PE = (1/4\pi\epsilon_o)q_1 q_2/r$ $1/4\pi\epsilon_o = k$ where $k = 9 \times 10^9 \, N \, m^2/C^2$	The potential energy between two point charges is related to the product of the charges ($q_1 q_2$) and is inversely proportional to the distance r between their centers. If the charges are both positive or both negative, then the PE is positive and decreases with increasing r. If the charges are of opposite sign, then PE is negative and increases with increasing r.
molecular rotational energy	$E_{rot} = L(L + 1) \, \hbar^2/2I$ where L = 0, 1, 2, etc.	Molecular rotational energy (E_{rot}) depends on the quantum number (L), the moment of inertia (I), and $\hbar$, where $\hbar = h/2\pi$.
molecular vibrational energy	$E_{vib} = (v + \frac{1}{2}) \, h \, f$ where v = 0, 1, 2, etc.	Energy levels for vibrational motion depend on the frequency of vibration (f) of the molecule and the vibrational quantum number (v).

CONCEPT SUMMARY

Chemical Bonds

The force that holds the atoms of a molecule together is referred to as a **bond**. The following is a summary of the two main types of chemical bonds: covalent and ionic.

Covalent Bond

In the case of diatomic molecules, such as H_2, O_2, and N_2, the outermost electrons are shared equally by both atoms. The type of bond in which the electrons are shared is called a **covalent bond**.

The cloud model from quantum mechanics is useful in attempting to explain chemical bonding. A simple molecule to consider is hydrogen. When two hydrogen atoms are at a distance, the electron clouds repel and the positively charged nuclei repel. There is no unbalanced force between the atoms.

As the atoms approach, the positively charged nucleus of one atom attracts the electron cloud of the other atom and the shapes of the electron clouds become distorted. The nuclear charge is concentrated and as a result the attraction of one nucleus for the electron cloud of the other atom is greater than the repulsion between the clouds.

As the electron clouds overlap, the overlapping regions cause the repulsion between the clouds to be further reduced. However, as the atoms come closer the repulsion between the positively charged nuclei increases until the forces balance. The distance between the nuclei when the balance point is reached is called the bond length.

Polar Covalent Bond

The chemical bond formed by molecules such as H_2, O_2, and N_2 is a pure covalent bond. This is because the electrons which form the chemical bond are shared equally by the atoms which form the molecule. When the atoms involved are from different elements, then the electrons which form the bond are not shared equally and the bond is not a pure covalent bond.

The water molecule contains two atoms of hydrogen and one atom of oxygen. Hydrogen has one proton in its nucleus while oxygen has eight. The result is that oxygen's large nuclear charge tends to pull the electron cloud of the hydrogen toward it so that the region near the oxygen atom is negatively charged while the region near each hydrogen is positively charged. Because one end of the molecule is charged positively while the other end is charged negatively, the molecule is called a **polar molecule**. This type of a bond is known as **polar covalent**.

TEXTBOOK QUESTION 3. Does the H_2 molecule have a permanent dipole moment? Does O_2? Does H_2O? Explain.

ANSWER: Both H_2 and O_2 have a pure covalent bond. The electrons which form the bond between the atoms are equally shared by the atoms; therefore, neither molecule has a permanent dipole moment.

H_2O consists of two hydrogen atoms and one oxygen atom. As described on page 29-3, oxygen's large nuclear charge tends to pull the electron cloud of each hydrogen toward it with the result that the region near the oxygen atom is negatively charged while the region near each hydrogen atom is positively charged. Overall, the water molecule is electrically neutral. However, one end of the molecule is charged positively while the other end is charged negatively and the molecule does have a permanent dipole moment.

Ionic Bond

An **ionic bond** is formed when one or more electrons are transferred from one atom to another. The bond formed is based on the electrostatic attraction of the negatively charged ion for the positively charged ion.

An example of a compound exhibiting ionic bonding is sodium chloride, NaCl. The nucleus of the sodium atom contains 11 protons while that of chlorine contains 17 protons. The nuclear charge of the chlorine exerts a greater force on the outer electron of the sodium ion than does the sodium nucleus. The result is the transfer of the outer electron of the sodium ion to the chlorine atom. The sodium ion (Na⁺) exerts a force of electrostatic attraction on the chlorine (Cl⁻). Because the force is between two ions, the bond is called an ionic bond.

Strong vs. Weak Bonds

Energy is required in order to break a chemical bond which holds the atoms of a molecule together. The energy required to break a bond is called the **bond energy** or **binding energy**. The binding energy for covalent and ionic bonds is usually in the range of 2 eV to 5 eV. The bonds which hold atoms of a molecule together are called "**strong**" bonds.

The term "**weak**" **bond** or **Van der Waals bond** usually refers to electrostatic attraction between molecules. An example of a weak bond is between two dipoles and such a bond is often called a **dipole-dipole bond**. When one of the atoms in a dipole-dipole bond is hydrogen, then the bond is usually referred to as a hydrogen bond. Another type of weak bond is a dipole-induced dipole bond. This type of bond results when a polar molecule with a permanent dipole moment induces a dipole moment in an electrically balanced, nonpolar molecule.

The strength of a weak bond is in the range of 0.04 to 0.3 eV. In a biological cell the average kinetic energy of molecules is in the same range. A weak bond can be broken during molecular collisions and therefore weak bonds are not permanent. Strong bonds are almost never broken by molecular collisions and are therefore relatively permanent. They can be broken by chemical action in a biological cell with the aid of an enzyme.

Potential Energy Diagrams for Molecules

As discussed in Chapter 17, the potential energy between two point charges separated by a distance r is given by

$$PE = (1/4\pi\epsilon_o)q_1 q_2/r \quad \text{where} \quad 1/4\pi\epsilon_o = k \quad \text{and} \quad k = 9 \times 10^9 \text{ N m}^2/\text{C}^2$$

If the charges are both positive or both negative, then the PE is positive and decreases with increasing r as shown in figure A. If the charges are of opposite sign, then PE is negative and increases with increasing r as shown in figure B.

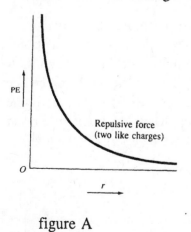

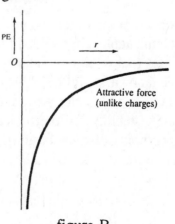

figure A   figure B

EXAMPLE PROBLEM 1. The spacing between adjacent sodium (Na$^+$) ions and chlorine (Cl$^-$) ions in a sodium chloride crystal is approximately 0.28 nm. a) Determine the electrostatic potential energy between adjacent Na$^+$ and Cl$^-$ ions. Assume that the sodium ion carries a charge of +1.0e while the chlorine ion carries a charge of -1.0e. Express your answer in both joules and electron volts. b) What is the magnitude of the electrostatic attractive force acting between the ions?

Part a. Step 1.	Solution: (Section 29-2)
Calculate the magnitude of the electrostatic potential energy in joules.	$PE = (1/4\pi\epsilon_o)q_1q_2/r$ But $q_1\ q_2 = (1.6 \times 10^{-19}\ C)(-1.6 \times 10^{-19}\ C) = -2.6 \times 10^{-38}\ C^2$. $PE = (9.0 \times 10^9\ N\ m^2/kg^2)(2.6 \times 10^{-38}\ C^2)/(0.28 \times 10^{-9}\ m)$ $PE = -8.4 \times 10^{-19}\ J$
Part a. Step 2. Determine the energy in eV.	$PE = (-8.4 \times 10^{-19}\ J)(1.0\ eV/1.6 \times 10^{-19}\ J)$ $PE = -5.2\ eV$
Part b. Step 1. Determine the magnitude of the electrostatic force attraction between the two ions.	$F = (1/4\pi\epsilon_o)q_1\ q_2/r^2$ $= (9 \times 10^9\ N\ m^2/kg^2)(-2.6 \times 10^{-38}\ C^2)/(0.280 \times 10^{-9}\ m)^2$ $F = -3.0 \times 10^{-9}\ N$

Covalent Bond

The potential energy function of a covalent bond, e.g., H_2, is shown in figure C. As discussed on page 29-3, as the two atoms approach, they tend to attract and share their valence electrons. The value of the PE decreases to a minimum value at a certain optimum distance between their nuclei (r_o). This distance is known as the bond length. However, if the distance between the nuclei becomes less than r_o, then the nuclei repel and the PE increases. In figure C, r_o is the approximate point of greatest stability for the molecule and the approximate point of lowest energy. The energy at this point is called the **binding energy**. The binding energy is the amount of energy required to separate the two atoms to infinity, at infinity PE = 0.

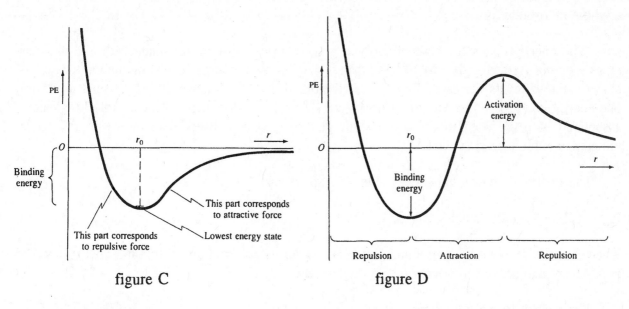

figure C   figure D

Activation Energy

For many molecules, the force between the atoms as they approach is repulsive. In order for the atoms to form a molecule, additional energy called the **activation energy** must be added to force them together. Figure D represents the PE curve for a situation where the activation energy must be considered.

Ionic Bond

The figure at right is a PE diagram for a typical ionic bond, in this case NaCl. As discussed on page 29-4, the sodium atom tends to lose its 3s electron to the chlorine atom. The result is a force of attraction between Na^+ and Cl^-. As shown in the diagram, the equilibrium distance between the ions is 0.24 nm and at this point the PE is a minimum and corresponds to a bonding energy of 4.2 eV.

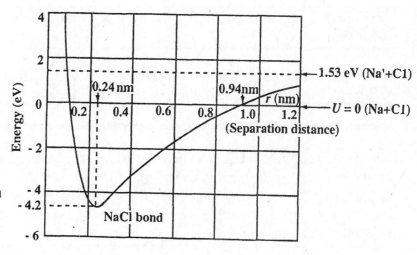

Molecular Spectra

As stated in the text, "When atoms combine to form molecules, the energy levels of the outer electrons are altered because they now interact with each other. Additional energy levels also become possible because the atoms can vibrate with respect to each other, and the atom as a whole can rotate."

"The energy levels for both vibrational and rotational motion are quantized, and are very close together (typically 10^{-1} to 10^{-3} eV apart). Each atomic energy level thus becomes a set of closely spaced levels corresponding to the vibrational and rotational motions. Transitions from one energy level to another appear as many very closely spaced lines." The simple line spectra associated with the Bohr model of the hydrogen atom is not observed in molecules. Instead, molecules exhibit band spectra.

The quantized rotational energy levels are given by

$$E_{rot} = L(L + 1)\; \hbar^2/2I \quad \text{where} \quad L = 0, 1, 2, \text{etc.}$$

where I is the moment of inertia and $\hbar = h/2\pi$. Transitions from one rotational energy level to another are subject to the selection rule $\Delta L = \pm 1$.

The energy levels for vibrational motion are given by

$$E_{vib} = (v + \tfrac{1}{2})\; h\; f$$

where v is the vibrational quantum number, $v = 0, 1, 2$, etc. and f is the classical frequency of vibration of the molecule. Transitions from one vibrational energy level to another are subject to the selection rule $\Delta v = \pm 1$.

TEXTBOOK QUESTION 5. The energy of a molecule can be divided into four categories. What are they?

ANSWER: The categories are electrostatic potential energy, translational kinetic energy, rotational kinetic energy, and vibrational kinetic energy.

EXAMPLE PROBLEM 2. It is determined that a photon of wavelength 0.00698 m causes the O_2 molecule to make a transition from the lowest rotational energy state (L = 0) to the first excited state (L = 1). Determine the a) energy of the photon, b) difference in energy between the rotational energy states, c) moment of inertia of the O_2 molecule, and d) bond length of the molecule. Note: the mass of an oxygen atom is 2.67×10^{-26} kg.

Part a. Step 1.	Solution: (Section 29-4)
Determine the energy of the photon.	$E = h\, f$ but $f = c/\lambda$ Therefore,
	$E = hc/\lambda$
	$E = (6.63 \times 10^{-34} \text{ J s})(3 \times 10^8 \text{ m/s})(0.00698 \text{ m}) = 2.85 \times 10^{-23}$ J

Part b. Step 1;	The photon caused the transition from $L = 0$ to $L = 1$; therefore, the energy difference between the two states equals the energy of the photon.
Determine the difference in energy between the states where $L = 0$ and $L = 1$.	$$\Delta E = E_1 - E_0 = 2.85 \times 10^{-23} \text{ J}$$

Part c. Step 1.	The formula for the quantized rotational energy level is
Determine the moment of inertia of the O_2 molecule.	$$E_{rot} = L(L + 1)\, \hbar^2/2I$$
	If $L = 0$, $E_0 = 0(0 + 1)\, \hbar^2/2I = 0$ J.
	If $L = 1$, $E_1 = 1(1 + 1)\, \hbar^2/2I = 2\, \hbar^2/2I = \hbar^2/I$
	$$\Delta E = E_1 - E_0 = \hbar^2/I - 0 \text{ J} = \hbar^2/I$$
	$$2.85 \times 10^{-23} \text{ J} = \hbar^2/I$$
	and $I = \hbar^2/(2.85 \times 10^{-23} \text{ J})$
	$$I = [(6.63 \times 10^{-34} \text{ J s})/2\pi]^2/(2.85 \times 10^{-23} \text{ J})$$
	$$I = 3.91 \times 10^{-46} \text{ J s}^2 = 3.91 \times 10^{-46} \text{ kg m}^2$$

Part d. Step 1.	Both oxygen atoms have the same mass and each atom can be considered to be a point mass. The center of mass of the molecule lies midway between the two atoms. The moment of inertia about the center of mass is given by
Determine the bond length of the molecule.	$$I = m_1 r_1^2 + m_2 r_2^2$$
	But $m_1 = m_2 = m$ and $r_1 = r_2 = r$. Therefore,
	$$I = 2\, m\, r^2$$
	$$3.91 \times 10^{-46} \text{ kg m}^2 = 2\,(2.67 \times 10^{-26} \text{ kg})\, r^2$$
	$$r^2 = 7.3 \times 10^{-21} \text{ m}^2$$
	$$r = 8.6 \times 10^{-11} \text{ m}$$
	The bond length is the distance between the oxygen atoms.
	bond length $= 2r = 1.71 \times 10^{-10} \text{ m}$

Bonding in Solids

In chapter 16, the ability of a solid to conduct an electrical current resulted in it being classified as a conductor, semiconductor, or insulator. This classification can now be discussed in terms of what is referred to as the **band theory of solids**.

As stated in the text, "If a large number of atoms come together to form a solid, then each of the original atomic levels becomes a band" As shown below, "The energy levels are so close together in each band that they seem essentially continuous."

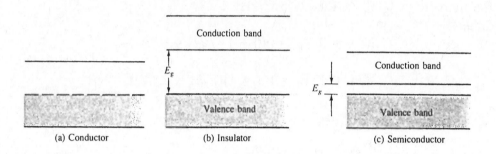

(a) Conductor   (b) Insulator   (c) Semiconductor

"In a good conductor, e.g., a metal, the highest energy band (valence band) containing electrons is only partially filled. As a result, many electrons are relatively free to move throughout the volume of the metal and the metal can carry an electric current."

"In a good insulator, the highest band (valence band) is completely filled with electrons. The next highest energy band, called the conduction band, is separated from the valence band by a large energy gap (E_g) of 5 to 10 eV." "At room temperature, molecular kinetic energy available due to collisions is only about 0.04 eV, so almost no electrons can jump from the valence to the conduction band." "When a potential difference is applied across the material, there are no available states accessible to the electrons, and no current flows. Hence the material is a good insulator."

"The bands for a pure semiconductor, such as silicon or germanium, are like those for an insulator, except that the unfilled conduction band is separated from the filled valence band by a much smaller energy gap (E_g), typically on the order of 1 eV. At room temperature, there will be a few electrons that can acquire enough energy to reach the conduction band and so a very small current can flow when a voltage is applied. At higher temperatures, more electrons will have enough energy to jump the gap. This effect can often more than offset the effects of more frequent collisions due to increased disorder at higher temperature, so that the resistivity of semiconductors can decrease with temperature."

EXAMPLE PROBLEM 3. The longest wavelength radiation absorbed by a certain semiconductor is 2350 nm. Determine the energy gap (E_g) for this semiconductor in joules and in electron volts.

Part a. Step 1. Determine the energy gap in joules.	Solution: (Section 29-6) E_g is the minimum energy required to cause an electron to jump from the valence band to the conduction band. The energy of the wavelength radiation must equal E_g. $E_g = E_{photon} = hc/\lambda$ $\quad = (6.63 \times 10^{-34} \text{ J s})(3 \times 10^8 \text{ m/s})/[(2350 \text{ nm})(1.0 \times 10^{-9} \text{ m/1nm})]$ $E_g = 8.46 \times 10^{-20} \text{ J}$
Part a. Step 2. Determine the energy in eV.	$E_g = (8.46 \times 10^{-20} \text{ J})(1.0 \text{ eV}/1.6 \times 10^{-19} \text{ J})$ $E_g = 0.529 \text{ eV}$

EXAMPLE PROBLEM 4. The energy gap for germanium is about 0.72 eV. Determine the a) lowest frequency photon which will cause an electron to jump from the valence band to the conduction band, and b) wavelength of the photon found in part a.

Part a. Step 1. Express the energy in joules.	Solution: (Section 29-6) $E_g = (0.72 \text{ eV})(1.6 \times 10^{-19} \text{ J}/1.0 \text{ eV})$ $E_g = 1.15 \times 10^{-19} \text{ J}$
Part a. Step 2. Determine the frequency of the photon.	$E_g = E_{photon} = hf$ $1.15 \times 10^{-19} \text{ J} = (6.63 \times 10^{-34} \text{ J s}) f$ $f = 1.74 \times 10^{14} \text{ Hz}$
Part b. Step 1. Determine the wavelength of the photon.	$c = f\lambda$ $3.0 \times 10^8 \text{ m/s} = (1.74 \times 10^{14} \text{ Hz}) \lambda$ $\lambda = 1.72 \times 10^{-6} \text{ m}$

Semiconductors and Doping

The electronic configuration of the valence electrons of silicon is $3s^2\, 3p^2$ and for germanium $4s^2\, 4p^2$, which means that each element has four outer electrons and is a relatively poor conductor of electricity. Silicon and germanium are examples of semiconductors.

However, if a small amount of an impurity such as arsenic ($4s^2\, 4p^3$) is introduced into the crystal structure of germanium, then arsenic's fifth electron is not bound and is free to move about. As a result, the "**doped**" semiconductor becomes highly conducting. An arsenic-doped germanium crystal is called an **n-type** semiconductor because electrons (negative charge) carry the current.

If a small amount of gallium ($4s^2\, 4p^1$) is added to the germanium crystal, then an empty place or **hole** is introduced because gallium has only three outer electrons. An electron from the germanium atom can jump into this hole but as a result the hole moves to a new location. Because most of the atoms of the crystal are germanium, this new location is invariably next to a germanium atom which is now positively charged because it has lost an electron. An electron can jump from another germanium atom to fill the previous hole and thus the hole can move through the crystal. The flow of electricity in this instance is called a hole current. A germanium crystal doped with gallium is called a **p-type** semiconductor since it is the positive holes which "appear" to carry the current.

TEXTBOOK QUESTION 17. A silicon semiconductor is doped with phosphorous. Will these atoms be donors or acceptors? What type of semiconductor will this be?

ANSWER: The outer electronic configuration of phosphorous is $2s^2\, 3p^3$. If a small amount of phosphorous is introduced into the crystal structure of silicon, then the fifth electron of the phosphorous atom is not bound and is free to move through the solid. Phosphorous is a donor atom and a semiconductor consisting of silicon doped with phosphorous is an n-type.

Semiconductor Diodes

A semiconductor **diode** is produced when a p-type and an n-type semiconductor are joined. This combination is called a **pn junction diode.** At the junction, a few electrons from the n-type diffuse into holes in the p-type and an internal difference in potential develops between the sections.

As shown at the top of the next page, if a battery is connected as shown in the left diagram, the diode is said to be **forward biased.** If the battery voltage is large enough to overcome the internal potential difference, a current will flow through the diode.

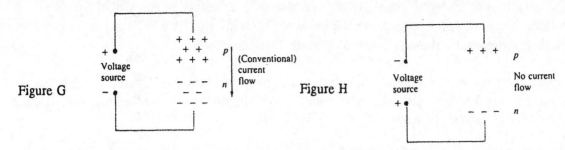

Figure G

Figure H

If the battery is connected as shown in the right diagram, the diode is **reverse biased**. This causes the holes and electrons to be separated, which means that the negative charge tends to be separated from the positive charge. The result is a diode that is essentially non-conducting. A graph of current versus voltage for a typical diode is shown in Fig. 29-28 of the text. As a reference, Fig. 29-28 is shown below.

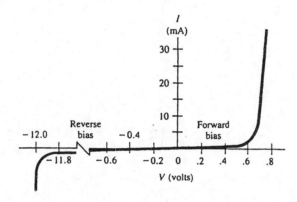

Since a pn junction diode allows current to flow only in one direction, it will allow current from an ac source to flow only in one direction through the circuit. Therefore, it can be used to change an ac current to a dc current. This is called rectification and in the simple circuit shown below, the diode is acting as a half-wave rectifier.

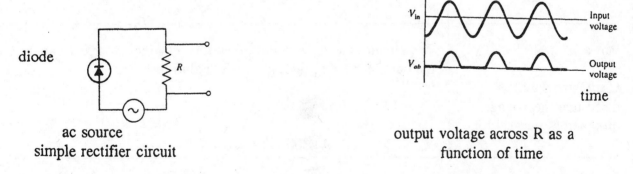

diode

ac source
simple rectifier circuit

output voltage across R as a
function of time

It should be noted that the symbol for a semiconductor diode is ⟶▶⟶ . The diode allows current to flow in the direction of the arrow but not in the reverse direction.

TEXTBOOK QUESTION 11. Compare the resistance of a pn junction diode when connected in forward bias to when connected in reverse bias.

ANSWER: A diode is forward biased when connected to a battery as shown in figure G. A battery voltage large enough to overcome the internal potential difference can cause a current to flow through the diode and the resistance to current flow is small.

A diode is reverse biased when connected to a battery as shown in figure H. This causes the holes and electrons to be separated. The result is that the diode is essentially nonconducting and the electrical resistance to current flow is very large.

EXAMPLE PROBLEM 5. A silicon diode whose current voltage characteristics are in the figure shown below is connected in series with a voltage source and a 300-Ω resistor. What voltage is needed to cause a 20-mA current to flow through the circuit?

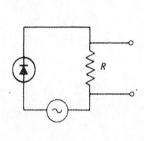
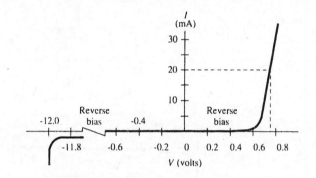

Part a. Step 1.	Solution: (Section 29-8)
Use the diagram to determine the approximate voltage drop across the diode.	Using the figure it can be determined that a voltage of approximately 0.75 volt is needed to produce a 20 mA current.

Part a. Step 2.

Use Ohm's law to determine the voltage drop across the resistor.

The circuit is a series circuit; therefore, the diode current equals the current flowing through the 300 ohm resistor.

$V_R = I R = (20 \text{ mA})(300 \ \Omega)$

$V_R = (20 \times 10^{-3} \text{ A})(300 \ \Omega) = 6.0 \text{ volts}$

Part a. Step 3.

Use Kirchhoff's voltage rule to determine the battery voltage.

The resistor is in series with the diode and the voltage source,

$V_{battery} = V_{diode} + V_R$

$\qquad = 0.75 \text{ V} + 6.0 \text{ volts}$

$V_{battery} = 6.75 \text{ V}$

PROBLEM SOLVING SKILLS

For problems related to electrostatic potential energy:

1. Determine the charge on each ion and the distance between the ions.
2. Use $PE = (1/4\pi\epsilon_o)q_1q_2/r$ to solve for the potential energy.

For problems related to rotational energy states:

1. Determine the difference in energy between the rotational energy states.
2. Use the energy difference to determine the moment of inertia of the molecule.
3. Use the moment of inertia to determine the bond length.

For problems related to the energy gap between the valence band and conduction band for a semiconductor or insulator:

1. If the energy for the electron to travel from the valence band to the conduction band is provided by a photon, determine the energy of the photon. The energy of a photon can be determined by using $E = h f$ or $E = h c/\lambda$.
2. The energy difference (energy gap) between the valence band and conduction band is equal to the lowest frequency (or longest wavelength) photon capable of causing the electron's transition.

For problems related to a series circuit containing a voltage source, a diode, and an additional resistor:

1. Use a figure giving the current-voltage characteristics of the diode to determine the approximate voltage drop across the diode.
2. Use Ohm's law to determine the voltage drop across the series resistor.
3. Use Kirchhoff's voltage rule to determine the voltage of the source.

SOLUTIONS TO SELECTED TEXT PROBLEMS

TEXTBOOK PROBLEM 1. Estimate the binding energy of a KCl molecule by calculating the electrostatic potential energy when the K^+ and Cl^- ions are at their stable separation of 0.28 nm. Assume each has a charge of 1.0e.

Part a. Step 1.	Solution: (Sections 29-1 to 29-3)
Calculate the magnitude of the electrostatic potential energy in joules.	$PE = k\, q_1q_2/r$
	But $q_1\, q_2 = (1.6 \times 10^{-19}\ C)(-1.6 \times 10^{-19}\ C) = -2.6 \times 10^{-38}\ C^2$.
	$PE = [(9 \times 10^9\ N\ m^2/C^2)(-2.6 \times 10^{-38}\ C^2)]/(0.28 \times 10^{-9}\ m)$
	$PE = -8.3 \times 10^{-19}\ J$

Part a. Step 2.	PE = $(-8.3 \times 10^{-19}$ J$)[(1.0$ eV$/1.6 \times 10^{-19}$ J$)]$
Determine the energy in eV.	PE = -5.2 eV

TEXTBOOK PROBLEM 4. Binding energies are often measured experimentally in kcal per mole, and then the binding energy in eV per molecule is calculated from that result. What is the conversion factor in going from kcal per mole to eV per molecule? What is the binding energy of KCl (= 4.43 eV) in kcal per mole?

Part a. Step 1. Convert from kcal/mole to eV/molecule.	Solution: (Sections 29-1 to 29-3) Note: 1kcal = 4186 J $[(1$ kcal/mol$)(4186$ J/kcal$)(1$ mol$/6.02 \times 10^{23}$ molecule$)]/[(1$ eV$)/(1.6 \times 10^{-19}$ J$)]$ = 0.0434 eV/molecule

- -

Part a. Step 2. Determine the binding energy of KCl in kcal/mole.	$(4.43$ eV/molecule$)[(1$ kcal/mole$)/(0.0434$ eV/molecule$)]$ = 102 kcal/mole

---

TEXTBOOK PROBLEM 7. The so-called "characteristic rotational energy," $\hbar^2/2I$, for N_2 is 2.48×10^{-4} eV. Calculate the N_2 bond length.

Part a. Step 1 Determine the moment of inertia of the N_2 molecule.	(Solution: Section 29-4) $\Delta E = \hbar^2/2I$ where $\hbar = h/2\pi = (6.62 \times 10^{-34}$ J s$)/(2\pi) = 1.05 \times 10^{-34}$ J s $I = \hbar^2/[2(\Delta E)]$ $I = (1.05 \times 10^{-34}$ J s$)^2/[2(2.48 \times 10^{-4}$ eV$)(1.6 \times 10^{-19}$ J$)/(1$ eV$)]$ $I = (1.05 \times 10^{-34}$ J s$)^2/(7.94 \times 10^{-23}$ J$)$ $I = 1.39 \times 10^{-46}$ kg m^2

- -

Part a. Step 2. Determine the bond length of the molecule.	Both nitrogen atoms have the same mass and each atom can be considered to be a point mass. The center of mass of the molecule lies midway between the two atoms. The moment of inertia about the center of mass is given by $I = m_1 \, r_1^2 + m_2 \, r_2^2$ ● ● $\|\leftarrow r_1 \rightarrow \cdot \leftarrow r_2 \rightarrow \|$ But $m_1 = m_2 = m$ and $r_1 = r_2 = r$. Therefore, $I = 2 \, m \, r^2$ Note: the atomic mass of a nitrogen atom = 14 amu 1.39×10^{-46} kg m^2 = 2 (14 amu)[(1.66 $\times$ 10^{-27} kg)/(1 amu)] r^2 $r^2 = 2.99 \times 10^{-21}$ m^2 $r = 5.47 \times 10^{-11}$ m The bond length is the distance between the nitrogen atoms. bond length = 2r = 2(5.47 $\times$ 10^{-11} m) = 1.09 $\times$ 10^{-10} m

TEXTBOOK PROBLEM 17. Calculate the longest-wavelength photon that can cause an electron in silicon (E_g = 1.1 eV) to jump from the valence band to the conduction band.

Part a. Step 1. Express the energy in joules.	Solution: (Section 29-6) E_g = (1.1 eV)(1.6 $\times$ 10^{-19} J/1.0 eV) E_g = 1.76 $\times$ 10^{-19} J
Part a. Step 2. Determine the minimum frequency of the photon.	The photon's energy must be equal to or greater than the energy gap. $E_{photon} \geq E_g$ and $E_{photon} = hf$ 1.76 $\times$ 10^{-19} J = (6.63 $\times$ 10^{-34} J s) f f = 2.65 $\times$ 10^{14} Hz

Part a. Step 3.	$c = f\lambda$
Determine the wavelength of the longest-wavelength photon.	3.0×10^8 m/s $= (2.65 \times 10^{14}$ Hz$)$ λ $\lambda = 1.13 \times 10^{-6}$ m

---

TEXTBOOK PROBLEM 19. The energy gap E_g in germanium is 0.72 eV. When used as a photon detector, roughly how many electrons can be made to jump from the valence to the conduction band by the passage of a 760-keV photon that loses all of its energy in this fashion?

Part a. Step 1.	Solution: (Section 29-6)
Express the photon energy in eV.	$E_{photon} = (760$ keV$)[(1000$ eV$)/1$ keV$)] = 7.60 \times 10^5$ eV

- -

Part a. Step 2.	E_g = energy required for one electron to jump to the conduction band
Determine the maximum number of electrons which can be made to jump to the conduction band.	n = maximum number of electrons which can make the jump $E_{photon} = n\, E_g$ 7.60×10^5 eV $= n\, (0.72$ eV$)$ $n \approx 1.1 \times 10^6$ electrons

---

TEXTBOOK PROBLEM 27. An ac voltage of 120 V rms is to be rectified. Estimate very roughly the average current in the output resistor R (25 kΩ) for (a) a half-wave rectifier (Fig. 29-29) and (b) a full-wave rectifier (Fig. 29-30) without capacitor.

Part a. Step 1.	Solution: (Section 29-8)
Determine the average current for a half-wave rectifier without a capacitor.	For a half-wave rectifier without a capacitor, the current is zero half of the time. $I_{average} = \tfrac{1}{2}\, V_{rms}/R$ $= \tfrac{1}{2}\, (120$ V$)/(25{,}000\ \Omega)$ $I_{average} = 2.4 \times 10^{-3}$ A $= 2.4$ mA

Part b. Step 1.

Determine the average for current for a full-wave rectifier without a capacitor.

For a full-wave rectifier without a capacitor, the current is positive all of the time.

$$I_{average} = V_{rms}/R$$

$$= (120 \text{ V})/(25,000 \text{ }\Omega)$$

$$I_{average} = 4.8 \times 10^{-3} \text{ A} = 4.8 \text{ mA}$$

CHAPTER 30

NUCLEAR PHYSICS AND RADIOACTIVITY

OBJECTIVES

After studying the material of this chapter, the student should be able to:

- determine the number of neutrons in a nuclide of known atomic number and mass number.
- explain what is meant by an isotope of an element. State how isotopes of an element differ and state the properties they have in common.
- explain what is meant by the unified atomic mass unit. Calculate the energy equivalent in MeV of an atomic mass unit.
- given a table of nuclear masses, calculate the binding energy of a nucleus and the binding energy per nucleon.
- identify the three kinds of radiation emitted by radioactive substances. State which radiations are deflected by electric and magnetic fields.
- give the symbol used to represent each of the following: alpha particle, beta⁻ particle, beta⁺ particle, gamma ray.
- write a general equation to represent each of the following possible radiation decays: alpha decay, beta⁺ decay, beta⁻ decay, gamma decay.
- distinguish between the parent nucleus and daughter nucleus in a nuclear transmutation.
- calculate the disintegration energy (Q) for a given alpha decay.
- list the four conservation laws which apply to radioactive decays.
- write the equation which relates the half-life of a substance to its decay constant.
- write the equation for the law of radioactive decay. Explain the meaning of each symbol in the equation. Solve problems related to the law of radioactive decay.

KEY TERMS AND PHRASES

nucleus of an atom contains protons and neutrons. These particles are called **nucleons**.

1 unified atomic mass unit (u) = 1.660×10^{-27} kg.

atomic number is the number of protons contained in the nucleus.

atomic mass number is the total number of protons and neutrons in the nucleus.

isotopes are atoms which have the same number of protons but different number of neutrons in the nucleus.

binding energy is the energy required to break apart a nucleus into its constituent protons and neutrons.

strong nuclear force refers to attractive force which holds the nucleus together. The strong nuclear force acts between all nucleons, protons, and neutrons alike. This force is much greater than the force of electrostatic repulsion which exists between the protons. The strong nuclear force is a short-range force. It acts between nucleons if they are less than 10^{-15} m apart, but is essentially zero if the separation distance is greater than 10^{-15} m.

weak nuclear force is much weaker than the strong nuclear force and appears in a type of radioactive decay called **beta decay**.

radioactive decay results from the instability of certain nuclei. There are three different radiations produced by radioactive decay: **alpha, beta,** and **gamma**.

parent nucleus refers to the nucleus before radioactive decay.

daughter nucleus refers to the nucleus after radioactive decay.

alpha decay occurs when a helium nucleus is spontaneously emitted from the nucleus. An alpha particle (α) consists of two protons and two neutrons but no electrons.

disintegration energy (Q) is the total energy released during alpha decay.

beta decay occurs when an electron (e^-) and an antineutrino ($\bar{\nu}$) are spontaneously emitted from the nucleus. Beta decay (β^-) is observed in nuclei which have a high ratio of neutrons to protons.

electron capture occurs when the nucleus absorbs one of the inner orbital electrons in the atom to form a neutron.

transmutation is the changing of one element into a new element. Transmutation is the result of alpha or beta decay.

gamma decay occurs when a nucleus in an excited state drops to a lower energy state. In the process a photon called a gamma ray (γ) is emitted.

law of conservation of nucleon number states that the total number of nucleons before decay equals the total number of nucleons after decay.

half-life of an isotope is the time required for half of the radioactive nuclei present in the sample to decay.

radioactive decay series is a series of successive decays which starts with one parent isotope and proceeds through a number of daughter isotopes. The series ends when a stable, nonradioactive isotope is produced.

radioactive dating refers to a method of estimating the age of an object based on the object's half-life and the amount of the isotope present in the sample being analyzed.

SUMMARY OF MATHEMATICAL FORMULAS

atomic mass number	$^A_Z X$ $A = Z + N$	The nucleus of a chemical element is designated by $^A_Z X$. X is the chemical symbol of the element, Z is the atomic number, which is the number of protons contained in the nucleus, A is the atomic mass number, and N is the number of neutrons in the nucleus.
atomic radius	$r \approx (1.2 \times 10^{-15} \text{ m})(A^{\frac{1}{3}})$	The approximate atomic radius (r) of the nucleus increases with the mass number (A).
binding energy	$E = (\Delta m) c^2$ $E = (u)(931.5 \text{ MeV/u})$	The energy equivalent of the mass difference (Δm) is the binding energy (E) of the nucleus. The binding energy is determined by multiplying the mass difference, expressed in u, by the conversion factor 931.5 MeV/u.
alpha decay	$N(A, Z) \rightarrow N(A\text{-}4, Z\text{-}2) + {}^4_2 He$	An alpha decay (α) particle consists of a helium nucleus. N(A, Z) is the parent nucleus (N) with atomic number (Z) and mass number (A). N(A-4, Z-2) is the daughter nucleus and ${}^4_2 He$ is the alpha particle.

disintegration energy	$Q = (M_p - M_d - m_\alpha) c^2$	The total energy released during alpha decay is called the disintegration energy (Q). M_p is the mass of the parent nucleus, M_d is the mass of the daughter nucleus, and m_α is the mass of the alpha particle.
beta decay	$N(A, Z) \rightarrow N(A, Z +1) + {}_{-1}^{0}e + \bar{\nu}$	Beta decay occurs when an electron (e⁻) and an antineutrino ($\bar{\nu}$) are spontaneously emitted from the nucleus. The electron carries one negative charge and as a result the atomic number of the daughter nucleus is one greater than the atomic number of the parent nucleus. The antineutrino carries no electric charge and has zero rest mass.
electron capture	$N(A, Z) + {}_{-1}^{0}e \rightarrow N(A, Z-1) + \nu$	Electron capture occurs when the nucleus absorbs one of the inner orbital electrons in the atom to form a neutron. When electron capture occurs, a neutrino (ν) is spontaneously emitted from the nucleus.
gamma decay	$N^*(A, Z) \rightarrow N(A, Z) + \gamma$	Gamma decay occurs when a nucleus in an excited state drops to a lower energy state. In the process a photon called a gamma ray (γ) is emitted.
		Since the gamma ray is a photon, it carries no electric charge and has no rest mass. No change in nucleon number or atomic number occurs due to a gamma decay. $N^*(A, Z)$ is the nucleus in an excited state.

radioactive decay rate	$\Delta N = - \lambda N \Delta t$ or $\Delta N / \Delta t = - \lambda N$	The number of radioactive decays (ΔN) that occur in a short time interval (Δt) is proportional to the length of the time interval and the total number (N) of radioactive nuclei present. λ is the decay constant which is different for different isotopes. $\Delta N / \Delta t$ is the rate of decay or the activity of the isotope. The negative sign indicates that the number of radioactive nuclei present is decreasing.
law of radioactive decay	$N = N_o\, e^{-\lambda t}$	Based on the law of radioactive decay, the number of radioactive nuclei present (N) after time (t) depends on the number of radioactive nuclei present in the original sample (N_o) and the decay constant (λ) for the particular isotope.
half-life	$T_{1/2} = - 0.693 / \lambda$	The half-life ($T_{1/2}$) of an isotope is the time required for half of the radioactive nuclei present in the sample to decay. The half life is related to the decay constant (λ).

CONCEPT SUMMARY

Structure of the Nucleus

The nucleus of an atom contains protons and neutrons. These particles are called **nucleons**. The neutron is electrically neutral, while the proton carries a single positive charge of magnitude 1.6×10^{-19} C. The mass of each particle is $m_p = 1.673 \times 10^{-27}$ kg = 1.0073 u; $m_n = 1.675 \times 10^{-27}$ kg = 1.0087 u where 1 u = 1.660×10^{-27} kg = 1 unified atomic mass unit.

The nucleus of a chemical element is designated by $^A_Z X$. X is the chemical symbol of the element, e.g., H for hydrogen, Ca for calcium. Z is the **atomic number**, which is the number of protons contained in the nucleus. A is the **atomic mass number**. The atomic mass number is the total number of protons and neutrons in the nucleus.

Experiments indicate that the nuclei have a roughly spherical shape. The radius (r) of the nucleus increases with the mass number (A), and the approximate radius is given by
$r \approx (1.2 \times 10^{-15} \text{ m})(A^{1/3})$

Determine the approximate density of the nucleus of a C-12 atom in kg/m^3.

Part a. Step 1.	Solution: (Section 30-1)
Determine the radius of the nucleus.	The atomic mass number (A) of a C-12 atom is 12. The approximate radius of the nucleus may be determined as follows: $$r \approx (1.2 \times 10^{-15} \text{ m})(A^{\frac{1}{3}})$$ $$r \approx (1.2 \times 10^{-15} \text{ m})(12^{\frac{1}{3}}) \approx 2.7 \times 10^{-15} \text{ m}$$
Part a. Step 2.	$$V = (4/3) \pi r^3$$
Determine the approximate volume of the nucleus.	$$\approx (4/3)(3.14)(2.7 \times 10^{-15} \text{ m})^3$$ $$V \approx 8.6 \times 10^{-44} \text{ m}^3$$
Part a. Step 3.	$$m = (12 \text{ u})(1.66 \times 10^{-27} \text{ kg/u})$$
Determine the mass of the nucleus in kg.	$$m = 1.99 \times 10^{-26} \text{ kg}$$
Part a. Step 4.	$$\rho = m/V \approx (1.99 \times 10^{-26} \text{ kg})/(8.6 \times 10^{-44} \text{ m}^3)$$
Determine the approximate density of the nucleus.	$$\rho \approx 2.3 \times 10^{17} \text{ kg/m}^3$$

Isotopes

Atoms which have the same atomic number but different mass numbers are called **isotopes**. Isotopes of the same element have the same 1) atomic number, 2) number of electrons, 3) electronic configuration, and 4) chemical properties. Isotopes differ in the number of neutrons in the nucleus. For example, the two isotopes of lithium are Li-6 and Li-7. Each isotope contains three protons; however, one contains three neutrons while the other contains four neutrons.

TEXTBOOK QUESTION 1. What do different isotopes of a given element have in common? How are they different?

ANSWER: Isotopes are atoms which have the same number of protons but different number of neutrons in the nucleus. Isotopes of the same element have the same chemical properties and the neutral atoms of the same element contain the same number of electrons. For example, Cu-63 and Cu-65 are isotopes Both have 29 protons in the nucleus but Cu-63 has 34 neutrons while Cu-65 has 36 neutrons. As shown in Appendix B, the atomic mass number as well as atomic mass of isotopes of the same element differ.

TEXTBOOK QUESTION 5. Why are the atomic masses of many elements (see the periodic table) not close to whole numbers?

ANSWER: Isotopes of the same element contain the same number of protons but different number of neutrons. The weighted average of the percent abundance of the isotopes is used to determine the atomic mass of the element. For most elements this weighted average is not close to a whole number. For example, in Appendix B chlorine is listed as having 17 protons but one isotope (^{35}Cl) has 18 neutrons while ^{37}Cl has 20 neutrons. The percent abundance of ^{35}Cl is 75.77% while ^{37}Cl is 24.23%. The atomic mass is determined as follows:
$(0.7577)(35) + (0.2423)37 = 35.45$

Binding Energy and Nuclear Forces

The total mass of the nucleus is always less than the sum of the masses of the protons and neutrons of which it is composed. The energy equivalent of the mass difference (Δm) is the **binding energy** of the nucleus and can be determined by multiplying the mass difference, expressed in u, by the conversion factor 931.5 MeV/u.

The average binding energy per nucleon is defined as the total binding energy divided by the mass number (A). Figure 30-1 in the text shows that the average binding energy per nucleon increases until A = 15, levels off at about 8.8 MeV per nucleon until A = 60, and then slowly decreases.

The nucleus is held together by an attractive force which acts between all nucleons, protons and neutrons alike. This force is called the **strong nuclear force** and is much greater than the force of electrostatic repulsion which exists between the protons. The strong nuclear force is a short-range force. It acts between nucleons if they are less than 10^{-15} m apart, but is essentially zero if the separation distance is greater than 10^{-15} m.

A second type of nuclear force is the **weak nuclear force**. This force is much weaker than the strong nuclear force and appears in a type of radioactive decay called **beta decay**.

EXAMPLE PROBLEM 2. The exact atomic mass of an Ar-40 atom is 39.9624 u. Determine the total mass of its constituent particles and calculate the binding energy of the atom. The mass of an electron, proton, and neutron are 0.00055 u, 1.00728 u, and 1.00867 u, respectively.

Part a. Step 1.	Solution: (Section 30-2)
Determine the total mass of the constituent particles.	The atomic number is 18; therefore, the Ar atom has 18 electrons and 18 protons. The mass number is 40; therefore, Ar has 40 - 18 = 22 neutrons. electrons 18 x 0.00055 u = 0.00990 u protons 18 x 1.00728 u = 18.1310 u neutrons 22 x 1.00867 u = 22.1907 u _____ total mass of nucleons 40.3316 u
Part a. Step 2. Determine the difference in mass between Ar-40 and its constituent particles.	mass of nucleons 40.3316 u mass of Ar 39.9624 u _____ mass difference 0.3692 u
Part a. Step 3. Determine the binding energy.	$E = (\Delta m)c^2$ $E = (0.3692 \text{ u})(931.5 \text{ MeV/u}) = 343.9 \text{ MeV}$

EXAMPLE PROBLEM 3. a) Use figure 30-1 from the text to determine the binding energy per nucleon for Cu-65. b) Use this value to determine the total binding energy for the Cu-65 nucleus.

Part a. Step 1.	Solution: (Section 30-2)
Determine the binding energy per nucleon.	It is not possible to determine an exact value from figure 30-1. An approximate value for the binding energy per nucleon for Cu-65 is 8.8 MeV per nucleon.
Part b. Step 1. Determine the total binding energy.	8.8 MeV/nucleon x 65 nucleons = 570 MeV

EXAMPLE PROBLEM 4. a) Determine the disintegration energy released when the following reaction occurs

$$^{214}_{84}\text{Po} \quad \rightarrow \quad ^{210}_{82}\text{Pb} \quad + \quad ^{4}_{2}\text{He}$$

b) Determine the kinetic energy of each of the products.

Part a. Step 1.
 	
Determine the mass before and after the decay.
	Solution: (Section 30-4)
	
	before after
	
	$^{214}_{84}\text{Po}$ 213.9952 u $^{210}_{82}\text{Pb}$ 209.9842 u
	
	                                                      $^{4}_{2}\text{He}$ 4.0026 u
	
	                            total mass of products 213.9868 u

Part a. Step 2. Determine the mass difference.	mass difference = 213.9952 u - 213.9868 u mass difference = 0.0084 u
Part a. Step 3. Determine the energy released.	(0.0084 u)(931.5 MeV/u) = 7.82 Mev
Part b. Step 1. Use the law of conservation of momentum to express the final velocity of the alpha particle in terms of the final velocity of the lead nucleus. Note: assume that the the initial momentum of the system was zero.	In any reaction, both energy and momentum must be conserved. Assuming that the polonium nucleus was initially at rest, then the initial momentum was zero and the total momentum after the decay must also be zero. This means that the alpha particle's momentum must be equal to but opposite that of the lead nucleus. initial momentum = final momentum $0 = m_\alpha \, v_\alpha + m_{Pb} \, v_{Pb}$ $m_\alpha \, v_\alpha = - m_{Pb} \, v_{Pb}$ and $v_\alpha = -(m_{Pb}/m_\alpha) \, v_{Pb}$ $v_\alpha = -(209.9842 \text{ u}/4.0026 \text{ u}) \, v_{Pb}$ $v_\alpha = - 52.46 \, v_{Pb}$

Part b. Step 2. Determine the ratio of the KE of the lead nucleus to the KE of the alpha particle.	$KE_{Pb}/KE_\alpha = (\frac{1}{2}\, m_{Pb}\, v_{Pb}^2)/(\frac{1}{2}\, m_\alpha\, v_\alpha^2)$ but $v_\alpha = -52.46\, v_{Pb}$ $KE_{Pb}/KE_\alpha = [\frac{1}{2}\, (209.9842\, u)(v_{Pb})^2]/[\frac{1}{2}\, (4.0026\, u)(-52.46\, v_{Pb})^2]$ Upon simplifying, $KE_{PB}/KE_\alpha = 0.01906$
Part b. Step 3. Determine the kinetic energy of the alpha particle and the daughter nucleus.	The total kinetic energy of the daughter nucleus and the alpha particle is 7.82 MeV. $KE_\alpha + KE_{Pb} = 7.82$ MeV but $KE_{Pb} = 0.01906\, KE_\alpha$ and substituting gives $KE_\alpha + 0.01906\, KE_\alpha = 7.82$ MeV $1.0196\, KE_\alpha = 7.82$ MeV $KE_\alpha = 7.67$ MeV $KE_{Pb} = 7.82$ MeV $- 7.67$ MeV $= 0.15$ MeV

Radioactive Decay Mode

Certain nuclei are unstable and undergo **radioactive decay**. There are three different radiations produced by radioactive decay: **alpha, beta**, and **gamma**.

Alpha Decay

An **alpha decay** particle consists of a helium nucleus. Thus, an alpha particle (α) consists of two protons and two neutrons but no electrons. An alpha decay can be represented by the following equation:

$N(A, Z) \rightarrow N(A\text{-}4, Z\text{-}2) + {}^{4}_{2}He$

$N(A, Z)$ is the **parent nucleus (N)** with atomic number (Z) and mass number (A).

$N(A\text{-}4, Z\text{-}2)$ is the **daughter nucleus** and ${}^{4}_{2}He$ is the alpha particle.

When alpha decay occurs, the atomic number of the element decreases by two while the mass number decreases by four. Alpha decay occurs in large nuclei in which the strong nuclear force is not strong enough to hold the nucleus together. It occurs when the mass of the parent nucleus is greater than the mass of the daughter plus the mass of the alpha particle, i.e., $M_p > M_d + \alpha$. The mass difference appears in the form of kinetic energy of the daughter nucleus and the alpha particle (but mainly in the alpha particle).

The total energy released during alpha decay is called the disintegration energy Q where $Q = (M_p - M_d - m_\alpha)\, c^2$

Beta Decay

Beta⁻ decay occurs when an electron (e⁻) and an antineutrino $(\bar{\nu})$ are spontaneously emitted from the nucleus. Beta decay (β-) is observed in nuclei which have a high ratio of neutrons to protons. The electron is not a nucleon; therefore, there is no change in the mass number of the daughter nucleus. However, the electron carries one negative charge and as a result the atomic number of the daughter nucleus is one greater than the atomic number of the parent nucleus. Beta⁻ decay is represented by the following equation:

$$N(A, Z) \rightarrow N(A, Z+1) + {}_{-1}^{0}e + \bar{\nu}$$

The **antineutrino** $(\bar{\nu})$ carries no electric charge and has zero rest mass.

Certain unstable isotopes have a low neutron to proton ratio. In such a situation, a positron and a neutrino (ν) may be emitted from the nucleus. The charge on a positron (${}_{+1}^{0}e$) is opposite that of the electron but the mass of both particles is the same. The positron is represented as e⁺ or β+ and is the antiparticle to the electron. As in the case of the antineutrino, the neutrino carries no electric charge and has zero rest mass. The equation for positron decay is as follows:

$$N(A, Z) \rightarrow N(A, Z-1) + {}_{+1}^{0}e + \nu$$

Electron Capture

Electron capture occurs when the nucleus absorbs one of the inner orbital electrons in the atom to form a neutron. When electron capture occurs, a neutrino is spontaneously emitted from the nucleus. Electron capture may occur if the neutron-proton ratio is low. Electron capture is represented by the following equation:

$$N(A, Z) + {}_{-1}^{0}e \rightarrow N(A, Z-1) + \nu$$

Transmutation

A new element is formed when alpha or beta decay occurs. The changing of one element (parent nucleus) into a new element (daughter nucleus) is called **transmutation**.

Gamma Decay

Gamma decay occurs when a nucleus in an excited state drops to a lower energy state. In the process a photon called a gamma ray (γ) is emitted. The process is analogous to photon emission when an orbital electron drops from a higher energy level to a lower energy level.

The nucleus may be in the excited state due to a violent collision with another particle or a previous radioactive decay leaves the nucleus in an excited state.

Since the gamma ray is a photon, it carries no electric charge and has no rest mass. Therefore, no change in nucleon number or atomic number occurs due to a gamma decay. A gamma decay can be represented as follows:

$$N*(A, Z) \rightarrow N(A, Z) + \gamma \quad \text{where} \quad N*(A, Z) \text{ is the nucleus in an excited state.}$$

ANSWER:

PART a) $^{24}_{11}$Na(β-) is an example of β- decay. The general equation for this type of decay is

$$N(A, Z) \rightarrow N(A, Z+1) + {}^{0}_{-1}e + \nu \quad \text{therefore,}$$

$$^{24}_{11}\text{Na} \rightarrow {}^{24}_{12}\text{Mg} + {}^{0}_{-1}e + \bar{\nu}$$

PART b) $^{22}_{11}$Na(β+) is an example of beta^{+} decay. The general equation for this type of decay is

$$N(A, Z) \rightarrow N(A, Z-1) + {}^{0}_{+1}e + \nu \quad \text{therefore,}$$

$$^{22}_{11}\text{Na} \rightarrow {}^{22}_{10}\text{Ne} + {}^{0}_{+1}e + \nu$$

PART c) $^{210}_{84}$Po(α) is an example of alpha decay. The general equation for this type of decay is

$$N(A, Z) \rightarrow N(A-4, Z-2) + {}^{4}_{2}\text{He} \quad \text{therefore,}$$

$$^{210}_{84}\text{Po} \rightarrow {}^{206}_{82}\text{Pb} + {}^{4}_{2}\text{He}$$

___

Conservation Laws

In addition to conservation of energy, linear momentum, angular momentum, and electric charge, a radioactive decay also obeys the law of **conservation of nucleon number**. This law states that the total number of nucleons before decay equals the total number of nucleons after decay.

___

EXAMPLE PROBLEM 5. Cite the type of reaction and then complete the following.

a) $^{234}_{90}$Th $\rightarrow$ X $+ {}^{0}_{-1}e + \bar{\nu}_e$ b) $^{7}_{4}$Be $+ {}^{0}_{-1}e \rightarrow$ X $+ \nu_e$

c) $^{210}_{84}$Ra $\rightarrow$ X $+ {}^{4}_{2}$He

Part a. Step 1.	Solution: (Sections 30-4, 30-5, and 30-7)
Cite the type of reaction, then use the law of conservation of charge and nucleon number to solve for X.	This reaction is a beta decay. The product nucleus must have an atomic number of 90 - (-1) = 91 in order to satisfy the law of conservation of charge. It must have a mass number of 234 in order to conserve nucleon number. Using the periodic table, it can be determined that X is protactinium; thus X is $^{234}_{91}$Pa

Part b. Step 1. Cite the type of reaction and repeat part a.	This is an example of electron capture. The atomic number of the product nucleus is 4 + (-1) = 3. The mass number is 7 + 0 = 7. Using the periodic table, the unknown nucleus is lithium; thus X is $_3^7\text{Li}$
Part c. Step 1. Cite the type of reaction and repeat part a.	This is an example of alpha decay. The atomic number of the product nucleus is 84 - 2 = 82. The mass number is 210 - 4 = 206. The unknown nucleus is lead; thus X is $_{82}^{206}\text{Pb}$

Half-life and Rate of Decay

The number of radioactive decays (ΔN) that occur in a short time interval (Δt) is proportional to the length of the time interval and the total number (N) of radioactive nuclei present.

$$\Delta N = - \lambda N \Delta t \quad \text{or} \quad \Delta N/\Delta t = - \lambda N$$

λ is the decay constant, which is different for different isotopes. $\Delta N/\Delta t$ is the rate of decay or the activity of the isotope. The negative sign in the equation indicates that the number of radioactive nuclei present is decreasing.

Based on the law of radioactive decay, the number of radioactive nuclei present is given by the equation

$$N = N_o \, e^{-\lambda t}$$

N_o is the number of radioactive nuclei present in the original sample, i.e., the number at $t = 0$ s. N is the number of radioactive nuclei present at time t and $e = 2.718$.

The rate of decay of any isotope is usually given by its **half-life**. The half-life ($T_{½}$) of an isotope is the time required for half of the radioactive nuclei present in the sample to decay. The relationship between the half-life and the decay constant is given by

$$T_{½} = - 0.693/\lambda$$

The graph shown at right indicates the number of undecayed nuclei (parent nuclei) present as a function of time where the time is expressed in terms of half-lives. This type of curve is known as an **exponential decay curve**.

Note: in this diagram, the time is given in terms of the half-life of an isotope of carbon, carbon-14.

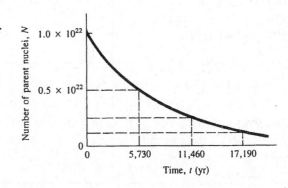

ANSWER: After two months the sample will not have completely decayed, in fact, ¼ of the original amount remains. The half-life of the isotope is one month, which means after one month half of the original sample remains. After another month has passed, half of the sample which remained after one month remains. If the initial amount is N, then after one month ½N remains. After two months, ½(½N) = ¼N is present.

Part a. Step 1. Determine the half-life in seconds.	Solution: (Section 30-8) (42.0 min)(60 s/1 min) = 2520 s
Part a. Step 2. Determine the decay constant for this element.	$\lambda = 0.693/T_{1/2}$ $= 0.693/2520$ s $\lambda = 2.75 \times 10^{-4}$/s
Part b. Step 1, Determine the number of half-lives that have passed in 2.8 h.	(2.8 hours)(60 min/1 hour) = 168 min (168 min)(1 half-life/42.0 min) = 4 half-lives
Part b. Step 2. Determine the number of moles of the original sample that is present after 4 half-lives.	The number of moles of the original present after 4 half-lives should be $(½)^4$ moles of the original; thus $(½)^4(2.00 \times 10^{-6}$ g)(1 mole/49 g) = 2.55×10^{-9} moles
Part b. Step 3. Determine the number of nuclei of Cr-49 present after 2.8 h.	N = $(2.55 \times 10^{-9}$ moles)(6.02×10^{23}/mole) N = 1.54×10^{15} nuclei

ALTERNATE SOLUTION: Use the law of radioactive decay to solve part b.

Part b. Step 1. Determine the number of moles of Cr-49 in the original sample.	$(2.0 \times 10^{-6} \text{ g})(1 \text{ mole}/49 \text{ g}) = 4.1 \times 10^{-8}$ moles
Part b. Step 2. Determine the number of nuclei present in the original sample.	$N_o = (4.1 \times 10^{-8} \text{ moles})(6.02 \times 10^{23}/\text{mole})$ $N_o = 2.46 \times 10^{16}$ nuclei
Part b. Step 3. Use the law of radio-active decay to solve for the number of Cr-49 nuclei remaining after 2.8 h.	$N = N_o e^{-\lambda t}$ where $\lambda = 2.75 \times 10^{-4}/\text{s}$, t = 2.8 h = 1.01×10^4 s, and $\lambda t = (2.75 \times 10^{-4}/\text{s})(1.01 \times 10^4 \text{ s}) = 2.77$ $N = (2.46 \times 10^{16} \text{ nuclei}) e^{-2.77} = (2.46 \times 10^{16} \text{ nuclei})(0.063)$ $N = 1.54 \times 10^{15}$ nuclei

Decay Series

A radioactive decay often results in a daughter nucleus that is also radioactive. A **radioactive decay series** is a series of successive decays which starts with one parent isotope and proceeds through a number of daughter isotopes. The series ends when a stable, nonradioactive isotope is produced.

Radioactive Dating

If the half-life of a radioactive isotope is known, an estimate of the age of an object oftentimes can be made. For example, the ratio of carbon-14 to carbon-12 in a living object is relatively constant. However, a living object stops absorbing carbon-14 when it dies. Therefore, by knowing the half-life of carbon-14 (5700 years) and the object's $^{14}C/^{12}C$ ratio, an estimate of the object's age can be made.

If a rock contains uranium, geologists are able to estimate the age of the rock by determining the amount of U-238 in the rock relative to the amount of daughter nuclei that are present.

EXAMPLE PROBLEM 7. a) A sample of wood from a living tree is found to contain 100 grams of pure carbon. If carbon contains 1 part in 10^{12} of carbon-14, determine the activity of the sample. b) A wood artifact from an archeological site is found to contain 100 grams of pure carbon and produces 142.8 decays/min. Determine the age of the artifact. Note: the half-life of carbon-14 is 5700 years and the decay constant is $3.8 \times 10^{-12}/\text{s}$.

Part a. Step 1. Determine the number of carbon atoms in the sample.	Solution: (Sections 30-8, 30-9, and 30-11) $(100 \text{ g})(6.02 \times 10^{23} \text{ atoms}/12 \text{ g}) = 5.02 \times 10^{24}$ atoms
Part a. Step 2. Determine the number of carbon-14 nuclei in the sample.	$(5.02 \times 10^{24} \text{ atoms})(1 \text{ carbon-14 nucleus}/10^{12} \text{ atoms})$ $= 5.02 \times 10^{12}$ carbon-14 nuclei
Part a. Step 3. Determine the activity in the sample.	$-\Delta N/\Delta t = \lambda \ \Delta N$ $\qquad = (3.8 \times 10^{-12}/\text{sec})(5.02 \times 10^{12} \text{ nuclei})$ $-\Delta N/\Delta t = 19.1$ decays/sec
Part b. Step 1. Determine the number of decays per second.	$(142.8 \text{ decays/min})(1 \text{ min}/60 \text{ s}) = 2.38$ decays/s
Part b. Step 2. Determine the ratio of the decays/sec in the artifact to the decays/sec in the sample.	$(2.38 \text{ decays/sec})/(19.1 \text{ decays/sec}) = 0.125$ or ⅛
Part b. Step 3. Determine the age of the artifact.	The rate of activity in the artifact is ⅛ that of the sample from a living tree. Therefore, three half-lives have passed since the artifact was made. Since the half-life of carbon-14 is 5700 years, the age of the artifact is 3 x 5700 years = 17,100 yrs.

PROBLEM SOLVING SKILLS

For problems involving nuclear density:

1. Use the equation $r \approx (1.2 \times 10^{-15} \text{ m})(A^{1/3})$ to determine the approximate radius of the nucleus.
2. Use the equation $V = (4/3) \pi r^3$ to determine the approximate volume of the nucleus.
3. Convert the mass of the nucleus from unified atomic mass units to kg. Note: the mass of the nucleus expressed in atomic mass units can be determined by noting the mass number.
4. Use the equation $\rho = m/V$ to determine the approximate density.

For problems related to the binding energy of the nucleus:

1. Use a table of nuclear masses to determine the total mass of the constituent particles which make up the nucleus. Note also the exact atomic mass of the nucleus.
2. Determine the difference between the mass of the nucleus and the mass of the constituent particles.
3. Use the equation $\Delta E = (\Delta m) c^2$ to determine the binding energy.

For questions related to a particular type of nuclear decay when the parent nucleus is known:

1. Take note whether the decay is alpha, beta, or gamma.
2. Write down the general equation for the particular type of decay.
3. Use the conservation laws to determine the atomic number and mass number of the daughter nucleus.
4. Use a periodic table to identify the daughter element.

For problems related to the calculation of the disintegration energy of an alpha decay:

1. Use a table of nuclear masses to determine the mass of the parent nucleus, daughter nucleus, and alpha particle in atomic mass units.
2. Determine the difference between the mass of the parent nucleus and the mass of the daughter nucleus + alpha particle. Express this difference in atomic mass units.
3. Determine the disintegration energy by multiplying the mass difference expressed in atomic mass units by 931.5 MeV/u.
4. If the problem asks for the kinetic energy of the daughter nucleus and alpha particle:
    a. Use the law of conservation of momentum to determine the ratio of the velocity of the alpha particle to that of the daughter nucleus.
    b. Determine the ratio of the kinetic energy of the alpha particle to the daughter nucleus.
    c. Knowing the total kinetic energy of the daughter nucleus and alpha particle, and the ratio of the two kinetic energies, algebraically solve for the kinetic energy of each particle.

For problems related to radioactive decay:

1. Determine the number of radioactive nuclei in the sample.
2. Use $\Delta N/\Delta t = - \lambda N$ to determine the activity of the sample.

3. If the problem involves carbon dating, then determine the ratio of decays per second in the artifact to the number of decays per second in the sample.

4. Knowing the half-life of C-14, determine the age of the artifact.

For problems related to half-life, decay constant, and the law of radioactive decay.

1. If either the half-life or decay constant is given, use $T = 0.693/\lambda$ to determine either the half life or decay constant.

2. Use the law of radioactive decay $N = N_o e^{-\lambda t}$ to determine the number of nuclei remaining after a certain time has passed.

SOLUTIONS TO SELECTED TEXTBOOK PROBLEMS

> **TEXTBOOK PROBLEM 4.** (a) What is the approximate radius of a Cu-64 nucleus? (b) Approximately what is the value of A for a nucleus whose radius is 3.9×10^{-15} m?

Part a. Step 1.	Solution: (Section 30-1)
Determine the radius of the nucleus.	The mass number (A) of the copper atom is 64. The approximate radius of the nucleus may be determined as follows:
	$r \approx (1.2 \times 10^{-15} \text{ m})(A^{\frac{1}{3}})$
	$r \approx (1.2 \times 10^{-15} \text{ m})(64^{\frac{1}{3}}) \approx 4.8 \times 10^{-15}$ m
Part b. Step 1.	$r \approx (1.2 \times 10^{-15} \text{ m})(A^{\frac{1}{3}})$ rearranging gives
Use the same equation to determine the mass number.	$A \approx [r/(1.2 \times 10^{-15} \text{ m})]^3$
	$A \approx [(3.9 \times 10^{-15} \text{ m})/(1.2 \times 10^{-15} \text{ m})]^3 \approx 34$

> **TEXTBOOK PROBLEM 12.** Calculate the binding energy per nucleon for a N-14 nucleus.

Part a. Step 1.	Solution: (Section 30-2)
Use Appendix B to determine the mass of a neutron, proton, and a N-14 nucleus.	From Appendix B, $m_{neutron} = 1.008665$ u, $m_{proton} = 1.007825$ u
	$m_{N-14} = 14.003074$ u

Part a. Step 2. Determine the mass difference between the nucleons and the N-14 nucleus.	N-14 contains 7 neutrons and 7 protons. mass difference = 7(1.008665 u) + 7(1.007825 u) - 14.003074 u $\qquad$ = 7.060655 u + 7.054775 u - 14.003074 u mass difference = 0.112356 u
Part a. Step 3. Determine the total binding energy.	Note: the binding energy per nucleon = 931.5 MeV total binding energy = (0.112356 u)(931.5 MeV/u) total binding energy = 104 MeV
Part a. Step 4. Determine the binding energy per nucleon.	binding energy per nucleon = (total binding energy)/(# of nucleons) $\qquad$ = (104 MeV)/14 binding energy per nucleon = 7.48 MeV

TEXTBOOK PROBLEM 23. A U-238 nucleus emits an α particle with KE = 4.20 MeV. a) What is the daughter nucleus and b) what is the approximate atomic mass (in u) of the daughter atom? Ignore recoil of the daughter nucleus.

Part a. Step 1. Determine the atomic number and mass number of the unknown nucleus.	Solution: (Sections 30-3 to 30-7) An α particle has an atomic number of 2 and a mass number of 4. The U-238 nucleus has an atomic number of 92 and a mass number of 238. The atomic number of the unknown element (X) = 92 - 2 = 90 The mass number of the unknown element (X) = 238 - 4 = 234
Part a. Step 2. Use the periodic table at the back of the text to identify the unknown nucleus.	The unknown element is $_{90}^{234}\text{Th}$ or thorium 234.

Part b. Step 1.	$m_{U\text{-}238} = 238.050783$ u
Use Appendix B to determine the mass of U-238, and the α particle.	$m_\alpha = 4.002602$ u

---

Part b. Step 2.	$KE = (\Delta m)\, c^2$
Determine the mass difference between the U-238, α and the Th-234.	4.20 MeV $= (\Delta m)\, c^2$
	$\Delta m = (4.20\ \text{MeV})/c^2$
	$\qquad = (4.20\ \text{MeV})(10^6\ \text{eV/1 MeV})(1.6 \times 10^{-19}\ \text{J/eV})/(9 \times 10^{16}\ \text{m/s}^2)$
	$\Delta m = (7.47 \times 10^{-30}\ \text{kg})[(1\ u)/(1.66 \times 10^{-27}\ \text{kg}) = 0.00450$ u

---

Part b. Step 3.	$\Delta m = m_{U\text{-}238} - m_{Th\text{-}234} - m_\alpha$
Determine the mass of the daughter atom, i.e., Th-234.	0.00450 u $= 238.050783$ u $- m_{Th\text{-}234} - 4.002602$ u
	0.00450 u $= 234.048181$ u $- m_{Th\text{-}234}$
	$m_{Th\text{-}234} = 234.043681$ u

TEXTBOOK PROBLEM 33. a) What is the energy of the alpha particle emitted in the decay $^{210}_{84}Po \rightarrow\ ^{206}_{82}Pb + \alpha$? Take into account the recoil of the daughter nucleus.

Part a. Step 1.	Solution: (Sections 30-3 to 30-7)
Determine the mass before and after the decay.	before after
	$^{210}_{84}Po$ 209.982416 u $\qquad$ $^{206}_{82}Pb$ 205.974449 u
	$\qquad\qquad\qquad\qquad\qquad\qquad\qquad\qquad$ α $\qquad$ 4.002603 u
	total mass of products $\qquad\qquad$ 209.977052 u

Part a. Step 2.	mass difference $= 209.982416$ u $- 209.977052$ u
Determine the mass difference.	mass difference $= 0.005364$ u

30-20

Part a. Step 3. Determine the energy released.	$(0.005364 \text{ u})(931.5 \text{ MeV/u}) = 4.9966 \text{ Mev}$
Part a. Step 4. Use the law of conserva- tion of momentum to express the final velocity of the alpha particle in terms of the final velocity of the lead nucleus. Note: assume that the the initial momentum of the system was zero.	In any reaction, both energy and momentum must be conserved. Assuming that the polonium nucleus was initially at rest, then the initial momentum was zero and the total momentum after the decay must also be zero. This means that the alpha particle's momentum must be equal to but opposite that of the lead nucleus. initial momentum = final momentum $0 = m_\alpha \, v_\alpha + m_{Pb} \, v_{Pb}$ $m_\alpha \, v_\alpha = - \, m_{Pb} \, v_{Pb}$ and $v_\alpha = -(m_{Pb}/m_\alpha) \, v_{Pb}$ $v_\alpha = -(205.974449 \text{ u})/(4.002603 \text{ u}) \, v_{Pb}$ $v_\alpha = - \, 51.46 \, v_{Pb}$
Part a. Step 5. Determine the ratio of the KE_{Pb} nucleus to the KE_α particle.	$KE_{Pb}/KE_\alpha = (\frac{1}{2} \, m_{Pb} \, v_{Pb}{}^2)/(\frac{1}{2} \, m_\alpha \, v_\alpha{}^2)$ but $v_\alpha = - \, 51.46 \, v_{Pb}$ $KE_{Pb}/KE_\alpha = [\frac{1}{2} \, (205.9744 \text{ u})(v_{Pb})^2]/[\frac{1}{2} \, (4.0026 \text{ u})(-51.46 \, v_{Pb})^2]$ Upon simplifying, $KE_{PB}/KE_\alpha = 0.01943$
Part a. Step 6. Determine the kinetic energy of the alpha particle and the daughter nucleus.	The total kinetic energy of the daughter nucleus and the alpha particle is 4.9966 MeV. $KE_\alpha + KE_{Pb} = 4.9966 \text{ MeV}$ but $KE_{Pb} = 0.01943 \, KE_\alpha$ and substituting gives $KE_\alpha + 0.01943 KE_\alpha = 4.9966 \text{ MeV}$ $1.01943 \, KE_\alpha = 4.9966 \text{ MeV}$ $KE_\alpha = 4.90 \text{ MeV}$ and $KE_{Pb} = 0.0952 \text{ MeV}$

TEXTBOOK PROBLEM 36. A radioactive material produces 1280 decays per minute at one time, and 4.6 h later produces 320 decays per minute. What is its half-life?

Part a. Step 1. Determine the decay constant.	Solution: (Section 30-8 to 30-11) $\Delta N/\Delta t = (\Delta N/\Delta t)_0 \, e^{-\lambda t}$ 320 decays/ min = (1280 decays/min) $e^{-\lambda(4.6\text{ h})}$ $0.25 = e^{-\lambda(4.6\text{ h})}$ $\ln(0.25) = \ln e^{-\lambda(4.6\text{ h})}$ $-1.386 = -\lambda\ (4.6\text{ h}) \ln e$ but $\ln e = 1$ $\lambda = 0.301/\text{h}$
Part a. Step 2.	$T_{1/2} = (0.693)/\lambda$ $T_{1/2} = (0.693)/(0.301/\text{h})$ $T_{1/2} = 2.3$ hours

TEXTBOOK PROBLEM 43. The iodine isotope I-131 is used in hospitals for diagnosis of thyroid function. If 682 μg are ingested by a patient, determine the activity (a) immediately, (b) 1.0 h later when the thyroid is being tested, and (c) 6 months later. Use Appendix B.

Part a. Step 1. Use Appendix B to determine the half-life of I-131.	Solution: (Sections 30-8 to 30-11) From Appendix B, $T_{1/2} = 8.0207$ days Converting the half-life to seconds gives (8.0207 days)[(24 hours)/(1day)][(3600 s)/(1 hour)] = 6.93×10^5 s
Part a. Step 2. Determine the decay constant.	$\lambda = (0.693)/T_{1/2}$ $\lambda = (0.693)/(6.93 \times 10^5\text{ s}) = 1.00 \times 10^{-6}/\text{s}$
Part a. Step 3. Determine the number of nuclei in the initial sample. Note: 682 μg = 6.82×10^{-4} g	1 mole of I-131 = 130.906 g $N_0 = (6.82 \times 10^{-4}\text{ g})[(1.0\text{ mole})/(130.906\text{ g})][(6.02 \times 10^{23}\text{ atoms})/(1\text{ mole})]$ $N_0 = 3.14 \times 10^{18}$ nuclei

Part a. Step 4. Determine the number of decays at t = 0 s.	$\Delta N/\Delta t = -\lambda N e^{-\lambda t}$ $\quad = -(1.00 \times 10^{-6}/s)(3.14 \times 10^{18}\text{ nuclei})\, e^{-(1.00 \times 10\text{-}6/s)(0\ s)}$ $\Delta N/\Delta t = -3.14 \times 10^{12}$ decays/s
Part b. Step 1. Determine λt at t = 1.0 hour.	$1.0\ h = 3600\ s$ $\lambda t = (1.00 \times 10^{-6}/s)(3600\ s) = 0.00360$
Part b. Step 2. Determine the number of decays per second at 1.0 h.	$\Delta N/\Delta t = (\Delta N/\Delta t)_o\, e^{-\lambda t}$ $\quad = (3.14 \times 10^{12}\text{ decays/s})\, e^{-0.00360}$ $\Delta N/\Delta t = 3.13 \times 10^{12}$ decays/s
Part c. Step 1. Determine λt at 6 months.	Note: assume that 6 months is approximately 180 days. $t \approx (180\text{ days})[(24\text{ hours})/(1\text{day})][(3600\ s)/(1\text{ hour})]$ $t \approx 1.56 \times 10^7\ s$ $\lambda t \approx (1.00 \times 10^{-6}/s)(1.56 \times 10^7\ s) \approx 15.6$
Part c. Step 2. Determine the approximate numberof decays after 6.0 months.	$\Delta N/\Delta t = (\Delta N/\Delta t)_o\, e^{-\lambda t}$ $\quad \approx (3.14 \times 10^{12}\text{ decays/s})\, e^{-15.6}$ $\Delta N/\Delta t \approx 5.1 \times 10^5$ decays/s

CHAPTER 31

NUCLEAR ENERGY; EFFECTS AND USES OF RADIATION

OBJECTIVES

After studying the material of this chapter, the student should be able to:

- explain what is meant by a nuclear reaction.
- write the general equation for a nuclear reaction in both the long form and the short form. Explain what each symbol in the equation represents.
- given a problem involving a nuclear reaction which is written in the short form, determine the missing particle or nucleus.
- calculate the reaction energy (Q value) for a nuclear reaction and state whether the reaction is exothermic or endothermic.
- distinguish between nuclear fission and fusion and give an example of each process.
- explain what is meant by a self-sustaining chain reaction and how this reaction is kept under control in a nuclear reactor.
- explain what is meant by dosimetry.
- define the curie, roentgen, gray, and rad. Explain what is meant by the term rem.
- given the level of activity of a radioactive sample in curies, calculate the number of decays per second and the yearly dosage in rem absorbed by a person in contact with the sample. Determine if the dosage exceeds the recommended standard.
- describe three devices that can be used to detect the presence and level of activity of radiation.
- describe three devices that can be used to photograph the paths taken by elementary particles.

KEY TERMS AND PHRASES

nuclear reaction occurs when two nuclei collide and two (or more) nuclei are produced. In the process, the nucleus of one element is changed into a different element, and therefore a transmutation of elements has occurred.

exothermic or **exoergic reaction** occurs when energy is released in the reaction.

endothermic or **endoergic reaction** occurs when the total kinetic energy of the projectile particle and the target particle must be greater than or equal to a minimum amount for the reaction to occur. This minimum energy is called the **threshold energy**.

nuclear fission is the process in which a nucleus is split into approximately equal parts along with the release of neutrons and a large amount of energy.

nuclear chain reaction occurs when the neutrons released in a typical fission cause other nuclei to fission.

critical mass is the minimum amount of fissionable material required for a self-sustaining chain reaction to occur.

nuclear reactor is a device where the chain reaction is controlled and the energy is released gradually.

atomic bomb is a device in which the chain reaction is uncontrolled and the energy release occurs in a few moments of time.

nuclear fusion is the process by which small nuclei combine to form heavier nuclei.

thermonuclear, or **hydrogen,** bomb is a device in which uncontrolled fusion reactions release enormous amount of energy in a few moments of time.

dosimetry refers to measuring the quantity, or dose, of radiation that passes through a material.

1 **rad** is the amount of radiation which deposits 10^{-2} J of energy per kg of absorbing material.

rem (rad equivalent man) is a measure of the biological damage caused by radiation.

relative biological effectiveness (RBE) of a certain type of radiation is defined as the number of rad of x or γ radiation that produces the same biological damage as 1 rad of the given radiation.

Geiger counter is a gas-filled device with electrodes that detects electrons and ions produced by emissions of radioactive nuclei.

scintillation counter makes use of a solid, liquid, or gas which emits light when struck by ionizing radiation.

semiconductor detector makes use of a pn junction diode. The device detects ionizing radiation by the electrons and holes produced in the semiconductor.

bubble chambers detect the paths of elementary particles by photographing the vapor tracks left by the particles as they pass through a superheated liquid.

cloud chambers detect ionizing radiation by photographing the condensation track left in a supercooled gas.

SUMMARY OF MATHEMATICAL FORMULAS

nuclear reaction	$x + X \rightarrow y + Y$	A nuclear reaction occurs when two nuclei collide and two (or more) nuclei are produced. In the process, the nucleus of one element is changed into a different element, and therefore a transmutation of elements has occurred. x is the bombarding particle, sometimes called the bullet. X is the nucleus hit by the bullet and usually referred to as the target nucleus. y is a small particle or possibly a photon and is called the product particle. Y is the recoil nucleus.
reaction energy or Q-value	$Q = (M_a + M_X - M_b - M_Y) c^2$ $Q = (\Delta m)c^2$ or $Q = KE_b + KE_Y - KE_a - KE_X$	The reaction energy (or Q-value) is related to the mass defect (Δm) by Einstein's equation. a is a projectile particle or small nucleus that strikes nucleus X, producing nucleus Y and particle b. Q can also be written in terms of the change in kinetic energy.
nuclear fission	A typical fission reaction is $^{1}_{0}n + ^{235}_{92}U \rightarrow$ $^{90}_{38}Sr + ^{136}_{54}Xe + 10\,^{1}_{0}n$	Nuclear fission is the process in which a nucleus is split into approximately equal parts along with the release of neutrons and a large amount of energy. Fission occurs only in the nuclei of certain elements. These elements have a large nucleon number, e.g., U-235. The nucleus may fission if struck by a slow-moving neutron.

dosimetry curie	1 Ci = 3.70 x 10^{10} disintegrations per second 1 μCi = 10^{-6} Ci.	Dosimetry refers to measuring the quantity, or dose, of radiation that passes through a material. The activity of a radioactive isotope is measured in terms of curies (Ci) or microcuries (μCi).
dosimetry roentgen	One roentgen (1 R) is the amount of radiation that will produce 2.1 x 10^9 ion pairs/cm³ of air at STP. Today the roentgen is defined as that amount of X or γ radiation that deposits 0.878 x 10^{-2} J of energy per kg of air.	The roentgen (R) is a unit of dosage that was previously defined in terms of the amount of ionized air produced by the radiation.
dosimetry rad radiation equivalent man (rem) and relative biological effectiveness (RBE)	1 rad is the amount of radiation which deposits 10^{-2} J of energy per kg of absorbing material. rem = rad x RBE	The roentgen has been replaced by the rad. The rem (rad equivalent man) is the unit which refers to the biological damage caused by radiation. RBE is the relative biological effectiveness of a certain type of radiation. It is defined as the number of rad of x or γ radiation that produces the same biological damage as 1 rad of the given radiation. The RBE of x and γ rays is 1.0, β particles 1.0, α particles 10-20, slow neutrons 3-5, and fast neutrons and protons 10.

CONCEPT SUMMARY

Nuclear Reactions and Transmutation of Elements

A **nuclear reaction** occurs when two nuclei collide and two (or more) nuclei are produced. In the process, the nucleus of one element is changed into a different element, and therefore a transmutation of elements has occurred.

A nuclear reaction can be written as an equation as follows:

x + X → y + Y

x is the bombarding particle, sometimes called the bullet. X is the nucleus hit by the bullet and usually referred to as the target nucleus. y is a small particle or possibly a photon and is called the product particle. Y is the recoil nucleus.

For example, in the reaction $^{1}_{1}H + ^{23}_{11}Na \rightarrow ^{20}_{10}Ne + ^{4}_{2}He$

x is $^{1}_{1}H$, X is $^{23}_{11}Na$, y is $^{4}_{2}He$ and Y is $^{20}_{10}Ne$.

The proton $^{1}_{1}H$ may be written as p, while the alpha particle $^{4}_{2}He$ may be written as α.

TEXTBOOK QUESTION 1(d). Fill in the missing particles or nuclei:

$\alpha + ^{197}_{79}Au \rightarrow ? + d$ where d stands for deuterium.

ANSWER: The equation of the reaction is

$^{197}_{79}Au + ^{4}_{2}He \rightarrow ^{A}_{Z}X + ^{2}_{1}H$

conservation of nucleon number	conservation of charge
197 + 4 = A + 2	79 + 2 = Z + 1
201 = A + 2	81 = Z + 1
A = 199	Z = 80

Therefore, A = 199 and Z = 80. Using the periodic table, it is found that the atomic number of mercury is 80. Therefore, the resulting nuclide is $^{199}_{80}Hg$.

TEXTBOOK QUESTION 3. When $^{22}_{11}Na$ is bombarded by deuterons ($^{2}_{1}H$), an α particle is emitted. What is the resulting nuclide?

ANSWER: The reaction is written as follows:

$^{22}_{11}Na + ^{2}_{1}H \rightarrow ^{A}_{Z}X + ^{4}_{2}He$

conservation of nucleon number	conservation of charge
22 + 2 = A + 4	11 + 1 = Z + 2
24 = A + 4	12 = Z + 2
A = 20	Z = 10

Therefore, A = 20 and Z = 10. Using the periodic table, it is found that the atomic number of neon is 10. Therefore, the resulting nuclide is

$^{20}_{10}Ne$.

EXAMPLE PROBLEM 1. a) Given the following reaction, determine the atomic number (Z) and mass number (A) of element X. b) Use the periodic table to determine element X.

$$_2^4He + {}_7^{14}N \rightarrow {}_Z^AX + {}_1^1H$$

Part a. Step 1.	Solution: (Section 31-1)
Use the law of conservation of nucleon number to solve for A.	conservation of nucleon number $4 + 14 = A + 1$ $18 = A + 1$ $A = 17$
Part a. Step 2. Use the law of conservation of charge to solve for Z.	conservation of charge $2 + 7 = Z + 1$ $9 = Z + 1$ $Z = 8$
Part b. Step 1. Use the periodic table to determine the element.	Z represents the atomic number of the element. The atomic number of oxygen is 8. Therefore, the resulting nuclide is $_8^{17}O$.

EXAMPLE PROBLEM 2. When $_3^7Li$ is bombarded by protons $_1^1H$ a neutron particle is emitted. What is the resulting nuclide?

Part a. Step 1. Write the reaction in the long form.	Solution: (Section 31-1) Writing the reaction gives $_1^1H + {}_3^7Li \rightarrow {}_Z^AX + {}_0^1n$
Part a. Step 2. Use the law of conservation of nucleon number to solve for A.	conservation of nucleon number $7 + 1 = A + 1$ $8 = A + 1$ $A = 7$

Part a. Step 3. Use the law of conservation of charge to solve for Z.	conservation of charge $3 + 1 = Z + 0$ $4 = Z + 0$ $Z = 4$
Part a. Step 4. Use the periodic table to determine the element.	Z represents the atomic number of the element. The atomic number of beryllium is 4. Therefore, the resulting nuclide is $^{7}_{4}Be$.

EXAMPLE PROBLEM 3. Fill in the missing particle or nucleus.

a) $^{24}_{12}Mg + d \rightarrow ? + \alpha,$ b) $^{12}_{6}C + d \rightarrow n + ?$

Part a. Step 1. Re-write the reaction.	Solution: (Section 31-1) d represents a deuteron $^{2}_{1}H$, while α represents an alpha particle $^{4}_{2}He$. $^{24}_{12}Mg + ^{2}_{1}H \rightarrow ? + ^{4}_{2}He$
Part a. Step 2. Apply the law of conservation of charge to determine the atomic number.	$12 + 1 = ? + 2$; thus the atomic number of the unknown nucleus is 11 and the element is sodium (Na).
Part a. Step 3. Apply the law of conservation of nucleon number and use the periodic table to identify the unknown particle.	$24 + 2 = ? + 4$; thus the mass number is 22. The unknown nucleus is $^{22}_{11}Na$.
Part b. Step 1. Re-write the reaction.	$^{12}_{6}C + ^{2}_{1}H \rightarrow ? + ^{1}_{0}n$

Part b. Step 2.	$6 + 1 = ? + 0$; therefore, the atomic number of the nucleus is 7 and the element is nitrogen (N).
Use conservation of charge.	
Part b. Step 3.	$12 + 2 = ? + 1$ and the mass number of the nucleus is 13. The
Use conservation of nucleon number to identify the unknown nuclide.	unknown nucleus is $^{13}_{7}N$.

Reaction Energy or Q Value

In any nuclear reaction, all the conservation laws hold, and therefore it is possible to determine if a particular reaction can occur. If, in a reaction, the rest mass of the products is less than the rest mass of the reactants, it is possible for the reaction to occur and energy be released to account for the missing mass. However, in some reactions, the rest mass of the products may be greater than that of the reactants. In such a case, the reaction will not occur unless the bombarding particle has sufficient kinetic energy. The reaction energy (or Q value) can be determined as follows:

$$Q = (M_a + M_X - M_b - M_Y) \, c^2$$

where a is a projectile particle or small nucleus that strikes nucleus X, producing nucleus Y and particle b. Q can also be written in terms of the change in kinetic energy as follows:

$$Q = KE_b + KE_Y - KE_a - KE_X$$

If $Q > 0$, then the reaction is exothermic or exoergic and energy is released in the reaction. If $Q < 0$, then the reaction is endothermic or endoergic and the total kinetic energy of the projectile particle and the target particle must be greater than or equal to a minimum amount for the reaction to occur. This minimum energy is called the **threshold energy**. The threshold energy, when added to Q, results in an energy great enough to allow the final products to have velocities which obey both the law of conservation of momentum as well as the law of conservation of energy.

EXAMPLE PROBLEM 4. Determine the Q-value of the following reaction. State whether the reaction is exothermic or endothermic.

$^{4}_{2}He + {}^{9}_{4}Be \rightarrow {}^{12}_{6}C + {}^{1}_{0}n$

Part a. Step 1.	Solution: (Section 31-1)
Determine the total mass of the reactants and the total mass of the products.	reactants products

reactants	products
$^{4}_{2}$He 4.002603 u	$^{12}_{6}$C 12.000000 u
$^{9}_{4}$Be 9.012183 u	$^{1}_{0}$n 1.008665 u
13.014786 u	13.0086665 u

Part a. Step 2.

Determine the mass difference between the products and the reactants.

mass difference = 13.014786 u - 13.0086665 u

mass difference = 0.006121 u

The products have less mass than the reactants; therefore, the reaction is exothermic.

Part a. Step 3.

Determine the value of Q in joules and MeV.

$Q = (M_a + M_X - M_b - M_Y) c^2$

$Q = (\Delta m) c^2$

$\Delta m = (0.006121 \text{ u})(1.66 \times 10^{-27} \text{ kg/u}) = 1.016 \times 10^{-29} \text{ kg}$

$Q = (1.016 \times 10^{-29} \text{ kg})(3.0 \times 10^8 \text{ m/s})^2 = 9.145 \times 10^{-13} \text{ J}$

$Q = (9.145 \times 10^{-13} \text{ J})(1 \text{ MeV}/1.6 \times 10^{-13} \text{ J}) = 5.70 \text{ MeV}$

Alternate solution:

$Q = (0.006121 \text{ u})(931.5 \text{ MeV/u}) = 5.70 \text{ MeV}$

Nuclear Fission

Nuclear fission is the process in which a nucleus is split into approximately equal parts along with the release of neutrons and a large amount of energy. Fission occurs only in the nuclei of certain elements. These elements have a large nucleon number, for example, $^{235}_{92}$U and $^{239}_{94}$Pu. Both of these nuclei may fission if struck by a slow-moving neutron. A typical fission reaction is

$$^{1}_{0}n + ^{235}_{92}U \rightarrow ^{90}_{38}Sr + ^{136}_{54}Xe + 10 \, ^{1}_{0}n$$

It can be shown that in this reaction, the sum of the rest masses of the reactants is greater than the sum of the rest masses of the products. The missing mass is converted to energy, $Q > 0$, and the amount of energy released can be found by using $E = (\Delta m)c^2$

$$^{235}_{92}U + ^{1}_{0}n \rightarrow ^{141}_{56}Ba + ^{92}_{36}Kr + 3\,^{1}_{0}n$$

Part a. Step 1.	Solution: (Section 31-2)	
Determine the total mass of the reactants and the total mass of the products.	reactants	products
	$^{235}_{92}U$ 235.0439 u	$^{141}_{56}Ba$ 140.9144 u
	$^{1}_{0}n$ 1.0087 u	$^{92}_{36}Kr$ 91.9263 u
	236.0526 u	$3\,^{1}_{0}n$ 3.0261 u
		235.8668 u

Part a. Step 2.	mass difference = 236.0526 u - 235.8668 u
Determine the mass difference.	mass difference = 0.1858 u

Part a. Step 3.	The reactants have a greater mass than the products; therefore, the reaction can occur spontaneously.
Determine the energy released in the reaction.	(0.1858 u)(931.5 MeV/u) = 173 MeV

Nuclear Chain Reaction

The neutrons released in a typical fission can be used to cause other nuclei to fission. If enough U-235 or Pu-239 is available, it is possible to create a self-sustaining **chain reaction**. The minimum amount of fissionable material required for a self-sustaining chain reaction is known as the critical mass. The critical mass depends on a number of factors, including: the moderator used to slow the neutrons since the probability of causing U-235 or Pu-239 to fission increases with slow-moving neutrons, the type of fuel used, e.g., U-235 or Pu-239, and whether or not the fuel is enriched. Only 0.7% of naturally occurring uranium is the fissionable U-235 and in order to have a self-sustaining chain reaction the percentage of U-235 must be increased.

In a **nuclear reactor**, the chain reaction is controlled and the energy is released gradually. In an "atomic bomb," the chain reaction is uncontrolled and the energy release occurs in a few moments of time.

ANSWER: Neutrons, which are electrically neutral, are not repelled by the positive charges located in the nucleus of the atom. As a result they can approach the nucleus to the point where the strong nuclear force plays a dominant role.

Nuclear Fusion

Nuclear fusion is the process by which small nuclei combine to form heavier nuclei. In the reaction, large amounts of energy are released. The force of electrostatic repulsion keeps the nuclei from combining, and therefore the fusion of light nuclei does not occur spontaneously. In order for fusion to happen, it is necessary for the nuclei to collide while traveling at high velocity. Such velocities do occur if the nuclei are part of a hot gas in which temperature approximates 100 million degrees Kelvin. Temperatures of this magnitude are found in the interior of stars, and thus fusion accounts for the enormous energy released by the Sun and other stellar objects.

Uncontrolled fusion reactions have been achieved in the form of the thermonuclear, or hydrogen, bomb. This is because the temperature required for the fusion is achieved by detonating a fission or atomic bomb. The fission bomb creates the necessary temperatures and pressures for the fusion process to occur. However, the difficulty of achieving and sustaining the very high temperatures needed for fusion has thus far prevented the construction of a controlled fusion reactor.

TEXTBOOK QUESTION 20. Light energy emitted by the Sun and stars comes from the fusion process. What conditions in the interior of stars make this possible?

ANSWER: The Sun has a very large mass and because of the large gravitational force exerted by this mass it has a very high density. Besides the large gravitational force and high density, the temperature in the Sun's interior is very high. As a result of these conditions, the strong nuclear force of attraction is able to overcome Coulomb repulsion between adjacent nuclei and nuclear fusion occurs.

EXAMPLE PROBLEM 6. a) Determine the energy released when the following fusion reaction occurs:

$$^1_1H \ + \ ^3_1H \ \rightarrow \ ^4_2He$$

b) Determine the amount of energy in joules produced per u of tritium consumed in the reaction.

c) Determine the amount of energy in joules produced per kg of tritium consumed in the reaction.

Part a. Step 1.	Solution: (Section 31-3)
Determine the total mass of the reactants and the total mass of the products.	reactants products $${}_{1}^{1}\text{H} \quad 1.007825 \text{ u} \qquad\qquad {}_{2}^{4}\text{He} \quad\quad 4.002613 \text{ u}$$ $${}_{1}^{3}\text{H} \quad 3.016049 \text{ u}$$ $$\overline{\quad 4.023874 \text{ u}\quad} \qquad\qquad\qquad \overline{\quad 4.002613 \text{ u}\quad}$$

Part a. Step 2.	mass difference = 4.023874 u - 4.002613 u
Determine the mass difference and the energy released per reaction.	mass difference = 0.021261 u $$(0.021261 \text{ u})(931.5 \text{ MeV/u}) = 19.80 \text{ MeV}$$

Part b. Step 1.	$(19.80 \text{ MeV})(1.6 \times 10^{-13} \text{ J/MeV}) = 3.169 \times 10^{-12} \text{ J}$
Express the energy released per reaction in terms of joules.	

Part b. Step 2.	$(3.169 \times 10^{-12} \text{ J})/(3.016049 \text{ u}) = 1.050 \times 10^{-12} \text{ J/u}$
Determine the energy released per atomic mass unit of tritium consumed in the reaction.	

Part c. Step 1.	$1 \text{ u} = 1.66 \times 10^{-27} \text{ kg}$
Determine the energy released per kg of tritium consumed in the reaction.	$(1.050 \times 10^{-12} \text{ J/u})(1 \text{ u}/1.66 \times 10^{-27} \text{ kg}) = 6.33 \times 10^{14} \text{ J/kg}$

EXAMPLE PROBLEM 7. Suppose the reaction discussed in problem 6 could be used to produce electrical power in a nuclear power plant of the future. Determine the amount of tritium that would be consumed each day if the output from the plant is to be 100,000 watts and the plant is 33.3% efficient.

Part a. Step 1. Determine the input power required to generate 100,000 watts of output power.	Solution: (Section 31-3) efficiency = output/input x 100% 33.3% = 100,000 watts/input x 100% input = (100,000 watts)(100%)/(33.3%) input = 300,000 watts
Part a. Step 2. Determine the energy in joules of input energy during a 24 hour period.	Note: 24 hours = 86,400 second and 1 watt = 1 J/s. 300,000 J/s x 86400 s = 2.59×10^{10} J
Part a. Step 3. Determine the amount of tritium consumed.	In part c of problem 6 it was determined that 6.33×10^{14} joules of energy is produced by each kg of tritium consumed. m = $(2.59 \times 10^{10}$ J$)/(6.33 \times 10^{14}$ J/kg$)$ m = 4.09×10^{-5} kg = 0.0493 g

Dosimetry

Dosimetry refers to measuring the quantity, or dose, of radiation that passes through a material. There are several units used for this purpose. The activity of a radioactive isotope is measured in terms of curies (Ci) or microcuries (μCi) where 1 Ci = 3.70×10^{10} disintegrations per second and 1 μCi = 10^{-6} Ci.

The **roentgen (R)** is a unit of dosage that was previously defined in terms of the amount of ionized air produced by the radiation. One roentgen (1 R) is the amount of radiation that will produce 2.1×10^9 ion pairs/cm^3 of air at STP. Today the roentgen is defined as that amount of x or γ radiation that deposits 0.878×10^{-2} J of energy per kg of air. The roentgen has been replaced by the **rad** where 1 rad is the amount of radiation which deposits 10^{-2} J of energy per kg of absorbing material.

Units have been developed which refer to the biological damage caused by radiation. The **rem** (rad equivalent man) is such a unit and can be used for all types of radiation.

rem = rad x RBE

RBE is the relative biological effectiveness of a certain type of radiation. It is defined as the number of rad of x or γ radiation that produces the same biological damage as 1 rad of the given radiation. The RBE of x and γ rays is 1.0, β particles 1.0, α particles 10-20, slow neutrons 3-5, and fast neutrons and protons 10. While we are all subject to about 0.13 rem/year of background radiation from our environment, large doses of radiation can cause radiation sickness leading to death. However, the length of time over which the body receives the dose as well as the size of the dose are important.

EXAMPLE PROBLEM 8. At one time it was possible to purchase a watch that contained the radium isotope Ra-226. This isotope emits 4.76 MeV alpha particles which struck a special material in the paint used for the dial of the watch. This material gave off visible light when struck by the alpha particles. Assume that the dial contained 5.00 microcuries of the isotope and that the wearer had a mass of 70.0 kg and wore the watch 24 hours per day. Determine the a) number of decays per second, and b) yearly dosage in rem to the wearer if 10% of the disintegrations interact with the person's body and deposit all of their energy. Assume an RBE of 10 for the alpha particles. c) The maximum dosage recommended by the U.S. government is 5.0 rem/year. Does the answer to part b exceed this recommendation?

Part a. Step 1. Determine the number of decays per second.	Solution: (Section 31-5) 1.0 Ci = 3.7 x 10^{10} decays/s; thus $(5.00 \times 10^{-6}$ Ci$)[(3.7 \times 10^{10}$ decays/s$)/(1$ Ci$)] = 1.85 \times 10^{5}$ decays/s
Part b. Step 1. Determine the amount of energy absorbed by the person's body each second in MeV. Express this energy in joules.	Only 10% (i.e., 0.10) of the disintegrations deposit their energy in the person's body. $(0.10)(1.85 \times 10^{5}$ decays/s$) = (1.85 \times 10^{4}$ decays/s$)$ $(1.85 \times 10^{4}$ decays/s$)(4.76$ MeV/decay$) = 8.81 \times 10^{4}$ MeV $(8.81 \times 10^{4}$ MeV$)(1.6 \times 10^{-13}$ J/MeV$) = 1.41 \times 10^{-8}$ J/s
Part b. Step 2. Determine the amount of energy absorbed per kg of mass by the person's body.	$(1.41 \times 10^{-8}$ J/s$)/(70$ kg$) = 2.01 \times 10^{-10}$ J/kg s

Part b. Step 3. Convert the answer in step 2 to rad/s.	Note: 1 rad = 10^{-2} J/kg $(2.01 \times 10^{-10}$ J/kg s$)(1$ rad/10^{-2} J/kg$)$ = 2.01×10^{-8} rad/s
Part b. Step 4. Determine the number of rads absorbed by the body in 1 year.	1 year = 3.15×10^{7} seconds $(2.01 \times 10^{-8}$ rad/s$)(3.15 \times 10^{7}$ s/year$)$ = 0.634 rad/year
Part b. Step 5. Determine the dosage in rem/year.	dosage in rem = rad x RBE $\quad\quad\quad\quad$ = (0.634 rad/year)(10) dosage in rem = 6.34 rem/year
Part c. Step 1. Does the dosage exceed recommended limit?	Since the maximum recommended dosage for a person is 5.0 rem/year, the 6.34 rem/year due to the watch is in excess of the recommended limit set by the government.

Detection of Radiation

A number of ways have been developed to measure radiation dose. A film badge is worn by people who work around sources of radiation. When developed, the darkness of the film is a measure of the radiation dose received. The **Geiger counter, scintillation counter,** and **semiconductor detector** all detect the presence and level of activity of radiation. The paths of elementary particles can be photographed using devices such as **bubble chambers, cloud chambers**, and **spark chambers**.

PROBLEM SOLVING SKILLS

For problems which involve determination of a missing particle in a nuclear reaction:

1. If the reaction is given in the short form, re-write the reaction in the long form.
2. Apply the law of conservation of charge to determine the atomic number (Z) of the nuclide.
3. Apply the law of conservation of nucleon number to determine the mass number of the nuclide.
4. Use the periodic table to identify the unknown nuclide.

For problems involving the calculation of the Q-value of a nuclear reaction:

1. Determine the total mass of the reactants and the total mass of the products.
2. Determine the mass difference between the sum of the reactants and the sum of the products.
3. If the mass difference is expressed in kg, use $E = (\Delta m)c^2$ to determine the Q-value in joules. The Q-value may be expressed in MeV by using the conversion factor $1 \text{ MeV} = 1.6 \times 10^{-13} \text{J}$.
4. If the mass difference is expressed in atomic mass units (u), then the Q-value may be expressed in MeV by multiplying the mass difference by 931.5 MeV/u.
5. If the Q-value is positive, then the reaction is exothermic or exoergic and can occur spontaneously. If the Q-value is negative, then the reaction is endothermic or endoergic and energy must be added for the reaction to occur.

If the problem involves dosimetry:

1. Determine the number of decays which occur per second.
2. Determine the amount of energy absorbed per kg of mass per second.
3. Convert the energy absorbed per kg of mass per second to rads per second.
4. If required, determine the number of rads absorbed per year.
5. If the RBE value is given, then the yearly dosage in rem may be determined.

SOLUTIONS TO SELECTED TEXT PROBLEMS

TEXTBOOK PROBLEM 3. Is the reaction $^{238}_{92}U + n \rightarrow \, ^{239}_{92}U + \gamma$ possible with slow neutrons? Explain.

Part a. Step 1.	Solution: (Section 31-1)
Determine the total mass of the reactants and the total mass of the products.	reactants $\qquad$ products
	$^{238}_{92}U$ 238.050784 u $\qquad$ $^{239}_{92}U$ 239.054289 u
	$^{1}_{0}n \qquad$ 1.008666 u
	———————— $\qquad$ ————————
	$\qquad$ 239.05945 u $\qquad\qquad$ 239.054289 u

Part a. Step 2.	mass difference = 239.05945 u - 239.054289 u
Determine the mass difference between the products and the reactants.	mass difference = 0.005166 u
	The products have less mass than the reactants; therefore, the reaction is exothermic.

Part a. Step 3.	$Q = (M_a + M_X - M_b - M_Y) c^2$
Determine the value of Q in joules and MeV. Note: if Q > 0, then the reaction is exothermic and no threshold energy is required.	$Q = (\Delta m) c^2$
	$\Delta m = (0.005166\ u)(1.66 \times 10^{-27}\ kg/u) = 8.5756 \times 10^{-30}\ kg$
	$Q = (8.5756 \times 10^{-30}\ kg)(3.0 \times 10^8\ m/s)^2 = 7.718 \times 10^{-13}\ J$
	$Q = (7.718 \times 10^{-13}\ J)(1\ MeV/1.6 \times 10^{-13}\ J) = 4.82\ MeV$
	Alternate solution:
	$Q = (0.005166\ u)(931.5\ MeV/u) = 4.82\ MeV$
	Since Q > 0, the reaction is possible.

TEXTBOOK PROBLEM 18. How many fissions take place per second in a 200-MW reactor? Assume 200 MeV is released per fission.

Part a. Step 1.	Solution: (Section 31-2)
Convert 200 MW to joules per second.	$(200\ MW)[(1.0 \times 10^6\ watt)/(1\ MW)][(1\ J/s)/(1\ watt)] =$
	$2.0 \times 10^8\ J/s$

Part a. Step 2.	Let n = number of fissions per second
Determine the number of fissions per second.	power = n (energy per fission)
	$2.0 \times 10^8\ J/s = n\ (200\ MeV)[(1.6 \times 10^{-13}\ J)/(1.0\ MeV)]$
	$n = (2.0 \times 10^8\ J/s)/(3.2 \times 10^{-11}\ J)$
	$n = 6.25 \times 10^{18}$ fissions each second.

TEXTBOOK PROBLEM 25. What is the average kinetic energy of protons at the center of a star where the temperature is 10^7 K? [*Hint*: use Eq. 13-8.]

Part a. Step 1.	Solution: (Section 31-3)
Determine the average kinetic energy of the protons in joules.	$KE = 3/2 \; k \; T = (3/2)(1.38 \times 10^{-23} \; J/K)(10^7 \; K)$ $KE = 2.07 \times 10^{-16} \; J$
Part a. Step 2.	$KE = (2.07 \times 10^{-16} \; J)(1 \; eV/1.6 \times 10^{-19} \; eV)$
Express this energy in electron volts (eV).	$KE = 1290 \; eV$

TEXTBOOK PROBLEM 34. How much energy (J) is contained in 1.00 kg of water if its natural deuterium is used in the fusion reaction of Eq. 31-8a? Compare to the energy obtained from the burning of 1.0 kg of gasoline, about 5×10^7 J.

$${}_{1}^{2}H + {}_{1}^{2}H \rightarrow {}_{1}^{3}H + {}_{1}^{1}H \qquad (4.03 \; MeV) \qquad (Eq. \; 31\text{-}8a)$$

Part a. Step 1.	Solution: (Section 31-3)
Determine the number of moles of water containing deuterium in 1000 grams of water.	From Appendix B, relative abundance of deuterium is 0.015% or 1.5×10^{-4}. Also, 1 mole of H_2O containing deuterium has a MW of 18 grams. $[(1.5 \times 10^{-4})(1000 \; g)][(1 \; mol)/(18 \; g)] = 8.33 \times 10^{-3} \; mol$
Part a. Step 2.	The formula for water is H_2O. Therefore, there are two deuterium nuclei in each molecule.
Determine the number of deuterium nuclei in 1000 g of water.	$(8.33 \times 10^{-3} \; mol)(6.02 \times 10^{23} \; molecules/mol)(2 \; nuclei/molecule) =$ $1.00 \times 10^{22} \; nuclei$
Part a. Step 3.	As shown in equation 31-8, two deuterium nuclei are involved in each reaction.
Determine the energy contained in the 1.0 kg of water due to the deuterium.	$(1.00 \times 10^{22} \; nuclei)[(4.03 \; MeV/reaction)/(2 \; nuclei/reaction)] =$ $2.02 \times 10^{22} \; MeV$ $(2.02 \times 10^{22} \; MeV)(1.6 \times 10^{-13} \; J/MeV) = 3.23 \times 10^9 \; J$

Part a. Step 4.	$(3.23 \times 10^9$ J deuterium in water$)/(5 \times 10^7$ J gasoline$) \approx 65$
Compare the energy found in step a. 3 to the energy found in 1.0 kg of gasoline.	There is approximately 65 times more energy contained in 1.0 kg of water as compared to 1.0 kg of gasoline.

TEXTBOOK PROBLEM 46. What is the mass of a 1.00 μCi $^{14}_{6}C$ source?

Part a. Step 1.	Solution: (Section 31-5)
Determine the decay constant of C-14 in decays per second. Note: there are 3.156×10^7 seconds in 1 year.	$\lambda = (0.693)T_{1/2}$ where $T_{1/2} = 5730$ years
	$\quad = (0.693)/(5730 \text{ years}) = 1.21 \times 10^{-4}$ decays/year
	$\lambda = (1.21 \times 10^{-4}/\text{year})[(1 \text{ y})/(3.156 \times 10^7 \text{ s})] = 3.83 \times 10^{-12}$ decays/s
Part a. Step 2.	activity $= \lambda N$
Determine the number (N) of C-14 nuclei present in the sample.	$(1.0 \times 10^{-6}$ Ci$)[(3.7 \times 10^{10}$ decays/s$)/(1$ Ci$)] = (3.83 \times 10^{-12}$ decays/s$)N$
	$N = 9.66 \times 10^{15}$ nuclei
Part a. Step 3.	$(9.66 \times 10^{15}$ nuclei$)[(1 \text{ mol})/(6.02 \times 10^{23}$ nuclei$)] = 1.60 \times 10^{-8}$ mol
Determine the number of moles of C-14 present.	
Part a. Step 4.	$m = (1.60 \times 10^{-8}$ mol$)[(14 \text{ g})/(1 \text{ mol})] = 2.22 \times 10^{-7}$ g
Determine the mass of the sample.	$m = (2.22 \times 10^{-7}$ g$)[(1 \text{ μg})/(10^{-6} \text{ g})] = 0.222$ μg C-14

TEXTBOOK PROBLEM 53. One means of enriching uranium is by diffusion of the gas UF$_6$.

Calculate the ratio of the speeds of molecules of this gas containing $^{235}_{92}U$ and $^{238}_{92}U$, on which this process depends.

Part a. Step 1.	Solution: (Section 31-1)
Determine the molecular weight of each molecule.	for U-235: UF_6 235 u + 6(19 u) = 349 u
	for U-238: UF_6 238 u + 6(19u) = 352 u

Part a. Step 2.	$KE = \frac{1}{2} m v^2$ but $KE = 3/2\, k\, T$
Derive a formula for the rms speed of each molecule a function of temperature.	$\frac{1}{2} m v^2 = 3/2\, k\, T$
	$v^2 = 3\, k\, T/m$
	$v_{rms} = (3\, k\, T/m)^{\frac{1}{2}}$

Part a. Step 3.	$v_{235}/v_{238} = (3\, k\, T/m_{235})^{\frac{1}{2}}/(3\, k\, T/m_{238})^{\frac{1}{2}}$
Determine the ratio of the rms speeds of the molecules.	$= (m_{238}/m_{235})^{\frac{1}{2}} = [(352\ u)/(349\ u)]^{\frac{1}{2}} = (1.0086)^{\frac{1}{2}}$
	$v_{235}/v_{238} = 1.0043$

CHAPTER 32

ELEMENTARY PARTICLES

OBJECTIVES

After studying the material of this chapter, the student should be able to:

- describe how the Van de Graaff generator, cyclotron, synchrotron, and linear accelerator can be used to accelerate charged particles. Explain the advantages to be gained by using one machine over the others.
- describe each of the four forces found in nature. List the fundamental particle associated with each force.
- distinguish between a particle and its antiparticle.
- classify a particle according to its family group: photon, lepton, meson, baryon and list the particle associated with interactions within each family.
- write out the law of conservation of lepton number and the law of conservation of baryon number.
- list the quantities which must be conserved in order for a particular reaction or decay process to occur. Apply the conservation laws to determine if a particular reaction occurs.
- distinguish between a particle and a resonance.
- explain what is meant by strangeness and conservation of strangeness.
- give the names associated with the six types of quarks and describe properties associated with each quark.
- explain why it was necessary to introduce the concept of quark color.
- describe what is meant by the grand unified theory.

KEY TERMS AND PHRASES

Van de Graaff generator electrostatically accelerates charged particles through potential differences as high as 30 MV along a straight line.

cyclotron accelerates protons or electrons through a potential difference into a magnetic field directed perpendicular to their path. As their speed increases, the radius of the circle in which they travel increases until they reach the outer edge of the apparatus and either strike a target within the cyclotron or leave the cyclotron and strike an external target.

cyclotron frequency is the number of revolutions per second made by the charged particle.

synchrotron or synchrocyclotron was developed as a way to compensate for the mass increase with increasing speed. In this type of accelerator, the radius of the particle's path is held constant while the magnetic field increases as the particle's mass increases.

linear accelerator, or LINAC, accelerates a charged particle in steps along a straight line.

colliding beam accelerator causes two beams of charged particles moving in opposite directions to collide head on. The kinetic energy of the colliding particles is available for causing reactions or creating new particles.

electromagnetic force is explained by the wave-particle duality. It is possible to suggest that the electromagnetic force between charged particles is the result of an electromagnetic field produced by one particle that affects the second particle (wave theory) or by an exchange of photons (γ particles) between the charged particles (particle theory).

strong nuclear force is postulated to be caused by pi meson or pion.

weak nuclear force accounts for beta decay (β). The weak nuclear force is presumed to be due to particles known as W^+, W^-, and Z^o.

graviton is postulated to be responsible for the gravitational force.

positron is the antiparticle of the electron. The positron has the same mass as an electron but has the opposite charge. When the positron encounters an electron, the two annihilate each other and a gamma ray is produced.

antiparticle of the proton is the antiproton and the antiparticle of the neutron is the antineutron. A few particles, e.g., the photon and the neutral pion (π^o), have no antiparticle.

leptons (light particles) include the electron, muon, and two types of neutrino: the electron neutrino (ν_e) and muon neutrino (ν_μ) as well as their antiparticles.

hadrons are composed of two subgroups, the **baryons** (heavy particles) and the **mesons** (intermediate particles). The baryons include the proton and all heavier particles through the omega particle plus their antiparticles. The mesons include the pion, kaon, and eta particles plus their antiparticles.

conservation laws for reaction and decay processes to occur include energy (including mass), linear momentum, charge, angular momentum (including spin), baryon family number, electron-lepton family number, and muon-lepton family number.

resonances are super short lived particles which decay via the strong interaction. They do not travel far enough in a bubble chamber or a spark chamber to be detected and their existence is inferred because of decay products which can be detected.

strangeness and the **principle of conservation of strangeness** provide an explanation for the observation of certain reactions and also why other reactions do not occur. They also provide an explanation for the "long" lifetimes, 10^{-10} s to 10^{-8} s, for certain particles which interact via the strong interaction.

quarks are composed of four particles which are named up (u), down (d), strange (s), and charm and carry a fractional charge, either ⅓ or ⅔ the charge on the electron. The quarks have antiparticles called antiquarks. Two more quarks called top (t) or truth and bottom (b) or beauty have been postulated.

color and flavor are properties associated with quarks. Each of the flavors has three colors; red, green, and blue, while the antiquarks are colored antired, antigreen, and antiblue.

gluons are particles which transmit the strong force. According to the theory there are eight gluons. They are massless and six of the gluons have color charge.

electroweak theory proposes that the weak force and the electromagnetic force are two different manifestations of a single, more fundamental, electroweak interaction.

grand unified theory suggests that at very short distances (10^{-31} m) and very high energy, the electromagnetic, weak, and strong forces are different aspects of a single underlying force. In this theory the fundamental difference between quarks and leptons disappears. Attempts are presently being made to incorporate the four forces found in nature - gravity, electromagnetic, weak, and strong forces - into a single theory.

SUMMARY OF MATHEMATICAL FORMULAS

cyclotron frequency	$f = 1/T$ $f = (qB)/(2\pi m)$	The cyclotron frequency (f) is the number of revolutions per second made by the charged particle.
type of force	relative strength	field particle
strong nuclear	1	mesons/gluons, very short distance ($\approx 10^{-15}$ m)
electromagnetic	10^{-2}	photon
weak nuclear	10^{-9}	$W\pm$ and Z^o, extremely short range ($\approx 10^{-15}$ m)
gravitational	10^{-38}	graviton

CONCEPT SUMMARY

Particle Accelerators

As discussed in Chapter 25, the sharpness or resolution of the details of an image is limited by the wavelength of the incident radiation. The de Broglie wavelength of particles such as electrons or protons is given by $\lambda = h/mv$, and particles which have greater momentum have a shorter wavelength. There are a number of different types of particle accelerators available to produce the high energy, short wavelength particles needed to investigate nuclear structure.

The **Van de Graaff generator** electrostatically accelerates charged particles through potential differences as high as 30 MV along a straight line. The **cyclotron** accelerates protons or electrons through a potential difference into a magnetic field directed perpendicular to their path. The charged particles move within two D-shaped cavities, called **dees**, and each time they move into the space between the dees a potential difference is applied which increases their speed. As their speed increases, the radius of the circle in which they travel increases until they reach the outer edge of the dee and either strike a target within the cyclotron or leave the cyclotron and strike an external target.

The **cyclotron frequency** is given by $f = 1/T = (qB)/(2\pi m)$.

f is the cyclotron frequency, which is the number of revolutions per second made by the charged particle. T is the period in seconds of the particle's motion. q is the charge in coulombs on the particle. B is the strength in tesla of the magnetic field which is directed perpendicular to the particle's path and m is the mass of the particle in kg.

The cyclotron frequency determines the frequency of the applied voltage needed to accelerate the charged particle. The frequency does not depend on the radius of the circle in which the particle is traveling but it does depend on the particle's mass. As the speed increases, the mass of the particle increases according to Einstein's formula $m = m_o/(1 - v^2/c^2)^{1/2}$. In order to continue to accelerate the particle, the cyclotron frequency must be reduced as the mass of the particle increases. A device known as a **synchrocyclotron** or **synchrotron** was developed for this purpose.

The synchrotron was developed as an alternative way to compensate for the mass increase with increasing speed. In this type of accelerator, the radius of the particle's path is held constant while the magnetic field increases as the particle's mass increases. The radius of the synchrotron may exceed 1.0 km and accelerate protons to energies exceeding 500 GeV.

In both the cyclotron and the synchrotron the charged particle travels in a circle and undergoes centripetal acceleration. The accelerated charges radiate electromagnetic energy and considerable energy is lost through radiation. This effect is called synchrotron radiation.

In a **linear accelerator**, or **LINAC**, a charged particle is accelerated in steps along a straight line. Linear accelerators have been constructed which have accelerated electrons to over 20 GeV.

No magnetic fields are used in a linear accelerator.

In a **colliding beam accelerator,** two beams of charged particles moving in opposite directions collide head on. If the particles have the same mass and energy, then the momentum before and after is zero. If the kinetic energy after impact is zero, then all of the initial kinetic energy of the colliding particles is available for causing reactions or creating new particles.

EXAMPLE PROBLEM 1. Determine the kinetic energy in GeV of a proton whose de Broglie wavelength is 1.00 fm.

Part a. Step 1.	Solution: (Section 32-1)
Use the de Broglie formula to determine the energy. Note: the energy of the proton is in the order of GeV. Assume that the proton's velocity is approximately that of the speed of light.	$E = h \, c/\lambda$ where $\lambda = 1.00$ fm $= 1.00 \times 10^{-15}$ m
	$\quad = [(6.63 \times 10^{-34} \text{ J s})(3.0 \times 10^{8} \text{ m/s})]/(1.00 \times 10^{-15} \text{ m})$
	$E = 1.99 \times 10^{-10}$ J
	$E = (1.99 \times 10^{-10} \text{ J})(1.00 \text{ eV}/1.6 \times 10^{-19} \text{ J}) = 1.24 \times 10^{9}$ eV
	$\quad = (1.24 \times 10^{9} \text{ eV})(1.00 \text{ GeV}/10^{9} \text{ eV})$
	$E = 1.24$ GeV

EXAMPLE PROBLEM 2. A cyclotron is used to accelerate protons to an energy of 30 MeV. The strength of the magnetic field is 1.00 T. Determine the a) radius of the proton's orbit when the proton's energy reaches 30 MeV, b) cyclotron frequency, and c) period of the proton's motion.

Part a. Step 1.	Solution: Section (32-2)
Determine the proton's velocity when its energy reaches 30 MeV.	$KE = \frac{1}{2} m v^2$
	$(30 \text{ MeV})(1.6 \times 10^{-13} \text{ J/MeV}) = (1.67 \times 10^{-27} \text{ kg}) v^2$
	$v^2 = 5.75 \times 10^{15}$ m²/s² and solving for v gives
	$v = 7.58 \times 10^{7}$ m/s
	At this speed, the proton's mass is no longer equal to its rest mass; however, the relativistic mass is approximately equal to the rest mass ($m = 1.03 \, m_o$) and the rest mass will be used.

Part a. Step 2.	$F = q v B \sin \theta$ but $\theta = 90°$ and $\sin 90° = 1$
Solve for the radius of the proton's orbit.	The motion is circular; therefore, $F = m v^2/r$, then
	$q v B = m v^2/r$ and solving for r gives
	$r = (m v)/(q B)$
	$r = [(1.67 \times 10^{-27}$ kg$)(7.58 \times 10^7$ m/s$)]/[(1.6 \times 10^{-19}$ C$)(1.00$ T$)]$
	$r = 0.791$ m or 79.1 cm
Part b. Step 1.	For a particle traveling in a circle at constant speed (v),
Determine the cyclotron frequency.	$v = (2 \pi r)/T = 2 \pi r f.$
	From step a. 2, $r = (m v)/(q B)$; therefore,
	$r = (m\, 2 \pi r f)/(q B)$ Solving for f gives
	$f = (q B)/(2 \pi m)$
	$f = [(1.6 \times 10^{-19}$ C $\times 1.00$ T$)]/[(2 \pi \times 1.67 \times 10^{-27}$ kg$)]$
	$f = 1.53 \times 10^7$ Hz
Part c. Step 1.	The period is inversely related to the frequency; therefore,
Determine the period of the proton's motion.	$T = 1/f = 1/(1.53 \times 10^7$ Hz$)$
	$T = 6.55 \times 10^{-8}$ s

EXAMPLE PROBLEM 3. A head-on collision occurs between two protons having equal kinetic energy. Determine the minimum kinetic energy of each proton in order to produce the following reaction: $p + p \rightarrow n + p + \pi^+$

Part a. Step 1.	Solution: (Section 32-3)	
Determine the total mass of the reactants and the total mass of the products. Note: in table 32-2 of the text, the rest mass is expressed in MeV/c², therefore, use MeV/c² in solving the problem.	reactants p 938.3 MeV/c² p 938.3 MeV/c² ――――― 1876.6 MeV/c² total mass of reactants (in MeV/c²)	products n 939.6 MeV/c² p 938.3 MeV/c² π⁺ 139.6 MeV/c² ――――― 2017.5 MeV/c² total mass of products (in MeV/c²)

Part a. Step 2.	
Determine the mass difference (expressed in MeV) between the products and the reactants.	mass difference = 2017.5 MeV/c² - 1876.6 MeV/c² mass difference = 140.9 MeV/c²

Part a. Step 3.	
Determine the kinetic energy of each of the colliding protons.	The mass of the products is greater than the mass of the reactants; therefore, the kinetic energy of the colliding protons must provide the energy necessary for the reaction to occur. Since the protons have equal kinetic energy, then each proton must have [140.9 MeV/c²](c²/2) = 70.5 MeV of kinetic energy

The Four Forces in Nature

Because of the wave-particle duality, it is possible to suggest that the electromagnetic force between charged particles is the result of an electromagnetic field produced by one particle that affects the second particle (wave theory) or by an exchange of photons (γ particles) between the charged particles (particle theory).

In 1935, Hidecki Yukawa postulated the existence of a particle that produces the strong nuclear force that holds the atomic nucleus together. This particle, now known as the pi meson or pion and represented by the symbol π, was discovered in 1947. The particle can be charged positively (π+) or negatively (π-) or be uncharged (πo). The mass of the π+ or π- particle is 140 MeV/c² or 273 times the rest mass of the electron while the mass of the πo particle is 135 MeV/c², which is 264 times the rest mass of the electron.

The **weak nuclear force** is used to account for beta decay (β). The weak nuclear force is presumed to be due to particles known as W^+, W^-, and Z^o. In 1983, the discovery of the W and Z particles was announced by a group of scientists led by Carlo Rubbia. The group used the high-energy accelerator at CERN.

The gravitational force is believed to be due to a particle known as the **graviton**. The graviton has not yet been observed. The electromagnetic force and the gravitational force are known as "long-range" forces, decreasing as the square of the distance between interacting particles. The strong nuclear force and the weak nuclear force are very short range forces, limited to distances of approximately 10^{-15} m. This distance is the approximate size of the atomic nucleus. The following table lists each type of force, relative strength of the particular force as compared to the strong nuclear force, and the name of the particle credited with producing the force.

type	relative strength	field particle
strong nuclear	1	mesons/gluons
electromagnetic	10^{-2}	photon
weak nuclear	10^{-9}	$W\pm$ and Z^o particle
gravitational	10^{-38}	graviton

TEXTBOOK QUESTION 7. Which of the four interactions (strong, electromagnetic, weak, gravitational) does an electron take part in? A neutrino? A proton?

ANSWER: The electron and proton both possess electric charge and therefore interact with other charged particles via the electromagnetic force and photon-photon exchange. The neutrino does not carry electric charge and therefore does not interact via the electromagnetic force. All three particles interact via the gravitational force and the weak nuclear force. The proton is a baryon and interacts via the strong nuclear force as well as the weak nuclear force. The electron and neutrino are leptons and do not take part in interactions via the strong force.

Particles and Antiparticles

The first **antiparticle**, the **positron**, was discovered in 1932. The positron has the same mass as an electron but has the opposite charge. When the positron encounters an electron, the two annihilate one another and a gamma ray is produced. The energy of the gamma ray is equal to the mass equivalent of the electron and positron plus any kinetic energy the two possessed at the time of interaction.

It is now known that most particles have antiparticles, e.g., proton (p)-antiproton ($\bar{p}$), neutron (n)-antineutron ($\overline{\text{n}}$). A few particles, e.g., the photon and the neutral pion (π^o), have no antiparticle.

TEXTBOOK QUESTION 3. What would an "antiatom," made up of the antiparticles to the constituents of normal atoms, consist of? What might happen if antimatter, made of such antiatoms, came in contact with our normal world of matter?

ANSWER: The antiatom would contain the antiparticles to those found in the normal atom. A normal atom contains electrons, protons, and neutrons. Therefore, the antiatom would contain positrons, antiprotons, and antineutrons. The conservation laws which apply to atoms would also apply to antiatoms. It is known that antimatter-matter interactions, such as electron-positron interactions, result in the annihilation of the particles involved with the production of their energy equivalent.

Particle Classification

Table 32-2 in the textbook lists a number of particles according to their family group and the way they interact. Photons interact only through the electromagnetic force, while **leptons** (light particles) interact only through the weak nuclear force. The leptons include the electron, **muon**, and two types of **neutrino**: the electron neutrino (v_e) and muon neutrino (v_μ) as well as their antiparticles.

The **hadrons** interact through the strong nuclear force. The hadrons are composed of two subgroups, the **baryons** (heavy particles) and the **mesons** (intermediate particles). The baryons include the proton and all heavier particles through the omega particle plus their antiparticles. The mesons include the pion, kaon, and eta particles plus their antiparticles.

Particle Interactions and Conservation Laws

In order for reaction and decay processes to occur, seven quantities must be conserved: energy (including mass), linear momentum, charge, angular momentum (including spin), baryon family number, electron-lepton family number, and muon-lepton family number.

Energy is conserved if the sum of the rest energies and total kinetic energy of the reactants equals the sum of the rest energies and total kinetic energy of the products. Linear momentum must be conserved. In the case of a decay of a nucleus at rest, linear momentum can be satisfied if the decay products travel in different directions such that the total final momentum equals zero.

The law of conservation of charge is satisfied if the sum of the charges on the reactants equals the sum of the charges on the products.

Angular momentum is satisfied if the sum of the spin angular momentum quantum numbers of the reactants equals that of the products. The spin angular momentum of each particle is a multiple of $h/2\pi$ or $\hbar$. For example, the spin angular momentum of an electron equals $\pm\frac{1}{2}(h/2\pi)$, where the + or - is used since the spin can be up (+) or down (-). The spin angular momentum quantum number for each particle listed in table 32-2 is as follows:

category	particle name	spin quantum number
leptons	electron neutrino	½
	muon neutrino	½
	electron	½
	muon	½
photons	photon	1
hadrons		
mesons	pion	0
	kaon	0
	eta	0
baryons	proton	½
	neutron	½
	lambda	½
	sigma	½
	xi	½
	omega	3/2

The law of conservation of baryon number states that in every interaction the total baryon number remains unchanged. The baryon number for a baryon is $B = + 1$ and for an antibaryon $B = - 1$. All non-baryons have a baryon number $B = 0$.

The law of conservation of lepton number must be applied separately to each lepton family. In the electron-lepton family the electron and electron neutrino (ν_e) are $L_e = + 1$, while the positron and electron antineutrino ($\bar{\nu}_e$) are assigned $L_e = - 1$. The electron-lepton family number must be conserved in order for a reaction to occur.

The law of conservation of muon-lepton family number states that the muon-lepton number (L_μ) must be conserved for a process to occur. The μ^- and ν_μ particles have numbers $L_\mu = +1$ while for μ^+ and μ^- the number is $L_\mu = -1$. All other particles have a lepton number equal to zero.

TEXTBOOK QUESTION 4. What particle in a decay signals the electromagnetic interaction?

ANSWER: The appearance of a photon (γ) indicates that an electromagnetic interaction has occurred.

EXAMPLE PROBLEM 4. Use the conservation laws to determine if the following reactions may occur. In each case demonstrate which conservation laws are satisfied and which are violated. In part a assume that the reactants have sufficient kinetic energy for the reaction to occur. In parts b and c assume that the nucleus which decays is at rest.

a) $p + p \rightarrow p + \bar{n} + n$

b) $\kappa \rightarrow \mu^- + \bar{\nu}_\mu$

c) $e^- \rightarrow \nu_e + \gamma$

Part a. Step 1.	Solution: (Section 32-5 and 32-6)
Apply the law of conservation of charge.	$p + p \rightarrow p + \bar{n} + n$ $+1 + +1 \rightarrow +1 + 0 + 0$ (not allowed) The left side has a net charge of +2 while the right side has a net charge of +1. The law of conservation of charge is violated.
Part a. Step 2. Apply the law of conservation of angular momentum.	$\pm\frac{1}{2} + \pm\frac{1}{2} \rightarrow \pm\frac{1}{2} + \pm\frac{1}{2} + \pm\frac{1}{2}$ (not allowed) By inspection, there is no way of combining the possible values so that the left side will equal the right side of the equation. The law of conservation of angular momentum is violated.

Part a. Step 3. Apply the law of conservation of baryon number.	$1 + 1 \rightarrow 1 + (-1) + 1$ (not allowed) The left side has a net result of +2 while the right side has a net value of +1. The law of conservation of baryon family number is violated.
Part a. Step 4. Apply the law of conservation of lepton number.	$0 + 0 \rightarrow 0 + 0 + 0$ No leptons are involved in this process. The law is not applicable.
Part a. Step 5. Apply the laws of conservation of energy and conservation of of momentum.	The law of conservation of energy would be satisfied if the reactants have sufficient kinetic energies. Also, it is possible for the reaction to satisfy the law of conservation of momentum. However, for a reaction to be allowed, every law must be satisfied. Therefore, this reaction is not allowed.
Part b. Step 1. Apply the law of conservation of charge.	$\kappa^- \rightarrow \mu^- + \bar{\nu}$ charge $\quad -1 \rightarrow -1 + 0$ (allowed)
Part b. Step 2. Apply the law of conservation of angular momentum.	spin $\quad 0 \rightarrow \pm\frac{1}{2} + \pm\frac{1}{2}$ (allowed)
Part b. Step 3. Apply the law of conservation of baryon number.	baryon number $\quad 0 \rightarrow 0 + 0$ (not applicable)
Part b. Step 4. Apply the law of conservation of lepton number.	muon-lepton number $\quad 0 \rightarrow +1 + (-1)$ (allowed)

Part b. Step 5. Apply the laws of conservation of energy and conservation of momentum.	energy 493.8 MeV → 105.7 MeV + 0 (allowed) The missing energy can be accounted for in the kinetic energy of the product particles. momentum (allowed) The product particles could be moving such that they have equal but opposite momenta. All of the conservation laws are satisfied; therefore, the reaction is allowed.
Part c. Step 1. Apply the law of conservation of charge.	$e^- \rightarrow \nu_e + \gamma$ charge -1 → 0 + 0 (not allowed)
Part c. Step 2. Apply the law of conservation of angular momentum.	spin $\pm\frac{1}{2}$ → $\pm\frac{1}{2}$ + 1 (allowed)
Part c. Step 3. Apply the law of conservation of baryon number.	baryon 0 → 0 + 0 (not applicable) number
Part c. Step 4. Apply the law of conservation of lepton number.	electron-lepton 1 → 1 + 0 (allowed) number
Part c. Step 5. Apply the laws of conservation of energy and conservation of momentum.	energy 0.51 MeV → 0 MeV + 0 MeV (allowed) The product particles have no rest energy. However, the missing energy can be accounted for in the kinetic energy of each product particle.

| momentum | (allowed) |

The product particles could be moving such that they have equal but opposite momenta.

The law of conservation of charge is not satisfied; therefore, the reaction cannot occur.

Resonances

In addition to the particles listed in table 32-2, there are a great many particles which decay in 10^{-23} seconds or less. Such super short lived particles are known as **resonances** and decay via the strong interaction. They do not travel far enough in a bubble chamber or a spark chamber to be detected and their existence is inferred because of decay products which can be detected.

EXAMPLE PROBLEM 5. a) The estimated lifetime of a certain resonance is 8.6×10^{-23} s. Estimate the width of the resonance in joules and MeV. b) Express the width of the resonance in terms of the mass of an electron.

Part a. Step 1. Use the uncertainty principle to determine the width of the resonance in joules.	Solution: (Section 32-7) If the particle's lifetime is uncertain by an amount Δt, then its rest energy will be uncertain by an amount given by $\Delta E = (h/2\pi)/\Delta t$ $= (6.63 \times 10^{-34} \text{ J s}/2\pi)/(8.6 \times 10^{-23} \text{ s})$ $\Delta E = 1.23 \times 10^{-12} \text{ J}$
Part a. Step 2. Express the answer in MeV.	$\Delta E = (1.23 \times 10^{-12} \text{ J})(1 \text{ MeV}/1.6 \times 10^{-13} \text{ J})$ $\Delta E = 7.7 \text{ Mev}$
Part b. Step 1. Express the width of the resonance in terms of the rest mass of an electron.	The rest energy of an electron is 0.51 Mev. Therefore, the width of this particular resonance is approximately (7.7 MeV)/(0.51 MeV) = 15 times the rest mass of an electron

Strange Particles

The production and decay of certain particles led to the introduction of a new quantum number called **strangeness** (S). Particles are assigned a number called the strangeness number (see table 32-2). The use of strangeness and the principle of conservation of strangeness provides an explanation for the observation of certain reactions and also why other reactions do not occur. It also provides an explanation for the "long" lifetimes, 10^{-10} s to 10^{-8} s, for certain particles which interact via the strong interaction. Strangeness is conserved in the strong interaction but not in the weak interaction. Thus conservation of strangeness is an example of a "partially conserved" quantity.

Quarks

The four known leptons (e^-, μ^-, ν_e, ν_μ) seem to be truly elementary particles. There is no evidence that they have internal structure; also they have no measurable size and do not decay.

Experiments indicate that hadrons, mesons, and baryons do have internal structure, have a definite size (10^{-15} m in diameter) and, with the exception of the proton, do decay. It was proposed in 1963 that the hadrons are combinations of three constituent particles known as **quarks**. The quarks are given names up (u), down (d), and strange (s) and have a fractional charge, either ⅓ or ⅔ of the charge on the electron. The quarks have antiparticles called **antiquarks** and the properties of both quarks and antiquarks are listed in table 32-3. All of the hadrons known in 1963 could be constructed from a combination of quarks and antiquarks.

The up and down quarks were named because of their spin directions while the strange quark was named because it is associated with the concept of strangeness. In 1964, the existence of a fourth quark, called charmed (c) was postulated. The c quark has a charm number C = +1 while the c antiquark is $\bar{C}$ = - 1. In 1974, a particle called the J/ψ or ψ was found that did not fit the three-quark scheme but could be accounted for on the basis of a four-quark theory.

In recent years the existence of two more quarks called top (t) or truth and bottom (b) or beauty has been postulated. In 1977, a new meson was detected which is believed to be a combination of a beauty quark and its antiquark ($b\bar{b}$).

Quantum Electrodynamics

An extension of the quark theory suggests that quarks have a property called **color** with the distinction between the different quarks called **flavor**. Each of the flavors has three colors; red, green, and blue, while the antiquarks are colored antired, antigreen, and antiblue.

The concept of color is used to explain the force which binds the quarks together in a hadron. Each quark carries a **color charge** and the strong force between the quarks is called the **color force**. This theory of a strong force is called **quantum chromodynamics**, or QCD, in order to indicate that the theory refers to the force between color charges.

The particles which transmit the strong force are called **gluons**. According to the theory there are eight gluons. They are massless and six of the gluons have color charge.

Grand Unified Theories

In the 1960's, a theory called the **electroweak theory** was proposed in which the weak force and the electromagnetic force are viewed as two different manifestations of a single, more fundamental, electroweak interaction. Attempts have been made to incorporate the electroweak force and the strong (color) force into a theory known as the **grand unified theory**. One such theory suggests that at very short distances (10^{-31} m) and very high energy, the electromagnetic, weak, and strong forces are different aspects of a single underlying force. In this theory the fundamental difference between quarks and leptons disappears.

Attempts are presently being made to incorporate the four forces found in nature - gravity, electromagnetic, weak, and strong forces - into a single theory.

TEXTBOOK QUESTION 16. Suppose there were a kind of "neutrinolet" that was massless, had no color charge or electrical charge, and did not feel the weak force. Could you say that this particle even exists?

ANSWER: Laymen, as well as scientists, depend on their five senses and the instruments which extend the senses to detect the existence of objects. As described in the question, the "neutrinolet" has no mass, has no color charge or electrical charge, and does not feel the weak force. Therefore, the particle does not interact via any one of the four forces previously discussed. At the present time, there would be no way to detect the existence of this particle. Even if such a particle exists in theory, a method would have to be found to either directly or indirectly detect it.

PROBLEM SOLVING SKILLS

For problems involving high-energy projectiles:

1. The kinetic energy of particles in the 1.0-GeV range is approximately equal to mc^2. Express the energy of the particle in joules.
2. The de Broglie wavelength of such a particle (or the energy if the wavelength is given) can be found by using $\lambda = h/(mv) \approx (hc)/(mc^2)$ where $mc^2 = E$, the particle's energy.

For problems involving the cyclotron:

1. Determine the particle's velocity. If the velocity is above 0.1c, then equations used for special

relativity must be applied.
2. Derive an equation for the radius of the orbit, the cyclotron frequency, and the period of the motion.
3. Use the equations derived in step 2 to solve the problem.

For problems involving particle interactions:

1. Apply each of the conservation laws.
2. Each of the conservation laws must hold for the interaction to be allowed.

For problems involving the width of a resonance:

1. Use the uncertainty principle, i.e., $(\Delta E)(\Delta t) = h/2\pi$, to determine the width of the resonance in joules.
2. Express the width of the resonance in MeV.
3. Express the width of the resonance in terms of the rest mass of an electron.

SOLUTIONS TO SELECTED TEXTBOOK PROBLEMS

TEXTBOOK PROBLEM 2. Calculate the wavelength of 35-GeV electrons.

Part a. Step 1.	Solution: (Section 32-1)
Determine the rest energy of an electron in joules.	$E_o = m_o c^2$ $= (9.1 \times 10^{-31} \text{ kg})(3 \times 10^8 \text{ m/s})^2$ $E_o = 8.19 \times 10^{-14} \text{ J}$
Part a. Step 2.	$KE = (35 \text{ GeV})[(10^9 \text{ eV})/(1 \text{GeV})][(1.6 \times 10^{-19} \text{ J})/(1 \text{ eV})]$
Determine the kinetic energy of an electron in joules.	$KE = 5.6 \times 10^{-9} \text{ J}$
Part a. Step 3.	$E_{total} = KE + E_o$
Determine the total energy of an electron in joules.	$= 5.6 \times 10^{-9} \text{ J} + 8.19 \times 10^{-14} \text{ J}$ $E_{total} \approx 5.6 \times 10^{-9} \text{ J}$

Part a. Step 4.	At 35 GeV, the speed of an electron $\approx$ c (the speed of light)
Use the de Broglie formula [λ = h/(m v)] to determine the wavelength.	λ = h/(m v) $\approx$ h/(m c) $\approx$ h c/(m c^2) where E = m c^2
	$\approx$ h c/E
	λ = [(6.63 x 10^{-34} J s)(3.0 x 10^8 m/s)]/(5.6 x 10^{-9} J)
	λ $\approx$ 3.6 x 10^{-17} m

TEXTBOOK PROBLEM 3. What strength of magnetic field is used in a cyclotron in which protons make 2.8 x 10^7 revolutions per second?

Part a. Step 1.	Solution: (Section 32-1 and 32-2)
Derive a formula for the cyclotron frequency.	The direction of motion of the protons is perpendicular to the magnetic field. As a result the protons travel in a circular orbit.
	The force causing the circular motion is F = q v B sin θ where θ = 90° and sin 90° = 1
	F = m a where a = v^2/r, then
	q v B = m v^2/r
	q B = m v/r where v = 2 π r/T = 2 π r f
	q B = m (2 π r f)/r
	f = (q B)/(2 π m)
Part a. Step 2.	and solving for B gives
Rearrange the formula derived in step a. 1 and solve for the magnetic field strength.	B = (m 2 π f)/(q)
	= [2 π (1.67 x 10^{-27} kg)(2.8 x 10^7 rev/s)]/(1.6 x 10^{-19} C)
	B = 1.83 T

Part a. Step 1.	Solution: (Section 32-2 to 32-6)
Determine the total KE of the electron-positron in joules.	$2(420 \text{ keV})[(1000 \text{ eV})/(1 \text{ keV})][(1.6 \times 10^{-19} \text{ J})/(1 \text{ eV})] =$ $1.34 \times 10^{-13} \text{ J}$

Part a. Step 2.	$E_{rest} = (m_o c^2)_{electron} + (m_o c^2)_{positron}$
Determine the rest energy of an electron-positron in joules.	$\qquad = 2 (9.1 \times 10^{-31} \text{ kg})(3 \times 10^8 \text{ m/s})^2$ $E_{rest} = 1.64 \times 10^{-13} \text{ J}$

Part a. Step 3.	$E_{total} = KE + E_{rest}$
Determine the total energy of the electron-positron in joules.	$\qquad = 1.34 \times 10^{-13} \text{ J} + 1.64 \times 10^{-13} \text{ J}$ $E_{total} = 2.98 \times 10^{-13} \text{ J}$

Part a. Step 3.	The speed of a photon = c (the speed of light)
Use the de Broglie formula to determine the wavelength.	$\lambda = h/(m \, v) = h/(m \, c) = h \, c/(m \, c^2) \quad$ where $E = m \, c^2$ $\qquad = h \, c/E \quad$ but each photon shares ½ the total energy $\qquad = [(6.63 \times 10^{-34} \text{ J s})(3.0 \times 10^8 \text{ m/s})]/[½(2.98 \times 10^{-13} \text{ J})]$ $\lambda = 1.33 \times 10^{-12} \text{ m}$

Part a. Step 1.	Solution: (Sections 32-7 to 32-11)
Use the uncertainty principle to estimate its lifetime. See Chapter 28, section 3.	$(\Delta E)(\Delta t) \geq h/(2 \pi) \quad$ and rearranging $\Delta t \geq h/[(2 \pi)(\Delta E)]$ $\qquad \geq (6.63 \times 10^{-34} \text{ J s})/[2 \pi(277 \text{ keV})(1.6 \times 10^{-16} \text{ J})/(1 \text{ keV})]$

$$\geq (6.63 \times 10^{-34} \text{ J s})/(2.78 \times 10^{-13} \text{ J})$$

$$\Delta t \geq 2.38 \times 10^{-21} \text{ s}$$

TEXTBOOK PROBLEM 37. Which if the following decays are possible? For those that are forbidden, explain which laws are violated.

(a) $\Xi^\circ \rightarrow \Sigma^+ + \pi^-$ (b) $\Omega^- \rightarrow \Sigma^\circ + \pi^- + \nu$

(c) $\Sigma^\circ \rightarrow \Lambda^\circ + \gamma + \gamma$

Part a. Step 1.	Solution: (Section 32-7 and 32-11)
Apply the law of conservation of charge.	$\Xi^\circ \rightarrow \Sigma^+ + \pi^-$
Note: Table 32-2 contains information needed to answer this question.	$0 \rightarrow +1 + -1$ (allowed) The left side has a net charge of zero while the right side has a net charge of zero. The law of conservation of charge is not violated.

Part a. Step 2.	$\pm\frac{1}{2} \rightarrow \pm\frac{1}{2} + 0$ (allowed)
Apply the law of conservation of angular momentum.	By inspection, it is possible to combine the possible values so that the left side will equal the right side of the equation. The law of conservation of angular momentum is not violated.

Part a. Step 3.	$1 \rightarrow 1 + 0$ (allowed)
Apply the law of conservation of baryon number.	The left side has a net result of +1 while the right side has a net value of +1. The law of conservation of baryon family number is not violated.

Part a. Step 4.	No leptons are involved in this process. The law is not applicable.
Apply the law of conservation of lepton number.	

Part a. Step 5. Apply the law of conservation of energy.	1314.9 MeV < 1189.4 MeV + 139.6 MeV The law of conservation of energy is not conserved. For a reaction to be allowed, every law must be satisfied. Therefore, this reaction is not allowed.
Part b. Step 1. Apply the law of conservation of charge.	$\Omega^- \rightarrow \Sigma^\circ + \pi^- + \nu$ -1 → 0 + -1 (allowed) The left side has a net charge of zero while the right side has a net charge of zero. The law of conservation of charge is not violated.
Part b. Step 2.. Apply the law of conservation of angular momentum.	±3/2 → ±½ + ±½ (not allowed) By inspection, it is not possible to combine the possible values so that the left side will equal the right side of the equation. The law of conservation of angular momentum is violated.
Part b. Step 3. Apply the law of conservation of baryon number.	1 → 1 + 0 + 0 (allowed) The left side has a net result of +1 while the right side has a net value of +1. The law of conservation of baryon family number is not violated.
Part b. Step 4. Apply the law of conservation of lepton number.	0 → 0 + 0 + 1 The left side has a net result of 0 while the right side has a net of value of +1. This law is violated.
Part b. Step 5. Apply the law of conservation of energy.	1672.5 MeV > 1314.9 MeV + 139.6 MeV The law of conservation of energy is conserved. For a reaction to be allowed, every law must be satisfied. Therefore, this reaction is not allowed.
Part c. Step 1. Apply the law of conservation of charge.	$\Sigma^\circ \rightarrow \Lambda^\circ + \gamma + \gamma$ 0 → 0 + 0 + 0 (allowed) The left side has a net charge of zero while the right side has a net charge of zero. The law of conservation of charge is not violated.

Part c. step 2.	$\pm\frac{1}{2} \rightarrow \pm\frac{1}{2} + \pm 1 + \pm 1$ (allowed)
Apply the law of conservation of angular momentum.	By inspection, it is possible to combine the possible values so that the left side will equal the right side of the equation. The law of conservation of angular momentum is not violated.
Part c. Step 3.	$+1 \rightarrow +1 + 0 + 0$ (allowed)
Apply the law of conservation of baryon number.	The left side has a net result of +1 while the right side has a net value of +1. The law of conservation of baryon family number is not violated.
Part c. Step 4. Apply the law of conservation of lepton number.	No leptons are involved in this process. The law is not applicable.
Part c. Step 5.	1314.9 MeV > 1115.7 MeV + 0 + 0
Apply the law of conservation of energy.	The law of conservation of energy is conserved. For a reaction to be allowed, every law must be satisfied. Therefore, this reaction is allowed.

TEXTBOOK PROBLEM 47. (a) How much energy is released when an electron and a positron annihilate each other? (b) How much energy is released when a proton and an antiproton annihilate each other? (All particles KE ≈ 0.)

Part a. Step 1. Determine the rest energy of a electron plus a positron in joules.	Solution: (Section 32-3 to 32-6) $E_{rest} = (m_o c^2)_{electron} + (m_o c^2)_{positron}$ $= 2 (9.1 \times 10^{-31} \text{ kg})(3 \times 10^8 \text{ m/s})^2$ $E_{rest} = 1.64 \times 10^{-13} \text{ J}$
Part a. Step 2. Determine the total energy of the electron-positron pair in joules.	The statement of the problem implies that the particles are initially at rest; therefore, KE = 0 $E_{total} = KE + E_{rest}$ $= 0 \text{ J} + 1.64 \times 10^{-13} \text{ J}$ $E_{total} = 1.64 \times 10^{-13} \text{ J}$

Part a. Step 3. Express the energy released in MeV.	$(1.64 \times 10^{-13} \text{ J})[(1 \text{ MeV})(1.6 \times 10^{-13} \text{ J})] = 1.02 \text{ MeV}$
Part b. Step 1. Determine the rest energy of a proton plus antiproton in joules.	$E_{rest} = (m_o c^2)_{proton} + (m_o c^2)_{antiproton}$ $= 2 (1.67 \times 10^{-27} \text{ kg})(3 \times 10^8 \text{ m/s})^2$ $E_{rest} = 3.0 \times 10^{-10} \text{ J}$
Part b. Step 2. Determine the total energy of the proton-antiproton pair in joules.	$E_{total} = KE + E_{rest}$ $= 0 \text{ J} + 3.0 \times 10^{-10} \text{ J}$ $E_{total} = 3.0 \times 10^{-10} \text{ J}$
Part b. Step 3. Express the energy released in MeV.	$(3.0 \times 10^{-10} \text{ J})[(1 \text{ MeV})(1.6 \times 10^{-13} \text{ J})] = 1875 \text{ MeV}$

CHAPTER 33

ASTROPHYSICS AND COSMOLOGY

OBJECTIVES

After studying the material of this chapter, the student should be able to:

- use parallax to determine the distance to nearby stars.
- distinguish between a star, a galaxy, and a super cluster.
- given the absolute luminosity and distance from the earth to a star, calculate the apparent brightness of the star.
- calculate the apparent and absolute magnitude of a star.
- draw a Hertzsprung-Russell diagram. Explain what is meant by the main sequence and locate the current position of our sun on the main sequence.
- describe the current theory of stellar evolution as applied to our Sun. Describe the possible "death" of a star.
- state Einstein's principle of equivalence.
- explain what is meant by gravity as curvature in space and time.
- determine the Schwarzchild radius of the black hole formed by a star of given mass.
- explain how the Doppler effect indicates that the universe is expanding. Use Hubble's law to determine the velocity of recession of a star a known distance from the earth.
- cite evidence which indicates that the universe evolved from an initial explosion called the Big Bang.
- use the standard cosmological model to describe the evolution of the universe from 10^{-43} s to the present time.
- distinguish between an open universe and a closed universe. Describe how the critical density can be used to determine whether the universe is open or closed.

KEY TERMS AND PHRASES

astrophysics is a branch of astronomy that applies the techniques and theories of physics to the study of celestial objects.

cosmology is the study of the organization and structure of the universe and its evolution.

parallax is a method for measuring the distance to nearby stars. Parallax employs the apparent shift of position of a star against the background of more distant stars as the Earth moves about the Sun.

parsec is a measure of stellar distance. The distance to a star which changes its apparent position in the sky through an angle of 1″ (1 second of arc) in the course of the year is one parallax second or parsec. 1 parsec = 3.26 light years.

redshift refers to the displacement of a star's spectra toward the red end of the spectrum. Based on the Doppler effect, the red shift indicates that the universe is expanding.

A **galaxy** contains millions to hundreds of thousands of millions of stars. The Milky Way Galaxy contains about 10^{11} stars.

absolute luminosity refers to the total power radiated in watts by a star or galaxy.

apparent brightness is the power crossing per unit area perpendicular to the path of the light at the Earth.

Hertzsprung-Russell (H-R) diagram is a graph of absolute magnitude of stars versus temperature with stars represented by a point on the diagram. Most stars fall along the diagonal band called the **main sequence**.

black hole is the residual mass of a star which is so dense that no matter or light can escape from its gravitational field.

Schwarzchild radius represents the event horizon of a black hole. This is the surface beyond which no signals can ever reach us.

Hubble's law states that galaxies are moving away from one another at speeds proportional to the distance between them.

Big Bang theory suggests that the expansion of the universe is due to an explosion which probably occurred 10 to 15 billion years ago.

standard cosmological model gives an explanation of how the universe evolved after the Big Bang.

SUMMARY OF MATHEMATICAL FORMULAS

redshift	$\lambda' = \lambda[(1 + v/c)/(1 - v/c)]^{\frac{1}{2}}$	The observed wavelength (λ') depends on the true wavelength (λ) as well as the velocity of recession (v) and the speed of light (c).

apparent brightness	$\ell = L/4\pi d^2$	The apparent brightness (ℓ) is the total power (L) crossing per unit area perpendicular to the path of the light at the Earth. d is the distance from the Earth to the star or galaxy and the total area of a sphere of radius d is $4\pi d^2$. Note: the total power radiated in watts by a star or galaxy is called the absolute luminosity (L).
apparent magnitude	$\ell_1/\ell_2 = 100(m_2 - m_1)/5$ $m_2 - m_1 = 2.5 \log_{10}(\ell_1/\ell_2)$	The apparent magnitude (m) is based on a logarithmic scale and is defined so that for two objects whose luminosity ℓ_1 and ℓ_2 differ by a factor of 100, their apparent magnitudes m_1 and m_2 differ by 5.
absolute magnitude	$M = m - 5 \log_{10}(d/10 \text{ parsecs})$	The absolute magnitude (M) of an object is defined as the apparent magnitude that the object would have at a distance of 10 parsecs. d is the distance from the Earth to the object.
Schwarzchild radius	$R = 2 G M/c^2$	Schwarzchild radius (R) represents the radius of the event horizon of a black hole. M is the object's mass, G is the gravitational constant, and c is the speed of light in a vacuum.

CONCEPT SUMMARY

Astrophysics is a branch of astronomy that applies the techniques and theories of physics to the study of celestial objects. **Cosmology** is the study of the organization and structure of the universe and its evolution.

Stellar Distances

One method for measuring the distance to nearby stars is **parallax**. Parallax employs the apparent shift of position of a star against the background of more distant stars as the Earth moves about the Sun. The distance to a star which changes its apparent position in the sky through an angle of 1″ (1 second of arc) in the course of the year is one parallax second or parsec. 1 parsec = 3.26 light years where 1 light year is the distance that light travels through a vacuum in one year (approximately 10^{13} km or 6 trillion miles).

As shown in the diagram, the sighting angle of a star relative to the plane of the Earth's orbit (θ) is measured at different times of the year. The distance from the Earth to the Sun (d) is known and trigonometry can be used to determine the distance (D) to the star.

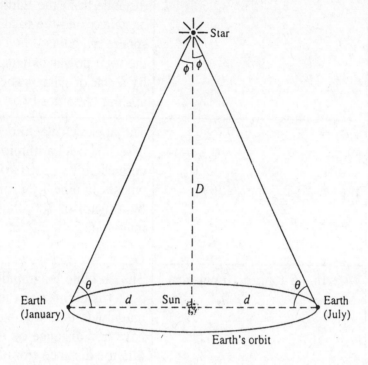

Parallax is useful for stars up to 100 light years (about 30 parsecs) from the Earth. The apparent brightness of more distant stars can give an approximate measurement of distance. Also useful is analysis of the shift of a star's spectra, the so-called redshift.

EXAMPLE PROBLEM 1. The brightest star in the night sky is Sirius, a star in the constellation Canis Major. Sirius is 8.6 light years from the Earth. Determine the a) parallax angle for Sirius, and b) distance to Sirius in parsecs.

Part a. Step 1.

Determine the parallax angle in radians.

Solution: (Section 33-1)

Using the above figure, note that $\tan \phi = d/D$. d is the radius of the Earth's orbit about the Sun; d = 1.5×10^8 km. D is the distance from the Earth to Sirius.

D = (8.6 Ly)(9.46 $\times 10^{12}$ km/Ly) = 8.14×10^{13} km

$\tan \phi$ = (1.5×10^8 km)/(8.14×10^{13} km)

$\tan \phi$ = 1.8×10^{-6}

	The angle is small; therefore, $\tan \phi \approx \phi$.
	$\phi = 1.8 \times 10^{-6}$ radians
Part a. Step 2.	$\phi = (1.8 \times 10^{-6} \text{ rad})(360°/2\pi \text{ rad})(3600''/1°)$
Express the angle in seconds of arc.	$\phi = 0.377''$ or 0.377 seconds of arc
Part b. Step 1.	1 parsec (pc) = $1/\phi$ where ϕ is expressed in seconds of arc
Determine the distance to Sirius in parsecs.	distance in parsecs = $1/(0.377'') = 2.63$ pc
	alternative solution:
	(8.6 Ly)(1 pc/3.26 Ly) = 2.63 pc

Redshift

Analysis of the spectra of stars and galaxies indicates a shift toward the red end of the spectrum. The Doppler effect suggests that this shift toward the red or **redshift** of light indicates that the star or galaxy is receding from the Earth and that the universe is expanding in size.

Based on equations derived from the Doppler effect, the observed wavelength (λ') is related to the true wavelength (λ) by the equation

$$\lambda' = \lambda[(1 + v/c)/(1 - v/c)]^{1/2}$$

where v = velocity of recession and c = speed of light.

The shift in the wavelength ($\Delta\lambda$) can be determined from the equation $\Delta\lambda = \lambda' - \lambda$.

EXAMPLE PROBLEM 2. a) Calculate the observed wavelength for the 656-nm line in the Balmer series for hydrogen in the spectrum of a galaxy which is receding at 0.1c from the Earth. b) Determine the change in the wavelength.

Part a. Step 1.	Solution: (Section 33-4)
Calculate the observed wavelength.	$\lambda' = \lambda[(1 + v/c)/(1 - v/c)]^{1/2}$

$$\lambda' = 656 \text{ nm } [(1 + 0.1c/c)/(1 - 0.1c/c)]^{\frac{1}{2}}$$

$$\lambda' = 656 \text{ nm } [(1 + 0.1)/(1 - 0.1)]^{\frac{1}{2}}$$

$$\lambda' = 656 \text{ nm } (1.1/0.9)^{\frac{1}{2}} = (656 \text{ nm})(1.11)$$

$$\lambda' = 725 \text{ nm}$$

Part b. Step 1.	$\Delta \lambda = \lambda' - \lambda$
Determine the shift in the wavelength.	$\quad = 725 \text{ nm} - 656 \text{ nm}$
	$\Delta \lambda = 69 \text{ nm}$ (redshift)

Stars and Galaxies

Our Sun is a single star in a **galaxy** known as the Milky Way Galaxy. A typical **galaxy** contains millions to hundreds of thousands of millions of stars. The Milky Way Galaxy contains about 10^{11} stars. It has a diameter of 100,000 light years and a thickness of about 6000 light years. Our Sun is located about halfway from the center to the edge, about 28,000 light years from the center. The Sun orbits the galactic center approximately once every 200,000 years at a speed of 250 km/s relative to the center of the galaxy.

Galaxies tend to be grouped in **galaxy clusters** with the clusters organized into clusters of clusters called **super-clusters**.

Luminosity and Magnitude

The total power radiated in watts by a star or galaxy is called the **absolute luminosity (L)**. The **apparent brightness** (ℓ) is the power crossing per unit area perpendicular to the path of the light at the Earth. Ignoring any absorption of the light as it travels through space,

$\ell = L/(4 \pi d^2)$ where d is the distance from the earth to the star or galaxy and the total area of area of a sphere of radius d is $4 \pi d^2$

The **apparent magnitude (m)** is based on a logarithmic scale and is defined so that for two objects whose luminosity ℓ_1 and ℓ_2 differ by a factor of 100, their apparent magnitudes m_1 and m_2 differ by 5; that is,

$\ell_1/\ell_2 = 100(m_2 - m_1)/5$ and $m_2 - m_1 = 2.5 \log_{10}(\ell_1/\ell_2)$

Visible stars, other than the Sun, range in apparent magnitude from approximately -1 (the brightest) to m = 6 (just visible to the naked eye).

The **absolute magnitude (M)** of an object is defined as the apparent magnitude that the object would have at a distance of 10 parsecs:

M = m - 5 $\log_{10}$(d/10 parsecs) where d is the distance from the Earth to the object

---

EXAMPLE PROBLEM 3. The apparent magnitude of Sirius is -1.47. Determine its absolute magnitude based on your answer to problem 1, part b.

Part a. Step 1.	Solution: Section 33-2
Determine the absolute magnitude.	The absolute magnitude (M) is related to the apparent magnitude (m) by the equation
	M = m - 5 $\log_{10}$(d/10)
	where d is the distance to Sirius in parsecs.
	M = -1.47 - 5 $\log_{10}$ (2.63/10)
	= -1.47 - 5 $\log_{10}$ 0.263
	= -1.47 - 5 (-0.579)
	M = -1.47 + 2.90 = + 1.43

Hertzsprung-Russell (H-R) Diagram

For most stars, the color is related to its absolute luminosity and therefore to its mass. This relationship is represented on a diagram called the Hertzsprung-Russell or H-R diagram. On the H-R diagram, shown on the next page, one axis represents the absolute magnitude while the other represents the temperature, and each star is represented by a point on the diagram.

Most stars fall along the diagonal band called the **main sequence**. The coolest stars, reddish in color, are located on the lower right. These stars are the least luminous and therefore must be of low mass. Stars on the upper left, bluish in color, are much more massive and have a higher luminosity. Stars that are not main-sequence type stars include red giants and white dwarfs.

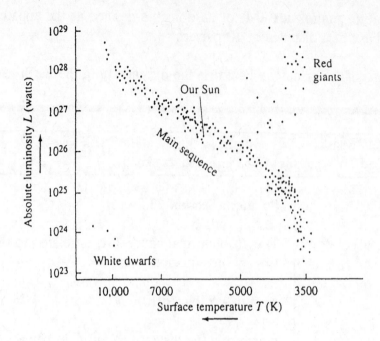

ANSWER: The H-R diagram relates only the surface temperature of a star to the absolute luminosity of the surface of the star. It does not directly reveal anything about the core. Theories on stellar evolution provide ideas on the core of a star and its place on the main sequence of the H-R diagram during its life cycle. Therefore, knowledge of a star's position on the H-R diagram might give some indirect knowledge of the stellar interior.

Stellar Evolution

Current theories suggest that **protostars** are formed from collapsing masses of hydrogen gas which exist in great clouds in space. During collapse, gravitational potential energy is transformed into kinetic energy, causing a heating of the gas. When the temperature reaches 10 million degrees kelvin, nuclear fusion begins. The dominant early reaction taking place in the core is the proton-proton cycle, in which four protons fuse to form a helium nucleus. After a time, the protostar becomes stable and can be placed on the main sequence of the H-R diagram. For a star with the mass of our Sun, it takes about 30 million years to reach the main sequence and it will remain there for approximately 10 billion years.

When the supply of hydrogen in the core is sufficiently depleted, the core contracts and the decrease in gravitational potential energy results in higher temperatures. This results in an increase in nuclear reactions, some of which result in the formation of more massive nuclei such as carbon. The outer part of the star expands and cools and the star's position on the H-R diagram moves from the main sequence to become a **red giant**. In about 5 billion years our Sun should become a red giant and its size could increase beyond the orbit of some if not all of the inner planets.

Stars with residual mass less than 1.4 solar masses will cool and become **white dwarfs**. The white dwarf will continue radiating energy until it eventually becomes a black dwarf, a dark, cold piece of ash.

More massive stars can continue to contract, due to their greater gravitational fields, and will eventually approach nuclear densities. Under sufficient pressure, the electrons combine with protons to form neutrons and the result may be a **neutron star**. Neutron stars are thought to have diameters in the order of ten kilometers. It is thought that **supernovae** are the result of the energy released in the final contraction of a neutron star.

It is theorized that an explosion in a supernovae can result in a rotating neutron star known as a **pulsar**. Pulsars are thought to be rotating neutron stars with extremely high magnetic fields which result in intense emission of radio waves in narrow beams. The result is similar to a lighthouse with a beam of light emitted at regular intervals as the light rotates.

If the star's residual mass is greater than two or three solar masses, it may contract still further and form a **black hole**, which is so dense that no matter or light can escape from its gravitational field.

General Relativity: Gravity and the Curvature of Space

Einstein's **principle of equivalence** states that "no experiment can be performed in a small region of space that could distinguish between a gravitational field and an equivalent uniform acceleration." As a result, there is an equivalence between gravitational mass and inertial mass.

Gravity is considered to be curvature in space-time. The curvature is greater near massive bodies. The theory requires the use of non-Euclidian geometries where the shortest distance between two points is not a straight line but possibly a curve called a geodesic. In such a space the sum of the angles of a triangle is not equal to 180°. The theory predicts that light rays passing near a massive object will be deflected. Confirmation came in 1919 when starlight passing the edge of the Sun during a solar eclipse was deflected by an amount consistent with the theory. In addition, the universe as a whole may be curved. If there is sufficient mass, the curvature of the universe is positive and the universe is closed and finite. Otherwise, the curvature is negative and the universe is open and infinite.

A black hole could produce extreme curvature of space-time. A black hole is the residual mass of a star which is so dense that no matter or light can escape from it. The **Schwarzchild radius** represents the **event horizon** of a black hole. This is the surface beyond which no signals can ever reach us. Inside the event horizon, at r = 0, there is, classically, an infinitely dense **singularity**. Whether such singularities can form is still unknown. Schwarzchild radius (R) is given by the equation $R = 2\,G\,M/c^2$ where G is the gravitational constant and c is the speed of light in a vacuum.

EXAMPLE PROBLEM 4. The largest planet in our solar system is Jupiter. The mass of Jupiter is 318 times the mass of the Earth. Determine the Schwarzchild radius for Jupiter.

Part a. Step 1.	Solution: (Section 33-3)
Determine the mass of Jupiter in kg.	$M = 318 \, M_e$
	$\quad = 318 \times 6.0 \times 10^{24}$ kg
	$M = 1.91 \times 10^{27}$ kg

Part a. Step 2.	$R = 2G \, M/c^2$
Determine the Schwarzchild radius for Jupiter.	$\quad = 2(6.67 \times 10^{-11} \text{ N m}^2/\text{kg}^2)(1.91 \times 10^{27} \text{ kg})/(3 \times 10^8 \text{ m/s})^2$
	$R = 2.83$ m
	This compares to a Schwarzchild radius of 2.95 km for the Sun and 8.9 mm for the Earth.

The Expanding Universe

Distant galaxies display a redshift of spectral lines. The Doppler effect is used to interpret the redshift, the interpretation indicates that the universe is expanding. The galaxies are moving away from one another at speeds (v) proportional to the distance (d) between them. **Hubble's law** describes the relationship as follows:

$v = H \, d$ where H is Hubble's constant

The value of Hubble's constant is 71 km/s/Mpc $\pm$ 4 km/s/Mpc or 22 km/s/Mly $\pm$ 1 km/s/Mly Note: 71 km/s/Mpc = 71 kilometers per second per million parsecs of distance.

The expansion of the universe is thought to be due to an explosive origin of the universe called the **Big Bang** which probably occurred 10 to 15 billion years ago.

There is a class of objects called **quasars** which do not seem to conform to Hubble's law. Quasars are quasi-stellar objects which are as bright as nearby stars but display very large redshifts. The large redshift indicates that they are very far away and because of their brightness they must be thousands of times brighter than normal galaxies.

ANSWER: You would see all other galaxies, including the Milky Way galaxy, receding from you. To use an analogy, imagine a deflated balloon covered with dots. As the balloon is inflated, your particular dot would recede from all of the other dots with a velocity directly proportional to the distance between the dots.

EXAMPLE PROBLEM 5. a) Use Hubble's law to estimate the recessional velocity of a galaxy 1.0 billion light years from the Earth. Note: let Hubble's constant $H = 22$ km/s per million light years of distance. b) Express the answer to part a as a fraction of the speed of light.

Part a. Step 1.	Solution: (Section 33-4)
Determine the recessional velocity of the galaxy.	$v = H\,d$
	$v = (22 \text{ km/s})/(1 \times 10^6 \text{ Ly}) \times 1.0 \times 10^9 \text{ Ly} = 22000 \text{ km/s}$
Part b. Step 1.	$(22000 \text{ km/s})(1000 \text{ m/1 km}) = 2.2 \times 10^7 \text{ m/s}$
Express the answer to part a as a fraction of the speed of light.	$(2.2 \times 10^7 \text{ m/s})/(3 \times 10^8 \text{ m/s}) = 0.073\,c = 7.3 \times 10^{-2}\,c$

The Big Bang and the Cosmic Microwave Background

In 1964, evidence to support the Big Bang theory was discovered. The discovery came in the form of radiation of wavelength 7.35 cm, which is in the microwave region of the spectrum. The intensity of this radiation was found to be independent of the time of day and came from all directions in the universe with equal intensity. The intensity of this radiation at $\lambda = 7.35$ cm corresponds to a blackbody radiation temperature of 2.725 K.

The **standard cosmological model** gives an explanation of how the universe evolved after the Big Bang. According to the model, starting at 10^{-43} s after the Big Bang, there was a series of phase transitions during which previously unified forces of nature condensed out one by one. The inflationary scenario assumes that during one of these phase transitions, the universe underwent a brief but rapid exponential expansion. Until about 10^{-35} s there was no distinction between quarks and leptons. Shortly thereafter the universe entered the hadron era. About 10^{-6} s after the Big Bang, the majority of hadrons disappeared, introducing the lepton era.

By the time the universe was 10 s old, the electrons too had mostly disappeared and the universe became radiation dominated. As the universe continued to expand, the radiation density decreased faster than the matter density. This eventually resulted in a matter-dominated universe. After about 500,000 years the universe was at a temperature of 3000 K and was cool enough for electrons to combine with nuclei and form atoms. After this time, the universe was cool enough for formation of stars and galaxies. In the 10 to 15 billion years since the Big Bang, the radiation has cooled to a temperature of 2.7 K which produces the 7.35-cm background radiation discovered in 1964.

TEXTBOOK QUESTION 19. Why were atoms, as opposed to bare nuclei, unable to exist until hundreds of thousands of years after the Big Bang?

ANSWER: As the universe expanded, the energy spread out over an increasingly larger volume and the temperature dropped. At temperatures higher than 3000 K the energy of the electrons and nuclei was too high for atoms to exist. It was necessary for the temperature of the universe to drop to approximately 3000 K before atoms could form.

The Future of the Universe

Whether the universe is open or closed depends on the matter density of the universe. If the average density is above a critical value known as the **critical density**, about 10^{-26} kg/m^3, then gravity will eventually stop the expansion and the universe will collapse back into a big crunch. If the average density is less than this value, the universe will continue to expand forever.

PROBLEM SOLVING SKILLS

For problems involving the parallax angle of a star and the stellar distance expressed in parsecs:

1. The parallax angle ϕ can be determined from the equation $\tan \phi = d/D$. d is the radius of the Earth's orbit about the Sun in km and D is the distance to the star in km.
   Note: 1 Ly = 9.46 x 10^{12} km
2. The parallax angle is small and therefore $\tan \phi \approx \phi$ when ϕ is expressed in radians.
3. The stellar distance in parsecs = $1/\phi$ where ϕ is expressed in seconds of arc. An alternative method to use is the conversion 1 pc = 3.26 Ly.

For problems involving the absolute magnitude of a star:

1. If necessary, convert stellar distances to parsecs.
2. The absolute magnitude (M) is related to the apparent magnitude (m) by the equation
   $M = m - 5 \log_{10} d/10$ where d is the distance to the star expressed in parsecs.

For problems involving the Schwarzchild radius:

1. Determine the mass of the object in kg.
2. The Schwarzchild radius (R) is related to the object's mass by the equation $R = G M/c^2$.

For problems involving Hubble's law:

1. Hubble's law states that the velocity (v) of a star/galaxy relative to the Earth is related to the distance to the star/galaxy by the equation $v = H d$. Express Hubble's constant in the appropriate units and solve the problem.

For problems involving the redshift of light from a star or galaxy receding from the Earth:

1. Express the velocity (v) of the galaxy as a fraction of the speed of light.
2. The observed wavelength (λ') is related to the true wavelength (λ) by the equation:
   $\lambda' = \lambda[(1 + v/c)/(1 - v/c)]^{1/2}$
3. The shift in the wavelength ($\Delta\lambda$) can be determined from the equation $\Delta\lambda = \lambda' - \lambda$.

SOLUTIONS TO SELECTED TEXTBOOK PROBLEMS

TEXTBOOK PROBLEM 4. A star is 36 pc away. What is its parallax angle? State (a) in seconds of arc, and (b) in degrees.

Part a. Step 1.	Solution: (Section 33-1 and 33-2)
Determine the parallax angle.	The stellar distance (D) in parsecs = $1/\phi$ where ϕ is expressed seconds of arc.
	$D = 1/\phi$ and $\phi = 1/D$
	$\phi = 1/(36 \text{ pc}) = 0.028$ seconds of arc or $0.028''$
Part b. Step 1.	There are 3600 second of arc in 1 degree of arc, i. e. $3600'' = 1°$
Convert seconds of arc to degrees of arc.	$(0.028'')[(1°)/(3600'')] = 7.7 \times 10^{-6}$ degrees of arc $= (7.7 \times 10^5)°$

TEXTBOOK PROBLEM 9. What is the relative brightness of the Sun as seen from Jupiter as compared to its brightness from Earth? (Jupiter is 5.2 times farther from the Sun than the Earth.)

Part a. Step 1. Use section 14-9 to determine the solar constant at the surface of the Earth.	Solution: (Sections 32-1 and 32-2) The solar constant refers the Sun's energy per second per square meter that strikes the Earth. From section 14-9, this value is approximately 1350 W/m².
Part a. Step 2. Determine the apparent brightness of the Sun as seen on Jupiter.	The apparent brightness (ℓ) at the surface of a planet depends on the energy radiated from the Sun, i.e., absolute luminosity (L) and the distance (d) from the Sun to the planet. $\ell_J = L/(4 \pi d_J^2)$ (Jupiter) and $\ell_E = L/(4 \pi d_E^2)$ (Earth) $\ell_J/\ell_E = [L/(4 \pi d_J^2)]/[L/(4 \pi d_E^2)]$ $\ell_J/\ell_E = (d_E^2)/(d_J^2) = (d_E/d_J)^2$ but $d_J = 5.2 d_E$ $\ell_J/(1350 \text{ W/m}^2) = [(d_E)/(5.2d_E)]^2$ $\ell_J = (0.19)^2 (1350 \text{ W/m}^2)$ $\ell_J \approx 50 \text{ W/m}^2$
Part a. Step 3. Determine the relative brightness of the Sun as observed on each planet.	$(\ell_J)/(\ell_E) \approx (50 \text{ W/m}^2)/(1350 \text{ W/m}^2) = 0.037$ The energy per second per square meter reaching Jupiter is only 3.7% as much as that striking the Earth.

TEXTBOOK PROBLEM 13. Calculate the density of a white dwarf whose mass is equal to the Sun's and whose radius is equal to the Earth's. How many times larger than the Earth's density is this?

Part a. Step 1. Refer to the text to determine the mass of the Sun and the mass and radius of the Earth.	Solution: (Section 33-1 and 33-2) From inside the front cover of the text: $m_{Sun} = 1.99 \times 10^{30}$ kg $r_{Earth} = 6.38 \times 10^3$ km $= 6.38 \times 10^6$ m $m_{Earth} = 5.97 \times 10^{24}$ kg

Part a. Step 2.	$\rho_{dwarf} = m_{dwarf}/V_{dwarf}$
Determine the density of the white dwarf.	$= (1.99 \times 10^{30} \text{ kg})/[(4/3) \pi (6.38 \times 10^{6} \text{ m})^{3}]$
	$= (1.99 \times 10^{30} \text{ kg})/(1.08 \times 10^{21} \text{ kg/m}^{3})$
	$\rho_{dwarf} = 1.83 \times 10^{9} \text{ kg/m}^{3}$

- -

Part a. Step 3.	$\rho_{Earth} = m_{Earth}/V_{Earth}$
Determine the density of the Earth.	$= (5.97 \times 10^{24} \text{ kg})/[(4/3) \pi (6.38 \times 10^{6} \text{ m})^{3}]$
	$= (5.97 \times 10^{24} \text{ kg})/(1.08 \times 10^{21} \text{ kg/m}^{3})$
	$\rho_{Earth} = 5.53 \times 10^{3} \text{ kg/m}^{3}$

- -

Part a. Step 4.	$\rho_{dwarf}/\rho_{Earth} = (1.83 \times 10^{9} \text{ kg/m}^{3})/(5.53 \times 10^{3} \text{ kg/m}^{3})$
Determine the ratio of the densities.	$\rho_{dwarf}/\rho_{Earth} = 3.3 \times 10^{5}$
	The density of the white dwarf is 330,000 times greater than the density of the Earth.

TEXTBOOK PROBLEM 26. Estimate the speed of a galaxy, and its distance from us, if the wavelength for the hydrogen line at 434 nm is measured on Earth as being 610 nm.

Part a. Step 1.	Solution: (Section 33-4)
Use the formula for the Doppler shift to determine the galaxy's speed.	$\lambda' = \lambda[(1 + v/c)/(1 - v/c)]^{\frac{1}{2}}$
	$610 \text{ nm} = 434 \text{ nm } [(1 + v/c)/(1 - v/c)]^{\frac{1}{2}}$
	$1.41 = [(1 + v/c)/(1 - v/c)]^{\frac{1}{2}}$
	$1.98 = (1 + v/c)/(1 - v/c)$
	$1.98 (1 - v/c) = 1 + v/c$
	$1.98 - 1.98 \, v/c = 1 + v/c$
	$0.98 = 2.98 \, v/c$

	0.328 = v/c
	v = 0.328 c

Part a. Step 2.	v = H d where H ≈ 71 km/s/Mpc
Use Hubble's law to determine its distance from the Earth.	$(0.328)(3 \times 10^8 \text{ m/s})[(1 \text{ km})/(1000 \text{ m})] \approx (71 \text{ km/s/Mpc}) \text{ d}$
	$\text{d} \approx (9.84 \times 10^4 \text{ km/s})/(71 \text{ km/s/Mpc})$
	$\text{d} \approx 1.39 \times 10^3 \text{ Mpc} \approx 4.5 \times 10^9$ light years or 4.5 billion light years

TEXTBOOK PROBLEM 44. What is the temperature that corresponds to the 1.8-TeV collisions at the Fermilab. To what era in cosmological history does this correspond? [Hint: see Fig. 33-25]

Part a. Step 1.	Solution: (Section 33-6)
Convert the collision energy to joules.	$(1.8 \text{ TeV})[(1 \times 10^{12} \text{ eV})/(1 \text{ TeV})][(1.6 \times 10^{-19} \text{ J})/(1 \text{ eV}) =$
	$2.9 \times 10^{-7} \text{ J}$

Part a. Step 2.	$KE = (3/2) \, k \, T$
Determine the temperature in degrees Kelvin.	$T = \frac{2}{3} \, KE/k$
	$T = (\frac{2}{3})(2.9 \times 10^{-7} \text{ J})/(1.38 \times 10^{-23} \text{ J/K})$
	$T = 1.4 \times 10^{16} \text{ K}$

Part a. Step 3.	Based on Fig. 33-26, this temperature corresponds to the hadron era.
Use Fig. 33-26 to determine which era in cosmological history this temperature corresponds.	

Part a. Step 1.	Solution: (Section 33-1 and 33-2)
Determine the total mass of the reactants and the total mass of the products.	reactants products

	reactants		products	
	$^{12}_{6}C$	12.000000 u	$^{24}_{24}Mg$	23.985042 u
	$^{12}_{6}C$	12.000000 u	γ	0.000000 u
		24.000000 u		23.985042 u

Part a. Step 2.	mass difference = 24.000000 u - 23.985042 u
Determine the mass difference between the products and the reactants.	mass difference = 0.014958 u

---

Part a. Step 3.	$Q = (M_a + M_X - M_b - M_Y) c^2$
Determine the value of Q in joules and MeV.	$Q = (\Delta m) c^2$
	$\Delta m = (0.014958 \text{ u})(1.66 \times 10^{-27} \text{ kg/u}) = 2.483 \times 10^{-29} \text{ kg}$
	$Q = (2.483 \times 10^{-29} \text{ kg})(3.0 \times 10^8 \text{ m/s})^2 = 2.234 \times 10^{-12} \text{ J}$
	$Q = (2.234 \times 10^{-12} \text{ J})(1 \text{ MeV}/1.6 \times 10^{-13} \text{ J}) = 13.97 \text{ MeV}$

Part b. Step 1.	$r = (1.2 \times 10^{-15} \text{ m}) A^{\frac{1}{3}}$
Use Eq. 30-1 to determine the radius of the carbon nucleus.	$= (1.2 \times 10^{-15} \text{ m}) (12)^{\frac{1}{3}}$
	$r = 2.75 \times 10^{-15} \text{ m}$

Part b. Step 2. Determine the total potential energy at closest approach.	The kinetic energy of the nuclei is converted to potential energy at closest approach. Note: at closest approach the distance between the centers of the carbon nuclei $= 2.75 \times 10^{-15}$ m $+ 2.75 \times 10^{-15}$ m $= 5.50 \times 10^{-15}$ m KE = PE PE $= k\ q_1\ q_2/r$ but $q_1 = q_2 = 6 \times 1.6 \times 10^{-19}$ C $= 9.6 \times 10^{-19}$ C $\quad = (9 \times 10^9$ N m^2/C$^2)(9.6 \times 10^{-19}$ C$)^2/(5.50 \times 10^{-15}$ m) PE $= 1.51 \times 10^{-12}$ J $(1.51 \times 10^{-12}$ J$)[(1$ MeV$)/(1.6 \times 10^{-13}$ J$)] = 9.4$ MeV
Part b. Step 3. Determine the KE of each nucleus.	$\text{KE}_{total} = \text{PE}_{total}$ but $\text{KE}_{total} = \text{KE}_1 + \text{KE}_2$ assuming the KE of each nucleus is the same, then the KE of each nucleus $= \frac{1}{2}(9.4$ MeV$) = 4.7$ MeV
Part c. Step 1. Determine the temperature of the nuclei in degrees Kelvin.	KE $= (3/2)\ k$ T where $k = 1.38 \times 10^{-23}$ J/K $(4.7$ MeV$)[(1.6 \times 10^{-13}$ J$)/(1$ MeV$)] = (3/2)(1.38 \times 10^{-23}$ J/K$)$ T 7.52×10^{-13} J $= (2.07 \times 10^{-23}$ J/K$)$ T T $= (7.52 \times 10^{-13}$ J$)/(2.07 \times 10^{-23}$ J/K$)$ T $= 3.6 \times 10^{10}$ K